AF589460

# Probability Theory and Stochastic Modelling

Volume 72

*Editors-in-Chief*

Søren Asmussen, Aarhus, Denmark
Peter W. Glynn, Stanford, CA, USA
Thomas Kurtz, Madison, WI, USA
Yves Le Jan, Paris, France

*Advisory Board*

Joe Gani, Canberra, Australia
Martin Hairer, Coventry, UK
Peter Jagers, Gothenburg, Sweden
Ioannis Karatzas, New York, NY, USA
Frank P. Kelly, Cambridge, UK
Andreas E. Kyprianou, Bath, UK
Bernt Øksendal, Oslo, Norway
George Papanicolaou, Stanford, CA, USA
Etienne Pardoux, Marseille, France
Edwin Perkins, Vancouver, Canada
Halil Mete Soner, Zurich, Switzerland

The **Stochastic Modelling and Probability Theory** series is a merger and continuation of Springer's two well established series Stochastic Modelling and Applied Probability and Probability and Its Applications series. It publishes research monographs that make a significant contribution to probability theory or an applications domain in which advanced probability methods are fundamental. Books in this series are expected to follow rigorous mathematical standards, while also displaying the expository quality necessary to make them useful and accessible to advanced students as well as researchers. The series covers all aspects of modern probability theory including

- Gaussian processes
- Markov processes
- Random Fields, point processes and random sets
- Random matrices
- Statistical mechanics and random media
- Stochastic analysis

as well as applications that include (but are not restricted to):

- Branching processes and other models of population growth
- Communications and processing networks
- Computational methods in probability and stochastic processes, including simulation
- Genetics and other stochastic models in biology and the life sciences
- Information theory, signal processing, and image synthesis
- Mathematical economics and finance
- Statistical methods (e.g. empirical processes, MCMC)
- Statistics for stochastic processes
- Stochastic control
- Stochastic models in operations research and stochastic optimization
- Stochastic models in the physical sciences

More information about this series at http://www.springer.com/series/13205

Makiko Nisio

# Stochastic Control Theory

## Dynamic Programming Principle

Makiko Nisio (emeritus)
Kobe University
Kobe, Japan

Osaka Electro–Communication University
Osaka, Japan

First edition published in the series ISI Lecture Notes, No 9, by MacMillan India Limited publishers, Delhi, © Makiko Nisio, 1981

ISSN 2199-3130 ISSN 2199-3149 (electronic)
ISBN 978-4-431-55122-5 ISBN 978-4-431-55123-2 (eBook)
DOI 10.1007/978-4-431-55123-2
Springer Tokyo Heidelberg New York Dordrecht London

Library of Congress Control Number: 2014953914

Mathematics Subject Classification: 93E20, 60H15

© Springer Japan 2015
This work is subject to copyright. All rights are reserved by the Publisher, whether the whole or part of the material is concerned, specifically the rights of translation, reprinting, reuse of illustrations, recitation, broadcasting, reproduction on microfilms or in any other physical way, and transmission or information storage and retrieval, electronic adaptation, computer software, or by similar or dissimilar methodology now known or hereafter developed. Exempted from this legal reservation are brief excerpts in connection with reviews or scholarly analysis or material supplied specifically for the purpose of being entered and executed on a computer system, for exclusive use by the purchaser of the work. Duplication of this publication or parts thereof is permitted only under the provisions of the Copyright Law of the Publisher's location, in its current version, and permission for use must always be obtained from Springer. Permissions for use may be obtained through RightsLink at the Copyright Clearance Center. Violations are liable to prosecution under the respective Copyright Law.
The use of general descriptive names, registered names, trademarks, service marks, etc. in this publication does not imply, even in the absence of a specific statement, that such names are exempt from the relevant protective laws and regulations and therefore free for general use.
While the advice and information in this book are believed to be true and accurate at the date of publication, neither the authors nor the editors nor the publisher can accept any legal responsibility for any errors or omissions that may be made. The publisher makes no warranty, express or implied, with respect to the material contained herein.

Printed on acid-free paper

Springer is part of Springer Science+Business Media (www.springer.com)

# Preface

The purpose of this book is to provide an introduction to stochastic controls theory, via the method of dynamic programming. The dynamic programming principle, originated by R. Bellman in 1950s, is known as the two stage optimization procedure. When we control the behavior of a stochastic dynamical system in order to optimize some payoff or cost function, which depends on the control inputs to the system, the dynamic programming principle gives a powerful tool to analyze problems. Exploiting the dependence of the value function (optimal payoff) on its terminal cost function, we will construct a nonlinear semigroup which allows one to formulate the dynamic programming principle and whose generator provides the Hamilton–Jacobi–Bellman equation. Here we are mainly concerned with finite time horizon stochastic controls. We also apply the semigroup approach to control-stopping problems and stochastic differential games, and provide with examples from the area of financial market models.

This book is organized as follows. Chapters 1–4 deal with completely observable finite-dimensional controlled diffusions. Chapters 5 and 6 are concerned with Hilbert space valued stochastic processes, related to partially observable control problems.

Chapter 1 is a review of stochastic analysis and stochastic differential equations with random coefficients for later uses. Chapter 2 deals with control problems with finite-time horizon. By a time-discretization method we construct a semigroup, associated with the value function, whose generator provides the Hamilton–Jacobi–Bellman equation. When the value function is smooth, it becomes a classical solution of the Hamilton–Jacobi–Bellman equation. However, it satisfies the equation in viscosity sense even if it is not smooth. Chapter 3 is concerned with viscosity solutions of nonlinear parabolic equation, including Hamilton–Jacobi–Bellman equations of stochastic controls and also stochastic optimal control-stopping problems. Chapter 4 presents zero sum, two-player, time-homogeneous, stochastic differential games and the Isaacs equations. We consider stochastic differential games by using progressive strategies. Then we construct semigroups associated with the upper and lower values, by using a semidiscretization method. These semigroups lead to the formulation of the dynamic programming principle and to the upper and lower Isaacs equations. The link between stochastic control and differential game is given via the risk sensitive control. Chapter 5 is a review on stochastic evolution equations on Hilbert spaces, in particular stochastic parabolic equations with colored Wiener noises. Basic definitions and results and Itô's formula are presented. Chapter 6 is concerned with control problems for Zakai equations. We again construct semigroups associated with the value functions. The dynamic programming principle and viscosity solutions of Hamilton–Jacobi–Bellman equations on Hilbert spaces are treated by using results obtained in the previous chapters. We show the connection between controlled Zakai equations and control of partially observable diffusions.

Kobe, Japan Makiko Nisio

# Acknowledgement

This book was planned as a new edition of *Stochastic Control Theory*, ISI Lecture Notes 9 (1981) following F. Delbaen's recommendation. I would like to acknowledge his recommendation together with valuable advice during preparation of the manuscript.

The author is greatly indebted to W. H. Fleming, who read carefully the manuscript and offered many valuable comments and suggestions, especially for Chap. 4, which led to a much improved version. Many thoughtful helps and encouragements had been given by experts and friends, particularly F. Asakura, Y. Fujita, H. Nagai, and T. Uratani who helped to improve the book at various stages. H. Morimoto assisted in writing the manuscript by carefully reading it and making valuable comments, especially on mathematical economics.

# Contents

# Notations

Let $a$ and $b$ be real numbers.
$a \vee b = \max\{a, b\}, \quad a \wedge b = \min\{a, b\}$
$a^+ = \max\{a, 0\}, \quad a^- = \min\{-a, 0\}$
$\delta_{i,j}$ = Kronecker symbol
$\mathbb{R}^d$ denotes $d$-dimensional Euclidean space
$x^i$ denotes the $i$-th coordinate of $x \in \mathbb{R}^d$
$x \cdot y = \sum_{i=1}^{d} x^i y^i, \quad |x|^2 = x \cdot x$
$\mathbb{R}_+^d = \{x \in \mathbb{R}^d; \quad x^i > 0, \quad i = 1, \ldots, d\}$
$S_r = \{x \in \mathbb{R}^d; \quad |x| \leq r\}$
$\mathbb{R}^d \otimes \mathbb{R}^m$ denotes the set of $d \times m$ matrices
$A_j^i = (i, j)$ entry of $A \in \mathbb{R}^d \otimes \mathbb{R}^m$
$A^\top$ = transpose of $A$
tr $A$ = trace of $A \in \mathbb{R}^d \otimes \mathbb{R}^d$
$|A|^2 = \mathrm{tr}(A^\top A) = \mathrm{tr}(AA^\top)$ for $A \in \mathbb{R}^d \otimes \mathbb{R}^m$
$I_d$ = $d$-dimensional unit matrix
$S^d$ = the set of symmetric $d \times d$ matrices
$S_+^d = \{A \in S^d; \quad A$ is nonnegative definite$\}$
$S_{++}^d = \{A \in S^d; \quad A$ is positive definite$\}$

$\mathcal{B}(U)$ = the $\sigma$-field generated by all open subsets of $U$
$L^p(\Omega, \mathcal{G}; \mathbb{R}^d)$ = the set of $\mathcal{G}$-measurable $d$-dimensional random variables with $E \mid \xi \mid^p < \infty$, for $p \geq 1$
$L^p([0, T] \times \Omega, (\mathcal{F}_t); \mathbb{R}^d)$ = the set of $(\mathcal{F}_t)$-progressively measurable $d$-dimensional processes with $\int_0^T E \mid X(t) \mid^p dt < \infty$, for $p \geq 1$
$\mathbf{\Gamma} = L^\infty([0, T] \times \Omega, (\mathcal{F}_t); \Gamma)$
$\sigma(\xi)$ = the $\sigma$-field generated by $\xi$
Let $\Sigma$ be a metric space.
$C(\Sigma)$ = the set of real valued continuous functions defined on $\Sigma$
$C_b(\Sigma) = \{\phi \in C(\Sigma); \quad \phi$ is bounded$\}$
$C_{bu}(\Sigma) = \{\phi \in C_b(\Sigma); \quad \phi$ is uniformly continuous$\}$

$C_p(\Sigma) = \{\phi \in C(\Sigma);\quad \phi$ is polynomial growing$\}$, when $\Sigma$ is a Banach space.
$C_K^\infty(\mathbb{R}^d) = \{\phi\, l \in C(\mathbb{R}^d);\quad \phi$ has compact support and continuous derivatives of any order$\}$
$C^{12}((0,T)\times\mathbb{R}^d) = \{\phi \in C((0,T)\times\mathbb{R}^d); \partial_t\phi,\ \partial_i\phi,\ \partial_{ij}\phi \in C((0,T)\times\mathbb{R}^d),$
$i,j = 1,\ldots,d\}$,
where $\partial_t = \frac{\partial}{\partial t},\quad \partial_i = \frac{\partial}{\partial x_i},\quad \partial_{ij} = \frac{\partial^2}{\partial x_i \partial x_j}$
$C^{12}([0,T)\times\mathbb{R}^d) = \{\phi \in C([0,T)\times\mathbb{R}^d) \cap C^{12}((0,T)\times\mathbb{R}^d);$
$\partial_t\phi,\ \partial_i\phi,\ \partial_{ij}\phi$ can be extended to continuous functions on
$C([0,T)\times\mathbb{R}^d),\quad i,j = 1,\ldots,d\}$
$C^{12}((0,T]\times\mathbb{R}^d)$ and $C^{12}([0,T]\times\mathbb{R}^d)$ are defined similarly
$\chi_\Lambda =$ the indicator function of the set $\Lambda$
$\partial_x\phi =$ gradient vector of $\phi$
$\partial_{xx}\phi =$ matrix of second order partial derivatives of $\phi = (\partial_{ij}\phi)_{i,j=1,\ldots,d}$
$\phi/\Lambda = \phi$ restricted to a set $\Lambda$

$H_0 = L^2(\mathbb{R}^d)$ with the usual inner product $(\cdot,\cdot)$ and norm $\|\cdot\|$
$\mathbf{I} =$ identity mapping
$\Delta =$ Laplacian operator
$H_p = \{\phi \in H_0;$ generalized derivatives of $\phi$ up to order $p$ belong to $H_0\},\quad p = 1,2,\ldots$
$\|\phi\|_{H_p} = \|(\mathbf{I}-\Delta)^{\frac{p}{2}}\phi\|$
$H_{-1} = \{$Borel functions $\phi;\quad (\mathbf{I}-\Delta)^{\frac{-1}{2}}\phi \in H_0\}$
$|||\phi||| = \|\phi\|_{H_1},\quad \|\phi\|_* = \|\phi\|_{H_{-1}} = \|(\mathbf{I}-\Delta)^{\frac{-1}{2}}\phi\|$
$\langle\phi,\psi\rangle =$ duality product between $\phi \in H_{-1}$ and $\psi \in H_1$
Let $\mathbb{H}$ and $\mathbb{Y}$ be Hilbert spaces.
$L^2(\mathbb{H};\ \mathbb{Y}) =$ Hilbert space of Hilbert–Schmidt operators from $\mathbb{H}$ into $\mathbb{Y}$
$\|\Phi\|_Q = \|\Phi Q^{\frac{1}{2}}\|_{L^2(\mathbb{H};\mathbb{Y})},\quad$ where $Q$ is a symmetric and nonnegative definite operator on $\mathbb{H}$

$\mathbb{M}^{2c}([0,T],\ (\mathcal{F}_t);\ \mathbb{H}) =$ set of continuous and square integrable $\mathbb{H}$-valued $(\mathcal{F}_t)$-martingales on $[0,T]$
$\langle M\rangle =$ quadratic variation process of $M$
$\langle M,N\rangle =$ quadratic variation process corresponding to $M$ and $N$
$D^pF(z) =$ $p$-th Fréchet derivative of $F$ at $z$
$C^{12}([0,T]\times\mathbb{H}) = \{F \in C([0,T]\times\mathbb{H});\quad \partial_t F,\ DF$ and $D^2F$ are continuous on $[0,T]\times\mathbb{H}\}$
$A \Rightarrow B =$ if $A$ then $B$

# Abbreviations

RHS = right-hand side
LHS = left-hand side
USC = upper semicontinuous
LSC = lower semicontinuous
ONB = orthonormal basis
w.r.t. = with respect to
a.e. = almost everywhere

# Chapter 1
# Stochastic Differential Equations

**Abstract** The purpose of this chapter is to overview elements of the theory of stochastic differential equations, based on Wiener processes, for use in the subsequent chapters. This theory was founded by K. Itô in 1942 (Itô, Zenkoku Shijo Sugaku Danwakai 244:1352–1400, 1942; Itô, On stochastic differential equations. Memoirs of the American Mathematical Society, vol 4. AMS, New York City, 1951). His aim was to construct Markov processes, governed by Kolmogorov's differential equations via Wiener processes, and to analyze their sample paths. After that, stochastic differential equations have been used to describe dynamical processes in random environments of various fields. Here we consider stochastic differential equations with random coefficients, because we aim at studying stochastic control problems.

The chapter is organized as follows. Section 1.1 is preliminaries. The basic definitions and results on stochastic processes are collected for later use. Stochastic differential equations and stochastic analysis will be introduced in Sect. 1.2. Section 1.3 deals with asset pricing problems as an application of previous results.

## 1.1 Review of Stochastic Processes

This section collects basic definitions and results on stochastic processes. Proofs of the results are mainly referred to standard and easily accessible books.

### *1.1.1 Random Variables*

**1. Measurable maps**

Let $\Omega$ be a non-empty set and $\mathcal{F}$ a $\sigma$-field of $\Omega$. We call the pair $(\Omega, \mathcal{F})$ a measurable space. Let $(\Omega, \mathcal{F})$ and $(\Omega', \mathcal{F}')$ be measurable spaces and $X : \Omega \mapsto \Omega'$ be a given map. For a $\sigma$-field $\mathcal{G}$ $(\subset \mathcal{F})$, $X$ is called $\mathcal{G}/\mathcal{F}'$-measurable (or $\mathcal{G}$-measurable in short), if $X^{-1}(B) \in \mathcal{G}$ for any $B \in \mathcal{F}'$. Sometimes $X$ is called measurable, when $\mathcal{G} = \mathcal{F}$.

© Springer Japan 2015

M. Nisio, *Stochastic Control Theory*, Probability Theory and Stochastic Modelling 72, DOI 10.1007/978-4-431-55123-2_1

**2. Random variables**

Let $(\Omega, \mathcal{F}, P)$ be a complete probability space, namely $(\Omega, \mathcal{F})$ is a measurable space, $P$ a probability measure on $\mathcal{F}$, and $\mathcal{N} := \{A \in \mathcal{F};\ P(A) = 0\}$ satisfies the condition

$$A \in \mathcal{N}, \quad B \subset A \Longrightarrow B \in \mathcal{N}. \tag{1.1}$$

(1) Let $(U, d)$ be a Polish space, i.e., $U$ is a separable complete metric space with a metric $d$, and let $\mathcal{B}(U)$ be the $\sigma$-field generated by the open subsets of $U$. $X : \Omega \mapsto U$ is called a $U$-valued random variable if $X$ is $\mathcal{F}/\mathcal{B}(U)$-measurable.

We identify two $U$-valued random variables $X$ and $Y$, whenever there is $N \in \mathcal{N}$, such that

$$X(\omega) = Y(\omega), \quad \forall \omega \notin N. \tag{1.2}$$

When (1.2) holds, we write

$$X = Y \quad P\text{-a.s. (almost surely).} \tag{1.3}$$

(2) Any $U$-valued random variable $X$ induces a probability measure $P_X$ on $\mathcal{B}(U)$ by

$$P_X(B) = P(X^{-1}(B)), \quad \forall B \in \mathcal{B}(U).$$

$P_X$ is called the probability distribution (or law) of $X$.

(3) When $U = \mathbb{R}^d$, we call $X$ a $d$-dimensional random variable.

**3. Expectations and conditional expectations**

(1) Let $X$ be a real random variable. The expectation (or mean) of $X$ is defined by

$$EX = \int_\Omega X(\omega)\, dP(\omega) = \int_{-\infty}^{\infty} x\, dP_X(x)$$

provided $\int |x|\, dP_X(x) < \infty$.

For a $d$-dimensional random variable $X = (X^1, \ldots, X^d)$, we put $EX = (EX^1, \ldots, EX^d)$.

(2) ***Conditional expectation***

Let $\mathcal{G}$ be a sub $\sigma$-field of $\mathcal{F}$. For a real integrable random variable $X$, we put

$$E[X; B] = \int_B X\, dP = E[\chi_B X], \quad B \in \mathcal{G}.$$

Since $E[X; \cdot]$ is a signed measure on $\mathcal{G}$, satisfying

$$E[X; B] = 0, \quad \text{if } P(B) = 0,$$

the Radon–Nikodým theorem yields a unique (up to indentification) $\mathcal{G}$-measurable integrable function $\xi$, such that

$$E[X;B] = \int_B \xi \, dP = E[\xi;B], \quad \forall B \in \mathcal{G}.$$

**Definition 1.1.** $\xi$ is called the conditional expectation of $X$ given $\mathcal{G}$, and is denoted by $E(X|\mathcal{G})$.

It holds that, if $\mathcal{G}_1 \subset \mathcal{G}_2 (\subset \mathcal{F})$, then

$$E(E(X|\mathcal{G}_2)|\mathcal{G}_1) = E(X|\mathcal{G}_1) \quad P\text{-a.s.} \tag{1.4}$$

For $\Lambda \in \mathcal{F}$, $E(\chi_\Lambda|\mathcal{G})$ is called the conditional probability of $\Lambda$ given $\mathcal{G}$, and denoted by $P(\Lambda|\mathcal{G})$.

**Definition 1.2.** Let $X$ be a $d$-dimensional random variable and $\mathcal{G}$ be a sub-$\sigma$-field of $\mathcal{F}$. We say $X$ and $\mathcal{G}$ are mutually independent, if

$$P(X \in A|\mathcal{G}) = P(X \in A) \quad P\text{-a.s.}, \quad \forall A \in \mathcal{B}(\mathbb{R}^d), \tag{1.5}$$

i.e.,

$$P((X \in A) \cap B) = P(X \in A)P(B), \quad \forall A \in \mathcal{B}(\mathbb{R}^d), \quad \forall B \in \mathcal{G}.$$

We say that $X$ and $Y$ are mutually independent if $X$ and $\sigma(Y)$ are mutually independent.

**4. Regular conditional probability**

Since $P(\Lambda|\mathcal{G})$ is defined uniquely up to sets of $P$-measure zero (possibly depending on $\Lambda$), $P(\cdot|\mathcal{G})(\omega)$ may not be a probability on $\mathcal{F}$ when $\omega$ is fixed. We next introduce a regular conditional probability.

**Definition 1.3.** $p : \mathcal{F} \times \Omega \mapsto [0,1]$ is called a regular conditional probability given $\mathcal{G}$, if

(a) $p(\cdot,\omega)$ is a probability measure on $(\Omega,\mathcal{F})$, $\forall \omega \in \Omega$,
(b) $p(A,\cdot)$ is $\mathcal{G}$-measurable, $\forall A \in \mathcal{F}$,
(c) $\forall A \in \mathcal{F}, \forall B \in \mathcal{G}$, $P(A \cap B) = \int_B p(A,\omega)\, dP(\omega)$.

Concerning the existence of regular conditional probability we have;

**Theorem 1.1.** *Suppose that $\Omega$ is a Polish space and $\mathcal{F} = \mathcal{B}(\Omega)$. Then there exists one and only one regular conditional probability given $\mathcal{G}$, i.e., if both of $p$ and $\hat{p}$ are regular conditional probabilities, then*

$$p(A,\cdot) = \hat{p}(A,\cdot), \quad \forall A \in \mathcal{F}, \quad P\text{-a.s.}$$

(see [SV79], p. 13, [IW81], p. 14 for details).

**5. Convergence theorems**

Let $X_n, n = 1, 2, \dots$ and $X$ be $d$-dimensional random variables.

**Definition 1.4.** (a) $X_n$ converges to $X$ in probability ($X_n \to X$ in prob), if $\lim_{n\to\infty} P(|X_n - X| > \varepsilon) = 0$, $\forall \varepsilon > 0$;
(b) $X_n$ converges to $X$ almost surely ($X_n \to X$ $P$-a.s.), if $\lim_{n\to\infty} |X_n - X| = 0$ $P$-a.s.;
(c) $X_n$ converges to $X$ in $p$-th mean ($X_n \to X$ in $L^p$), if $\lim_{n\to\infty} E|X_n - X|^p = 0$;
(d) $X_n$ converges to $X$ in law ($X_n \to X$ in law), if $\lim_{n\to\infty} Ef(X_n) = Ef(X)$, $\forall f \in C_b(\mathbb{R}^d)$.

For the law convergence, $X$ and $X_n, n = 1, 2, \dots$ may be defined on different probability spaces.

**Proposition 1.1.** *The following relations hold:*

(i) $X_n \to X$ *$P$-a.s.* $\Rightarrow X_n \to X$ *in prob* $\Rightarrow X_n \to X$ *in law.*
(ii) *If $X_n \to X$ in prob, then there is a subsequence which converges $P$-a.s.*

Now we state the basic theorems of the passage to the limit under the sign of conditional expectation.

Let $X_n$ be an integrable real random variable, $n = 1, 2, \dots$, and $\mathcal{G}$ be a sub-$\sigma$-field of $\mathcal{F}$.

### Monotone Convergence Theorem

Suppose that $X_n \geq 0$ and is increasing to $X(\leq \infty)$ $P$-a.s. Then

$$E(X_n|\mathcal{G}) \nearrow E(X|\mathcal{G})(\leq \infty) \quad P\text{-a.s.}$$

### Fatou's Lemma

Let $X_n \geq 0$, $n = 1, 2, \dots$. Then

$$E(\liminf_{n\to\infty} X_n|\mathcal{G}) \leq \liminf_{n\to\infty} E(X_n|\mathcal{G}) \quad P\text{-a.s.}$$

### Dominated Convergence Theorem

Suppose that $|X_n| \leq Y$, with integrable $Y, n = 1, 2, \dots$, and $X_n \to X$ $P$-a.s. Then

$$E(|X_n - X||\mathcal{G}) \longrightarrow 0 \quad P\text{-a.s.}$$

When $Y = C = \mathit{const}$, we call this the *bounded convergence theorem.*

**Convergence Theorem**

Let $X_n \to X$ $P$-a.s. Suppose that $X_n, n = 1, 2, \ldots$ are uniformly integrable given $\mathcal{G}$, that is, as $c \to \infty$

$$\sup_n E(|X_n|\chi(|X_n| > c)|\mathcal{G}) \longrightarrow 0 \quad \text{in prob.} \tag{1.6}$$

Then $E(|X_n - X||\mathcal{G}) \to 0$ $P$-a.s.

**Proposition 1.2.** *If there is $p > 1$ such that $\sup_n E(|X_n|^p|\mathcal{G}) < \infty$, $P$-a.s., then* (1.6) *holds.*

(see [LS01], Theorems 1.1–1.4).

## *1.1.2 Stochastic Processes*

### 1. Basic definitions

(1) Let $(\Omega, \mathcal{F}, P)$ be a complete probability space and $T$ a positive constant. A family of sub-$\sigma$-fields $(\mathcal{F}_t) := (\mathcal{F}_t; t \in [0, T])$ is called a filtration, if $\mathcal{F}_s \subset \mathcal{F}_t \subset \mathcal{F}$ for $0 \leq s < t$, and $\mathcal{F}_0$ contains all $P$-null sets.

If $\mathcal{F}_t$ is right continuous, namely $\mathcal{F}_t = \mathcal{F}_{t+} := \bigcap_{s>t} \mathcal{F}_s$, $\forall t \in [0, T)$, then the filtration$(\mathcal{F}_t)$ is said to satisfy the usual condition.

If $(\mathcal{F}_t)$ is a filtration, then $(\mathcal{F}_{t+})$ satisfies the usual condition.

**Definition 1.5.** We call a quadruple $(\Omega, \mathcal{F}, (\mathcal{F}_t), P)$ a filtered probability space when $(\mathcal{F}_t)$ satisfies the usual condition.

(2) A stochastic process $X = (X(t);\ t \in [0, T])$ is a family of ($d$-dimensional) random variables, defined on $(\Omega, \mathcal{F}, P)$. We identify two stochastic processes $X$ and $Y$, whenever there is a $P$-null set $N$ such that

$$X(t, \omega) = Y(t, \omega), \quad \forall t, \quad \text{for} \quad \omega \notin N.$$

We call $Y$ is a modification of $X$, if

$$P(X(t) = Y(t)) = 1, \quad \forall t.$$

For fixed $\omega$, the function $t \mapsto X(t, \omega)$ on $[0, T]$ is called a sample path.

If there is a $P$-null set $N$ such that $X(t, \omega)$ is continuous (or right continuous) in $t$ for $\omega \notin N$, then $X$ is called a continuous (or right continuous) process.

Put $\mathcal{F}_t^X$ = completion of the $\sigma$-field spanned by $(X(s); s \leq t)$.

(3) Let us introduce two kinds of measurability associated with the filtration $(\mathcal{F}_t)$.

**Definition 1.6.** Suppose $X : [0,T] \times \Omega \mapsto \mathbb{R}^d$ is $\mathcal{B}([0,T]) \times \mathcal{F}/\mathcal{B}(\mathbb{R}^d)$-measurable.

(a) $X$ is called $(\mathcal{F}_t)$-adapted, if $X(t)$ is $\mathcal{F}_t$-measurable for any $t \in [0,T]$.
(b) $X$ is called $(\mathcal{F}_t)$-progressively measurable, if the restriction of $X$ to $[0,t]$ is $\mathcal{B}([0,t]) \times \mathcal{F}_t/\mathcal{B}(\mathbb{R}^d)$-measurable, for any $t \in [0,T]$.

From the definition it is clear that if $X$ is $(\mathcal{F}_t)$-progressively measurable, then it is $(\mathcal{F}_t)$-adapted. When $X$ is continuous, the converse is true. We note that, when $X$ is $(\mathcal{F}_t)$-adapted, it has an $(\mathcal{F}_t)$-progressively measurable modification (see [My66], p. 18).

**2. Stopping times**

Let $(\Omega, \mathcal{F}, (\mathcal{F}_t), P)$ be a filtered probability space.

**Definition 1.7.** $\tau : \Omega \mapsto [0,T]$ is called an $(\mathcal{F}_t)$-stopping time, if

$$\{\omega; \tau(\omega) \leq t\} \in \mathcal{F}_t, \quad \forall t \in [0,T].$$

For any stopping time $\tau$, $\mathcal{F}_\tau$ is defined by

$$\mathcal{F}_\tau = \{A \in \mathcal{F}; A \cap (\tau \leq t) \in \mathcal{F}_t, \quad \forall t \in [0,T]\}. \tag{1.7}$$

$\mathcal{F}_\tau$ is a $\sigma$-field, which intuitively consists of information up to $\tau$.

*Example 1.1.* Let $X$ be a continuous $(\mathcal{F}_t)$-adapted $d$-dimensional process. For an open (or closed) set $A \subset \mathbb{R}^d$, the hitting time of $A$, defined as,

$$\tau(\omega) := \begin{cases} \inf\{t < T; X(t,\omega) \in A\} \\ T, \quad \text{if}\{\cdots\} = \text{empty} \end{cases}$$

is an $(\mathcal{F}_t)$-stopping time.

For $(\mathcal{F}_t)$-stopping times $\tau$ and $\sigma$, the following facts hold:

(i) $\mathcal{F}_\tau \cap \mathcal{F}_\sigma = \mathcal{F}_{\tau \wedge \sigma}$;
(ii) If $X$ is $(\mathcal{F}_t)$-progressively measurable, then $X(\tau)$ is $\mathcal{F}_\tau$-measurable;
(iii) Let $\xi$ be a random variable with $E|\xi| < \infty$. Then

$$E(E(\xi|\mathcal{F}_\sigma)|\mathcal{F}_\tau) = E(\xi|\mathcal{F}_{\sigma\wedge\tau}) \quad P\text{-a.s.}$$

For details, refer to [LS01], Section 1.3, [KS91], Section 1.1.2 and [IW81], 1.5.

**3. Continuous martingales**

(1) Let $X$ be a continuous $(\mathcal{F}_t)$-adapted real process.

**Definition 1.8.** $X$ is called a continuous $(\mathcal{F}_t)$-martingale (supermartingale or submartingale), if $E|X(t)| < \infty, \forall t$, and

$$\text{martingale property:} \qquad E(X(t)|\mathcal{F}_s) = X(s) \quad P\text{-a.s.} \quad \forall s \le t$$

$(E(X(t)|\mathcal{F}_s) \le X(s)$, or $E(X(t)|\mathcal{F}_s) \ge X(s)$ $P$-a.s. $\forall s \le t$, respectively).

For an $(\mathcal{F}_t)$-martingale $X$ (super- or submartingale), $EX(t)$ is constant (decreasing or increasing).

(2) Let us recall some basic properties of continuous martiangales, for later use.

(i) Let $X$ be a continuous $(\mathcal{F}_t)$-martingale.
Suppose that $f : \mathbb{R}^1 \mapsto \mathbb{R}^1$ is convex and $E|f(X(t))| < \infty, \forall t$. Then $(f(X(t)), t \in [0, T])$ is an $(\mathcal{F}_t)$-submartingale.
For instance, we can take $f(x) = |x|^p (p \ge 1)$.

(ii) ***Maximal inequality***
Let $X$ be a non-negative continuous $(\mathcal{F}_t)$-submartingale. Then

$$E\Big[\sup_{0 \le t \le T} X(t)^p\Big] \le \frac{p}{p-1} E[X(T)^p], \quad p > 1, \tag{1.8}$$

holds (see [LS01], Theorem 3.2, for the proof).

(iii) ***Optional sampling theorem***
Let $\sigma$ and $\tau$ be $(\mathcal{F}_t)$-stopping times and $X$ be a continuous $(\mathcal{F}_t)$-martingale. Then

$$E(X(\tau)|\mathcal{F}_\sigma) = X(\tau \wedge \sigma) \quad P\text{-a.s.}$$

(iv) ***Optional stopping theorem***
Let $\tau$ be an $(\mathcal{F}_t)$-stopping time and $X$ an $(\mathcal{F}_t)$-martingale. The stopped process $X^\tau(\cdot)$ is defined by $X^\tau(t) = X(\tau \wedge t)$. Then $X^\tau(\cdot)$ is an $(\mathcal{F}_{t\wedge\tau})$-martingale. $X^\tau(\cdot)$ is also an $(\mathcal{F}_t)$-martingale.

**4. Square integrable martingales**

Let $X$ be a continuous $(\mathcal{F}_t)$-martingale with $X(0) = 0$.

Suppose that $E|X(t)|^2 < \infty, \ \forall t$. Then $(X(t)^2; \ t \ge 0)$ is a continuous $(\mathcal{F}_t)$-submartingale.

Hence, by the Doob–Meyer decomposition theorem, there exists one and only one a continuous, integrable, and increasing $(\mathcal{F}_t)$-adapted process $(A(t); \ t \ge 0)$ with $A(0) = 0$, such that $(X(t)^2 - A(t); \ t \ge 0)$ is a continuous $(\mathcal{F}_t)$-martingale. $A(t)$ is denoted by $\langle X\rangle(t)$ and is called the quadratic variational process of $X$. $\langle X\rangle(t)$ is obtained in the following way. Let $0 = t_0 < t_1^N < \cdots < t_N^N = t$ be a partition of $[0, t]$ such that

$$\lim_{N\to\infty} \max_i |t_{i+1}^N - t_i^N| = 0.$$

Then, as $N \to \infty$.

$$\sum_i |X(t_{i+1}^N) - X(t_i^N)|^2 \longrightarrow \langle X \rangle(t) \quad \text{in prob} \tag{1.9}$$

(see [IW81], p. 36 and p. 69 for the proof).

Let $X$ and $Y$ be square integrable continuous $(\mathcal{F}_t)$-martingales with $X(0) = Y(0) = 0$. Then there exists one and only one continuous and integrable $(\mathcal{F}_t)$-adapted process $\langle X, Y \rangle(\cdot)$ with $\langle X, Y \rangle(0) = 0$ and bounded variation on $[0, T]$, $P$-a.s., and such that $(XY)(\cdot) - \langle X, Y \rangle(\cdot)$ is an $(\mathcal{F}_t)$-martingale. $\langle X, Y \rangle$ is called the quadratic variational process corresponding to $X$ and $Y$.

Indeed, $\langle X, Y \rangle(t) := \frac{1}{4}(\langle X + Y \rangle(t) - \langle X - Y \rangle(t))$ has the required properties.

Considering (1.9), we formally denote

$$(dX(t))^2 = d\langle X \rangle(t), \quad dX(t)dY(t) = d\langle X, Y \rangle(t).$$

Then

$$|d\langle X, Y \rangle(t)|^2 \le d\langle X \rangle(t) \cdot d\langle Y \rangle(t).$$

We have the following maximal inequality, for the quadratic variational process.

### Burkholder–Davis–Gundy Inequality

Let $X$ be a square integrable continuous $(\mathcal{F}_t)$-martingale, with $X(0) = 0$. Then, for $p \in (0, \infty)$ there exist positive constants $c_p$ and $C_p$, such that

$$c_p E\Big[\sup_{0 \le s \le t} |X(s)|^{2p}\Big] \le E[\langle X \rangle(t)^p] \le C_p E\Big[\sup_{0 \le s \le t} |X(s)|^{2p}\Big]. \tag{1.10}$$

In detail, $c_1 = \frac{1}{4}$, $C_1 = 1$,

$$\text{for } p > 1, \quad c_p = \left(\frac{2^{2p} p^{2p+1}}{(2p-1)^{2p-1}}\right)^{-p}, \quad C_p = (4p)^p,$$

$$\text{for } p \in (0, 1), \quad c_p = \left(\frac{16}{p}\right)^{-p}, \quad C_p = p^{-2p}$$

(see [IW81], p. 110 for the proof).

### 5. Local martingales

Let $X$ be a continuous $(\mathcal{F}_t)$-adapted real process. $X$ is called a continuous $(\mathcal{F}_t)$-local martingale, if there exists a nondecreasing sequence of $(\mathcal{F}_t)$-stopping times $(\tau_n, n = 1, 2, \ldots)$ such that $P(\lim_{n\to\infty} \tau_n = T) = 1$ and the stopped process $X^{\tau_n}$ is a continuous $(\mathcal{F}_t)$-martingale for each $n$.

Let $\tilde{\tau}_n$ be the exit time of $X$ from $[-n, n]$. Then $\sigma_n := \tau_n \wedge \tilde{\tau}_n$ is increasing to $T$ as $n \to \infty$, and $X^{\sigma_n}$ is a continuous square integrable $(\mathcal{F}_t)$-martingale. Since $\langle X^{\sigma_n}\rangle(t) = \langle X^{\sigma_{n+1}}\rangle(t)$ for $t < \sigma_n$, we can define $\langle X\rangle(t)$ by

$$\langle X\rangle(t) = \langle X^{\sigma_n}\rangle(t), \quad \forall t < \sigma_n.$$

$\langle X\rangle := (\langle X\rangle(t), t \in [0, T])$ does not depend on the choice of $\tau_n$, and is called the quadratic variational process of the local martingale $X$.

By $\mathbb{M}^c([0, T], (\mathcal{F}_t))$ and $\mathbb{M}^c_{loc}([0, T], (\mathcal{F}_t))$, we denote the set of continuous $(\mathcal{F}_t)$-martingales and continuous $(\mathcal{F}_t)$-local martingales, respectively.

We put

$$\mathbb{M}^{2c}([0, T],\ (\mathcal{F}_t)) = \{X \in \mathbb{M}^c([0, T], (\mathcal{F}_t));\ E|X(t)|^2 < \infty,\ \forall t \in [0, T]\}.$$

**6. $(\mathcal{F}_t)$-Wiener process**

Let $W = (W^1, \ldots, W^d)$ be a continuous $(\mathcal{F}_t)$-adapted $d$-dimensional process. $W$ is called a $d$-dimensional $(\mathcal{F}_t)$-Wiener process, if $W(0) = 0$ and $W(t) - W(s)$ is $N(0, (t-s)I_d)$-distributed and independent of $\mathcal{F}_s$, for any $s \le t$, that is

$$P(W(t) - W(s) \in A|\mathcal{F}_s) = \int_A (2\pi(t-s))^{-\frac{d}{2}} \exp\Big(-\frac{|y|^2}{2(t-s)}\Big)\, dy, \quad \forall A \in \mathcal{B}(\mathbb{R}^d).$$

The following properties hold:

(i) $W^i \in \mathbb{M}^{2c}([0, T], (\mathcal{F}_t)), i = 1, \ldots, d$.
(ii) $\langle W^i, W^j\rangle(t) = \delta_{ij}t$, where $\delta_{ij}$ = Kronecker symbol.
(iii) Let $X \in (X^1, \ldots, X^d)$ be a continuous $(\mathcal{F}_t)$-adapted process with $X(0) = 0$. If $X$ satisfies (i) and (ii), then $X$ is an $(\mathcal{F}_t)$-Wiener process.
(iv) $\mathcal{F}^W_t = \mathcal{F}^W_{t+}$, and $\mathcal{F}^W_0$ = trivial.
(v) Let us consider the time interval $[0, \infty)$, instead of $[0, T]$. Let $\tau$ be a bounded $(\mathcal{F}_t)$-stopping time and put $\tilde{\mathcal{F}}_t = \mathcal{F}_{t+\tau}$. Then $(W(t + \tau) - W(\tau); t \ge 0)$ is also an $(\tilde{\mathcal{F}}_t)$-Wiener process on $[0, \infty)$, independent of $\mathcal{F}_\tau$

(see [IW81], p. 74, [LS01], p. 192 for the proof).

**7. Continuous Markov processes**

(1) Let $p(\cdot)$ be a transition probability function, namely $p(\theta, x; t, \cdot)$ is a probability measure on $\mathcal{B}(\mathbb{R}^d)$, for any $0 \le \theta \le t$ and $x \in \mathbb{R}^d$, and $p(\theta, \cdot; t, A)$ is a Borel function on $\mathbb{R}^d$, satisfying $p(\theta, x; \theta, A) = \begin{cases} 1, x \in A, \\ 0, x \notin A. \end{cases}$

Moreover, the Chapman–Kolmogorov equation

$$p(\theta, x; t, A) = \int_{\mathbb{R}^d} p(\theta, x; s, dy) p(s, y; t, A), \quad \forall A \in \mathcal{B}(\mathbb{R}^d)$$

holds, for any $\theta \le s \le t$ and $x \in \mathbb{R}^d$.

(2) **Definition 1.9.** By a continuous Markov process with the transition probability function $p(\cdot)$, we mean a continuous $(\mathcal{F}_t)$-adapted process $X$, with

$$\text{(Markov property)} \quad P(X(t) \in A|\mathcal{F}_s) = p(s, X(s); t, A) \ P\text{-a.s.,}$$
$$\forall\, 0 \le s \le t \text{ and } A \in \mathcal{B}(\mathbb{R}^d).$$

Namely, if the condition "$X(s) = x$" is given then the probability distribution of the future of the process $(X(t); t > s)$ does not depend on its past $(X(\theta); \theta < s)$. We sometimes denote $p(s, x; t, A)$ by $P_{sx}(X(t) \in A)$, where the subscript $(s, x)$ indicates the initial condition "$X(s) = x$."

Let us define a family of linear operator $T_{\theta t}$ on $\mathbb{B}(\mathbb{R}^d)$ (= bounded Borel functions on $\mathbb{R}^d$), associated with the transition probability function $p(\cdot)$, i.e.,

$$\mathrm{T}_{\theta t}\phi(x) = \int_{\mathbb{R}^d} \phi(y)p(\theta, x; t, dy) = E_{\theta x}\phi(X(t)) \quad \text{for } \theta \le t.$$

$\mathrm{T}_{\theta t}$ is called the transition operator of the Markov process. .

Noticing that $\mathrm{T}_{\theta t}\phi \in \mathbb{B}(\mathbb{R}^d)$ and $\mathrm{T}_{tt} = \mathbf{I}$, we see

$$\mathrm{T}_{\theta s}(\mathrm{T}_{st}\phi)(x) = \mathrm{T}_{\theta t}\phi(x) \quad \text{for } \theta \le s \le t,$$

by the Chapman–Kolmogorov equation.

(3) ***Backward evolution operator*** $\mathcal{A}$

For a smooth function $\Phi$, we set

$$\mathcal{A}\Phi(\theta, x) = \lim_{h \to 0} \frac{1}{h}(E_{\theta x}\Phi(\theta + h, X(\theta + h)) - \Phi(\theta, x)),$$

whenever the RHS exists.

The operator $\mathcal{A}$ is called the backward evolution operator. By definition, the generator of $\mathrm{T}_{\theta,\theta+}$ acts by $\mathrm{G}_\theta\Psi = \lim_{s\to 0} \frac{1}{s}(\mathrm{T}_{\theta,\theta+s}\Psi - \Psi)$. Hence

$$\mathcal{A}\Phi(\theta, x) = \mathrm{G}_\theta\Phi(\theta, \cdot)(x) + \frac{\partial\Phi}{\partial\theta}(\theta, x).$$

When $X$ is given by a stochastic differential equation, we can compute $\mathcal{A}$ by using Itô's formula (see (2.178)).

### 1.1.3 *Itô Integrals*

We next define the Itô integral $\int \Phi(s)\, dW$. Let $W$ be a 1-dimensional $(\mathcal{F}_t)$-Wiener process. Since the paths of Wiener process are not differentiable, the integral cannot be defined in the ordinary way.

Let $L^0((\mathcal{F}_t))$ denote the set of $(\mathcal{F}_t)$-progressively measurable real process and

$$\begin{cases} L^2([0,T]\times\Omega,(\mathcal{F}_t)) = \left\{\Phi \in L^0((\mathcal{F}_t));\ E\int_0^T |\Phi(s)|^2\,ds < \infty\right\}, \\ L^2_{(\mathcal{F}_t)}([0,T]) = \left\{\Phi \in L^0((\mathcal{F}_t));\ P\left(\int_0^T |\Phi(s)|^2\,ds < \infty\right) = 1\right\}. \end{cases} \tag{1.11}$$

Setting $\|\Phi\|_T^2 = E[\int_0^T |\Phi(s)|^2\,ds]$, $L^2([0,T]\times\Omega,(\mathcal{F}_t))$ becomes a Hilbert space, with the norm $\|\ \ \|_T$.

Let us define $I(\Phi)(t,\omega) = \int_0^t \Phi(s,\omega)\,dW(s,\omega)$ in three steps.

**Step 1.** $\Phi$ is a simple process, say

$$\Phi(s,\omega) = \begin{cases} f_0(\omega), & \text{for } s = 0, \\ f_i(\omega), & \text{for } s \in (t_i, t_{i+1}], \quad i = 0,\dots,n. \end{cases}$$

where $0 = t_0 < t_1 < \cdots < t_n < t_{n+1} = T$, and $f_i$ is bounded and $\mathcal{F}_{t_i}$-measurable ($i = 0,\dots,n$). Define

$$\begin{aligned} I(\Phi)(t,\omega) &= \sum_{i=0}^{n} f_i(\omega)(W(t_{i+1}\wedge t,\omega) - W(t_i \wedge t,\omega)) \\ &= \sum_{i=0}^{j-1} f_i(\omega)(W(t_{i+1},\omega) - W(t_i,\omega)) + f_j(\omega)(W(t,\omega) - W(t_j,\omega)), \end{aligned}$$

for $t \in [t_j, t_{j+1}]$.

Thus, $I(\Phi) \in \mathbb{M}^{2c}([0,T],(\mathcal{F}_t))$ and satisfies

$$E[I(\Phi)(t)] = 0, \quad E[I(\Phi)(T)^2] = \|\Phi\|_T^2,$$

and

$$\langle I(\Phi)\rangle(t) = \int_0^t \Phi(s)^2\,ds.$$

**Step 2.** For $\Phi \in L^2([0,T]\times\Omega;(\mathcal{F}_t))$, there is a sequence of simple processes $(\Phi_n, n = 1,2,\dots)$, such that $\lim_{n\to\infty}\|\Phi - \Phi_n\|_T = 0$. Hence, as $n,m \to \infty$

$$E[|I(\Phi_n)(T) - I(\Phi_m)(T)|^2] = E[|I(\Phi_n - \Phi_m)(T)|^2] \longrightarrow 0. \tag{1.12}$$

From the Burkholder–Davis–Gundy inequality, it follows that there exists the limit, $\lim_{n\to\infty} I(\Phi_n)$, in $\mathbb{M}^{2c}([0,T],(\mathcal{F}_t))$ and is unique (up to equivalence of stochastic processes). Since this limit does not depend on the choice of $\Phi_n$, we can define $I(\Phi)$ by

$$I(\Phi)(t) = \lim_{n\to\infty} I(\Phi_n)(t) =: \int_0^t \Phi(s)\,dW(s).$$

Consequently, we have

$$I(\Phi) \in \mathbb{M}^{2c}([0,T],(\mathcal{F}_t)) \quad \text{with} \quad \langle I(\Phi)\rangle(t) = \int_0^t \Phi(s)^2 ds. \tag{1.13}$$

For any two $(\mathcal{F}_t)$-stopping times $\tau$ and $\sigma$, with $0 \le \sigma \le \tau$, $P$-a.s., we put

$$\int_\sigma^t \Phi(s)\,dW(s) = I(\Phi)(\tau) - I(\Phi)(\sigma). \tag{1.14}$$

The following properties hold:

(i) $E(\int_\sigma^\tau \Phi(s)\,dW(s)|\mathcal{F}_\sigma) = 0 \quad P$-a.s.;
(ii) $E(\int_\sigma^\tau \Phi(s)\,dW(s) \int_\sigma^\tau \Psi(s)\,dW(s)|\mathcal{F}_\sigma) = E(\int_\sigma^\tau \Phi(s)\Psi(s)\,ds|\mathcal{F}_\sigma) \quad P$-a.s.;
(iii) Put $\hat{\Phi}(s,\omega) = \Phi(s,\omega)\chi(s \le \sigma) := \Phi(s,\omega)\chi_{[0,\sigma]}(s)$. Then

$$\int_0^{t\wedge\sigma} \Phi(s)\,dW(s) = \int_0^t \hat{\Phi}(s)\,dW(s), \quad \forall\, t \in [0,T].$$

**Step 3.** For $\Phi \in L^2_{(\mathcal{F}_t)}([0,T])$, we define

$$\tau_n(\omega) = \begin{cases} \inf\{t < T; \quad \int_0^t \Phi(s,\omega) \ge n\}, \\ T, \quad \text{if}\{\cdots\} = \text{empty}. \end{cases} \tag{1.15}$$

Then $(\tau_n, n = 1,2,\dots)$ is a sequence of nondecreasing $(\mathcal{F}_t)$-stopping times satisfying $P(\lim_n \tau_n = T) = 1$. Set $\Phi_n(t,\omega) = \Phi(t,\omega)\chi(t \le \tau_n)$. Then $\Phi_n \in L^2([0;T]\times\Omega,(\mathcal{F}_t))$ and (iii) yields

$$\begin{aligned}\int_0^{t\wedge\tau_m} \Phi_n(s)\,dW(s) &= \int_0^t \Phi(s)\chi(s \le \tau_m)\chi(s \le \tau_n)\,dW(s) \\ &= \int_0^t \Phi_m(s)\,dW(s), \quad \text{for } m < n.\end{aligned}$$

By this consistency, the following is well-defined:

$$\int_0^t \Phi(s)\,dW(s) := \int_0^t \Phi_m(s)\,dW(s) \quad \text{for } t \le \tau_m. \tag{1.16}$$

Since $\int_0^{\tau_m\wedge\cdot} \Phi(s)\,dW(s) \in \mathbb{M}^{2c}([0,T],(\mathcal{F}_t)), m = 1,2,\dots,$ the Itô integral

$\int_0^t \Phi(s)\,dW(s)$ is defined as a continuous $(\mathcal{F}_t)$-local martingale (see [IW81], pp. 45–22, [KS98], pp. 129–141, [LS01], pp. 92–112, for details).

Replacing real one-dimensional processes by $m$-dimensional ones, we define $L^2([0,T]\times\Omega,(\mathcal{F}_t);\mathbb{R}^m)$ and $L^2_{(\mathcal{F}_t)}([0,T];\mathbb{R}^m)$ in the same way as in (1.11). Let $W$ be an $m$-dimensional $(\mathcal{F}_t)$-Wiener process. For $\Phi^i = (\Phi^i_1,\ldots,\Phi^i_m) \in L^2([0,T]\times\Omega,(\mathcal{F}_t);\mathbb{R}^m)$, we define $\int \Phi^i(s)\,dW(s)$ by

$$\int_0^t \Phi^i(s)\,dW(s) = \sum_{j=1}^m \int_0^t \Phi^i_j(s)\,dW^j(s).$$

**Definition 1.10.** $X$ is called a $d$-dimensional Itô process, if $X$ admits the representation

$$X(t) = X_0 + \int_0^t b(s)\,ds + \int_0^t \Phi(s)\,dW(s), \quad \forall t, \quad P\text{-a.s.,} \tag{1.17}$$

or the equivalent componentwise form

$$X^i(t) = X^i_0 + \int_0^t b^i(s)\,ds + \sum_{j=1}^m \int_0^t \Phi^i_j(s)\,dW^j(s), \quad i = 1,\ldots,d,$$

where

$X_0$ is an $\mathcal{F}_0$ - measurable $d$-dimensional random variable,
$b(\cdot) \in L^1([0,T]\times\Omega,(\mathcal{F}_t);\mathbb{R}^d)$,
$\Phi(\cdot) \in L^2([0,T]\times\Omega;\mathbb{R}^d\otimes\mathbb{R}^m)$,
$\int b(s)\,ds$ stands for the pathwise Lebesgue integral.

Sometimes we use for Itô process $X$ the differential representation

$$dX(t) = b(t)\,dt + \Phi(t)\,dW(t).$$

### *1.1.4 Itô's Formula*

Itô's formula is one of the most important tools in stochastic calculus.

Let $X$ be a $d$-dimensional Itô process, given by (1.17), and $c(\cdot)$ be a bounded $(\mathcal{F}_t)$-progressively measurable real process.

#### Itô's Formula

Let $F$ be a function of $C^{12}([0,T]\times\mathbb{R}^d)$ and $\tau$ be an $(\mathcal{F}_t)$-stopping time. Then $e^{-\int_0^t c(h)\,dh} F(t, X(t))$ is also a real Itô process and the following formula holds:

$$
\begin{aligned}
&e^{-\int_0^{t\wedge\tau} c(s)\,ds} F(t\wedge\tau, X(t\wedge\tau)) - F(0, X_0) \\
&= \int_0^{t\wedge\tau} e^{-\int_0^s c(h)\,dh} \Big\{ \partial_t F(s, X(s)) + b(s)\cdot\partial_x F(s, X(s)) \\
&\quad + \frac{1}{2}\mathrm{tr}\big(a(s)\partial_{xx} F(s, X(s))\big) - c(s) F(s, X(s)) \Big\}\, ds \\
&\quad + \int_0^{t\wedge\tau} e^{-\int_0^s c(h)\,dh} \partial_x F(s, X(s)) \Phi(s)\, dW(s), \quad \forall t,
\end{aligned}
\tag{1.18}
$$

where $a(t) = \Phi(s)\Phi(s)^\top$, $\partial_x F$ = gradient of $F$, $\partial_{xx} F$ = Hessian of $F$ (see [IW81], pp. 66–73, [LS01], pp. 123–128 for the proof).

Roughly speaking, the formal expansion

$$
d(e^{-\int_0^t c\,ds} F(t, X(t))) = e^{-\int_0^t c\,ds}\big(dF(t, X(t)) - c(t)F(t, X(t))\big)
$$

and the relation

$$
dF(t, X(t)) = \frac{\partial F}{\partial t} dt + \sum_{i=1}^{d} \frac{\partial F}{\partial x_i} dX^i + \frac{1}{2}\sum_{j=1}^{d} \frac{\partial^2 F}{\partial x_i \partial x_j} dX^i\, dX^j + o(dt)
$$

lead to Itô's formula, by using $dW^i(t)\, dW^j(t) = \delta_{ij}\, dt$.

We finally state Itô's formula with generalized derivatives.

**Itô–Krylov Formula ([Kr09], 2.10, Theorem 1)**

Suppose that

(a) $\Phi(\cdot)$ and $b(\cdot)$ are bounded,
(b) $a(t,\omega)$ is uniformly positive definite, i.e., there is a positive constant $\lambda_0$ such that

$$
y^\top a(t,\omega) y \ge \lambda_0 |y|^2, \quad \forall y \in \mathbb{R}^d, \quad \forall t\omega.
$$

Let $D$ be a bounded open set of $\mathbb{R}^d$. $\tau_D$ denotes the exit time of $X$ from $D$.

Then, for any $F \in W^{12}([0,T]\times D)$ and any $(\mathcal{F}_t)$-stopping time $\tilde{\tau}$, (1.18) holds when we replace $\tau$ with $\tau_D \wedge \tilde{\tau}$ and take generalized derivatives, $\frac{\partial F}{\partial t}$, $\frac{\partial F}{\partial x_i}$ and $\frac{\partial^2 F}{\partial x_i \partial x_j}$ $(i, j = 1, \ldots, d)$.

*Example 1.2 (Exponential martingales).* Let us define $q(t)$ by

$$
q(t) = \exp\Big(\int_0^t \Phi(s)\cdot dW(s) - \frac{1}{2}\int_0^t |\Phi(s)|^2\, ds\Big), \quad t \in [0, T],
$$

where $\Phi \in L^2([0,T]\times\Omega, (\mathcal{F}_t); \mathbb{R}^m)$.

Then $q(\cdot)$ is positive, continuous, and $(\mathcal{F}_t)$-adapted. Moreover,

(i) $q(\cdot)$ is a continuous $(\mathcal{F}_t)$-supermartingale with $q(0) = 1$.
(ii) $q(\cdot)$ is a continuous $(\mathcal{F}_t)$-martingale if and only if $Eq(T) = 1$.

When $q(\cdot)$ is a martingale, $q(\cdot)$ is called an exponential $(\mathcal{F}_t)$-martingale.

## 1.2 Stochastic Differential Equations

In this section, we will study stochastic differential equations (SDEs in short) related to stochastic control problems. Hence we are interested in two kinds of SDEs, those with random coefficients and those with deterministic Borel measurable coefficients. In Sect. 1.2.1 we consider Lipschitz continuous SDEs with random coefficient and prove the existence of a unique solution by using the successive approximation method. Section 1.2.2 deals with the Girsanov transformations of probability measures via exponential martingales, which provide a tool for studying SDEs with Borel measurable drift terms. We also recall the Bayes formula associated with Girsanov transformation. In Sect. 1.2.3 we overview the weak solutions of SDEs with deterministic Borel coefficients [Kr09]. Section 1.2.4 deals briefly with backward SDEs introduced by Pardoux and Peng [PP90].

### *1.2.1 Lipschitz Continuous SDEs with Random Coefficients*

Let

$$b : [0, T] \times \mathbb{R}^d \times \Omega \longmapsto \mathbb{R}^d$$

and

$$\alpha : [0, T] \times \mathbb{R}^d \times \Omega \longmapsto \mathbb{R}^d \otimes \mathbb{R}^m$$

be measurable and, for any $x \in \mathbb{R}^d, b(\cdot, x, \cdot)$ and $\alpha(\cdot, x, \cdot)$ be $(\mathcal{F}_t)$-progressively measurable.

We assume the following two conditions:

$(a_1)$ ***Equi-Lipschitz continuity***
$\exists l > 0$, such that

$$|b(t,x,\omega)-b(t,y,\omega)|+|\alpha(t,x,\omega)-\alpha(t,y,\omega)| \leq l|x-y|, \quad \forall x, y, \quad \forall (t,\omega).$$

$(a_2)$ ***Linear growth***
$\exists K > 0$, such that

$$|b(t,x,\omega)| + |\alpha(t,x,\omega)| \leq K(l + |x|), \quad \forall x, \quad \forall (t,\omega).$$

Let $\theta \in [0, T)$. Consider the $d$-dimensional SDE

$$dX(t) = b(t, X(t), \omega)\, dt + \alpha(t, X(t), \omega)\, dW(t), \quad t > \theta, \tag{1.19}$$

with the initial condition

$$X(\theta) = X_\theta \quad (\in L^p(\Omega, \mathcal{F}_\theta; \mathbb{R}^d) \text{ with } p \geq 1). \tag{1.20}$$

Although we customarily omit the probability parameter $\omega$, we put it in coefficients $b$ and $\alpha$ in order to stress the randomness of the coefficients.

**Definition 1.11.** $X = (X(t);\ t \in [\theta, T])$ is called a solution of (1.19)–(1.20), if $X$ is a continuous and $(\mathcal{F}_t)$-adapted $d$-dimensional process satisfying

$$X(t) = X_\theta + \int_\theta^t b(s, X(s), \omega)\, ds + \int_\theta^t \alpha(s, X(s), \omega)\, dW(s),\ \forall t \in [\theta, T],\ P\text{-a.s.}, \tag{1.21}$$

or, equivalently, the componentwise equation

$$X^i(t) = X^i_\theta + \int_\theta^t b^i(s, X(s), \omega)\, ds + \sum_{j=1}^m \int_\theta^t \alpha^i_j(s, X(s), \omega)\, dW^j(s),$$
$$\forall t \in [\theta, T], \quad i = 1, \dots, d, \quad P\text{-a.s.}$$

We say that the solution $X$ is unique if

$$P(X(t) = \hat{X}(t),\ \forall t \in [\theta, T]) = 1 \tag{1.22}$$

holds for any solution $X$ and $\hat{X}$.

When we want to stress the initial condition, we denote $X$ by $X_{\theta, X_\theta}$. $b(\cdot)$ and $\alpha(\cdot)$ are called the drift coefficient and the diffusion coefficient, respectively.

By using the successive approximation method, we have

**Theorem 1.2.** *Under conditions* $(a_1)$ *and* $(a_2)$, *problem* (1.19)–(1.20) *has a unique solution* $X$. *Moreover, the following properties hold:*

(i) *There is a constant* $K_p$, *which depends only on* $p$ *and* $K$, *such that*

$$E\Big[\sup_{\theta \leq s \leq T} |X(s)|^{2p}\Big] \leq K_p(1 + E|X_\theta|^{2p}), \tag{1.23}$$

$$E\Big[\sup_{t_1 \leq s \leq t_2} |X(s) - X(t_1)|^{2p}\Big] \leq K_p(1 + E|X_\theta|^{2p})(t_2 - t_1)^p$$
$$\textit{for any } \theta \leq t_1 < t_2 < T. \tag{1.24}$$

(ii) *There is a constant $\hat{K}_p$, which depends only on $p$, $K$ and $l$, such that*

$$E\Big[\sup_{\theta\le s\le T}|X(s)-\hat{X}(s)|^{2p}\Big]\le \hat{K}_p E|X_\theta-\hat{X}_\theta|^{2p}, \tag{1.25}$$

*where $\hat{X}$ is the solution of* (1.19) *with the initial condition $\hat{X}(\theta)=\hat{X}_\theta$, and*

$$E\Big[\sup_{\theta^*\le s\le T}|X(s)-X^*(s)|^{2p}\Big]\le \hat{K}_p(1+E|X_\theta|^{2p})(\theta^*-\theta)^p, \tag{1.26}$$

*where $X^*$ is the solution of* (1.19) *with the initial condition $X^*(\theta^*)=X_\theta$ and $\theta^*\ge\theta$.*

*Example 1.3 (Linear SDE).* Let us consider the SDE

$$\begin{aligned} dX(t) =&(B(t,\omega)X(t)+b_0(t,\omega))\,dt \\ &+\sum_{j=1}^m (A_j(t,\omega)X(t)+\alpha_j(t,\omega))\,dW^j(t), \end{aligned} \tag{1.27}$$

where $B, A_j : [0,T]\times\Omega\mapsto\mathbb{R}^d\otimes\mathbb{R}^d$, $b_0,\alpha_j : [0,T]\times\Omega\mapsto\mathbb{R}^d$, $j=1,\dots,m$ are bounded and $(\mathcal{F}_t)$-progressively measurable, and $W=(W^1,\dots,W^m)$ is an $m$-dimensional $(\mathcal{F}_t)$-Wiener process. Since $(a_1)$ and $(a_2)$ hold, SDE (1.27), together with the initial condition $X(\theta)=x$, admits a unique solution.

We will treat two cases.

(1) ***Linear Gaussian model***
Suppose that $B(\cdot), b_0(\cdot)$ and $\alpha_j(\cdot)$ are deterministic and $A_j(\cdot)=0$ $(j=1,\dots,m)$. Consider the SDE

$$\begin{cases} dX(t)=(B(t)X(t)+b_0(t))\,dt+\alpha(t)\,dW(t), & t\in[0,T], \\ X(\theta)=x. \end{cases} \tag{1.28}$$

In order to seek a formula for $X$, we consider the following matrix-valued differential equation:

$$\begin{cases} \dfrac{dZ}{dt}(t)=B(t)Z(t), & t\in(0,T], \\ Z(0)=I_d. \end{cases} \tag{1.29}$$

The solution is given by $Z(t)=\sum_{n=0}^{\infty}\frac{1}{n!}(\int_0^t B(s)\,ds)^n=\exp(\int_0^t B(s)\,ds)$ and its inverse matrix $Z^{-1}(t)$ is $\exp(-\int_0^t B(s)\,ds)$.

Consequently, for the solution $X$ of (1.28), we have

$$X(t) = Z(t)Z^{-1}(\theta)x + Z(t)\int_\theta^t Z^{-1}(s)b_0(s)\,ds + Z(t)\int_\theta^t Z^{-1}(s)\alpha(s)\,dW(s). \tag{1.30}$$

(2) ***1-dimensional linear SDE***
Let $b \in L^1([0,T]\times\Omega,(\mathcal{F}_t);\mathbb{R}^1)$ and $\alpha \in L^2([0,T]\times\Omega,(\mathcal{F}_t);\mathbb{R}^1\otimes\mathbb{R}^m)$ be given. Consider the linear SDE

$$\begin{cases} dX(t) = X(t)b(t,\omega)\,dt + X(t)\alpha(t,\omega)\,dW(t,\omega), & t \in (0,T], \\ X(0) = x_0\ (>0). \end{cases} \tag{1.31}$$

Although neither $(a_1)$ nor $(a_2)$ hold, Itô's formula yields

$$X(t) = x_0 \exp\Big(\int_0^t \alpha(s,\omega)\,dW(s) + \int_0^t \Big[b(s,\omega) - \frac{1}{2}|\alpha(s,\omega)|^2\Big]ds\Big) \tag{1.32}$$

is a unique solution (refer to [YZ99], Sections 3 and 4 in Chapter 6 for details on (1.27)).

### 1.2.2 Girsanov Transformations

Let $(\Omega,\mathcal{F},(\mathcal{F}_t),P)$ be a filtered probability space. Here we are concerned with transformations of probability distribution. So we write down the probability $P$ together with $(\mathcal{F}_t)$.

Consider an exponential $(\mathcal{F}_t,P)$-martingale

$$q(t) = \exp\Big(\int_0^t \Phi(s)\cdot dW(s) - \frac{1}{2}\int_0^t |\Phi(s)|^2\,ds\Big), \quad t \in [0,T], \tag{1.33}$$

with $\Phi \in L^2([0,T],(\mathcal{F}_t);\mathbb{R}^m)$.

Since $q(t) > 0$ $P$-a.s. and $Eq(T) = 1$, we can define a new probability $\hat{P}$ on $\mathcal{F}_T$ through the Radon–Nykodým derivative $q(T)$, namely

$$\hat{P}(\wedge) = E[q(T)\chi_\wedge], \quad \forall \wedge \in \mathcal{F}_T. \tag{1.34}$$

We denote $\hat{P}$ by $q \circ P$ and call the transformation, $P \mapsto q \circ P$ Girsanov transformation.

From the martingale property of $q(\cdot)$, we deduce the consistency property $\hat{P}(\Lambda) = E[q(t)\chi_\Lambda]$ for $\Lambda \in \mathcal{F}_t$.

The following proposition deals with the conditional expectation.

**Proposition 1.3 (Bayes formula).** *Let $Z$ be $\mathcal{F}_T$-measurable positive random variable with $EZ = 1$. Define a new probability $Q$ by*

$$Q(\Lambda) = E[Z\chi_\Lambda], \quad \forall \Lambda \in \mathcal{F}_T. \tag{1.35}$$

*Let $\zeta$ be $Q$-integrable and $\mathcal{F}_t$-measurable. Then, for $s < t$,*

$$E^Q(\zeta|\mathcal{F}_s) = E(E(Z|\mathcal{F}_t)\zeta|\mathcal{F}_s)/E(Z|\mathcal{F}_s) \quad Q\text{-a.s.}, \tag{1.36}$$

*where $E^Q$ denotes the expectation w.r.t. $Q$.*
*In particular, for $\hat{P} = q \circ P$,*

$$E^{\hat{P}}(\eta|\mathcal{F}_s) = E(\eta q(t)|\mathcal{F}_s)/q(s) \quad \hat{P}\text{-a.s.} \tag{1.37}$$

*for any $\mathcal{F}_t$-measurable random variable $\eta$ with $E^{\hat{P}}|\eta| < \infty$.*

*Proof.* Put $Z_t = E(Z|\mathcal{F}_t) > 0$. For $\Lambda \in \mathcal{F}_s$ $(s < t)$, we have

$$\begin{aligned}
E^Q[\zeta\chi_\Lambda] &= E[Z\zeta\chi_\Lambda] \\
&= E[E(Z|\mathcal{F}_t)\zeta\chi_\Lambda] \\
&= E\Big[Z_s\frac{E(Z_t\zeta|\mathcal{F}_s)}{Z_s}\chi_\Lambda\Big] \\
&= E\Big[E(Z|\mathcal{F}_s)\frac{E(Z_t\zeta|\mathcal{F}_s)}{Z_s}\chi_\Lambda\Big] \\
&= E\Big[Z\frac{E(Z_t\zeta|\mathcal{F}_s)}{Z_s}\chi_\Lambda\Big] \quad \Big(\text{since}\frac{E(Z_t\zeta|\mathcal{F}_s)}{Z_s}\chi_\Lambda \text{ is } \mathcal{F}_s\text{-measurable}\Big) \\
&= E^Q\Big[\frac{E(Z_t\zeta|\mathcal{F}_s)}{Z_s}\chi_\Lambda\Big],
\end{aligned} \tag{1.38}$$

which yields (1.36). □

Now we state a basic theorem.

**Theorem 1.3 (Girsanov's Theorem).** *Let $W$ be an $m$-dimensional $(\mathcal{F}_t, P)$-Wiener process and $\Phi \in L^2([0,T] \times \Omega, (\mathcal{F}_t); \mathbb{R}^m)$. Suppose that $q(t)$ given by (1.33) is an exponential $(\mathcal{F}_t, P)$-martingale. Then $\hat{W}(t) = (\hat{W}^1(t), \ldots, \hat{W}^m(t))$, given by*

$$\hat{W}^i(t) = W^i(t) - \int_0^t \Phi^i(s)\,ds, \quad i = 1, \cdots, m \tag{1.39}$$

*is a $(\mathcal{F}_t, q \circ P)$-Wiener process.*

For the proof, we show that $\hat{W}^i(t)$, $\hat{W}^i(t)^2 - t$, $\hat{W}^i(t)\hat{W}^j(t)$ $(i \neq j), i = 1, \ldots, m$ are all $(\mathcal{F}_t, q \circ P)$-martingales, by using the Itô and Bayes formulas (see [LS01], 6.3 for details).

We will give useful two criterions, under that $q(\cdot)$ becomes a martingale.

**Proposition 1.4 (Special case of Novikov's criterion).** *If*

$$E\exp\Big(\frac{1}{2}\int_0^T|\Phi(s)|^2\,ds\Big)<\infty, \tag{1.40}$$

*then $q(\cdot)$ is a martingale.*

(see [IW81], p. 142, [LS01], p. 229 for details).

**Proposition 1.5.** *Let $X$ be a continuous $(\mathcal{F}_t)$-adapted process. Suppose that $X$ and the $(\mathcal{F}_t, P)$-Wiener process $W$ satisfy*

$$X(t)=X_0+\int_0^t b(s,X(s),\omega)\,ds+\int_0^t \alpha(s,X(s),\omega)\,dW(s), \tag{1.41}$$

*where $E|X_0|^2<\infty$ and $b(\cdot,x,\cdot)$ and $\alpha(\cdot,x,\cdot)$ are $(\mathcal{F}_t)$-progressively measurable processes satisfying condition $(a_2)$. Assume that $\alpha$ is bounded.*

*Let $\psi : [0,T)\times\mathbb{R}^d\times\Omega\mapsto\mathbb{R}^m$ be a measurable map such that, for any $x\in\mathbb{R}^d$, $\psi(\cdot,x,\cdot)$ is $(\mathcal{F}_t)$-progressively measurable. Further assume that $\psi$ satisfies $(a_2)$. Define $M(t)$ by*

$$M(t)=\exp\Big(\int_0^t\psi(s,X(s),\omega)\cdot dW(s)-\frac{1}{2}\int_0^t|\psi(s,X(s),\omega)|^2\,ds\Big),\quad t\in[0,T]. \tag{1.42}$$

*Then, $M(\cdot)$ is an exponential $(\mathcal{F}_t,P)$-martingale.*

*Proof.* Since we can apply the same arguments as in [Be92], p. 77, to the random coefficients case, we will sketch the proof. First we claim that there is a constant $c_0$ such that

$$E[M(t)|X(t)|^2]\le c_0,\quad \forall t\in[0,T]. \tag{1.43}$$

Indeed, for $\varepsilon>0$, we compute $\frac{\zeta(t)}{1+\varepsilon\zeta(t)}$ $(=\frac{1}{\varepsilon}-\frac{1}{\varepsilon(1+\varepsilon\zeta(t))})$, where $\zeta(t)=M(t)|X(t)|^2$ $(\ge 0)$.

Since Itô's formula yields

$$\begin{aligned} d\Big(\frac{\zeta(t)}{1+\varepsilon\zeta(t)}\Big) &= \frac{1}{(1+\varepsilon\zeta(t))^2}M(t)\{|X(t)|^2\psi+2X(t)\alpha\}\,dW(t)\\ &\quad+\frac{1}{(1+\varepsilon\zeta(t))^2}M(t)\{2X(t)(b+\alpha\psi)+\mathrm{tr}(\alpha\alpha^\top)\}\,dt\\ &\quad-\frac{\varepsilon}{(1+\varepsilon\zeta(t))^3}\,d\langle\zeta\rangle(t) \end{aligned} \tag{1.44}$$

and

$$\text{2nd term in the RHS } \leq c_1\Big(\frac{\zeta(t)}{1+\varepsilon\zeta(t)} + M(t)\Big) \tag{1.45}$$

with a constant $c_1$ independent of $\varepsilon$ and $M(t)$ a supermartingale, we have

$$E\Big(\frac{\zeta(t\wedge\tau_n)}{1+\varepsilon\zeta(t\wedge\tau_n)}\Big) \leq E|X_0|^2 + c_1\int_0^t E\frac{\zeta(s\wedge\tau_n)}{1+\varepsilon\zeta(s\wedge\tau_n)}\,ds + c_1 T \tag{1.46}$$

where $\tau_n$ is the exit time of $M(\cdot)$ from $[0,n]$.

Thus, Gronwall's inequality and the bounded convergence theorem show that there is a constant $c_0$ independent of $\varepsilon$, such that

$$E\Big(\frac{\zeta(t)}{1+\varepsilon\zeta(t)}\Big) \leq c_0, \quad \forall t. \tag{1.47}$$

Now letting $\varepsilon \downarrow 0$ leads to (1.43), by the monotone convergence theorem.

Next, using the same arguments, we evaluate $E\left(\frac{M(T)}{1+\varepsilon M(T)}\right)$. By Itô's formula,

$$\frac{M(t)}{1+\varepsilon M(t)} = \frac{1}{1+\varepsilon} + \int_0^t \frac{dM(s)}{(1+\varepsilon M(s))^2} - \varepsilon\int_0^t \frac{d\langle M\rangle(s)}{(1+\varepsilon M(s))^3}.$$

Now take the expectation of both sites and apply (1.43) to the 3rd term of RHS. Letting here $\varepsilon \to 0$ yields $EM(t) = 1$, from which the proposition follows thanks to Example 1.2. □

*Example 1.4 (Transformation of the drift term).* Let $W$ be a $d$-dimensional $(\mathcal{F}_t, P)$-Wiener process and $b : [0,T]\times\mathbb{R}^d \mapsto \mathbb{R}^d$ a Borel function satisfying

$$|b(t,x)| \leq K(1+|x|), \quad \forall t, x. \tag{1.48}$$

We consider the SDE;

$$dX(t) = b(t, X(t))\,dt + dW(t), \quad t \in (\theta, T], \tag{1.49}$$

with the initial condition

$$X(\theta) = X_\theta (\in L^2(\Omega, \mathcal{F}_\theta, P)).$$

We hardly expect a solution adapted to $(W, X_\theta)$. However, we can construct a Wiener process $\hat{W}$ and a continuous $(\mathcal{F}_t)$-adapted process $\xi$ on a probability space $(\Omega, \mathcal{F}, (\mathcal{F}_t), \hat{P})$, so that (1.49) holds for $(\xi, \hat{W})$.

Indeed, let us consider the auxiliary SDE

$$\begin{cases} d\xi(t) = dW(t), \ t \in (\theta, T), \\ \xi(\theta) = X_\theta. \end{cases} \tag{1.50}$$

Here, $\xi(t) = X_\theta + (W(t) - W(\theta))$.

Define $\Phi \in L^2([0, T] \times \Omega, (\mathcal{F}_t); \mathbb{R}^d)$ and $q(\cdot)$ by

$$\Phi(s) = \begin{cases} b(s, \xi(s)), \ \theta \le s, \\ 0, \qquad\qquad s < \theta \end{cases} \tag{1.51}$$

and

$$\begin{aligned} q(t) &= \exp\Big(\int_0^t \Phi(s) \cdot dW(s) - \frac{1}{2}\int_0^t |\Phi(s)|^2\, ds\Big) \\ &= \begin{cases} \exp\Big(\int_\theta^t b(s, \xi(s)) \cdot dW(s) - \frac{1}{2}\int_\theta^t |b(s, \xi(s))|^2 ds\Big), \ \theta \le t, \\ 1, \qquad\qquad\qquad\qquad\qquad\qquad\qquad\qquad t < \theta, \end{cases} \end{aligned} \tag{1.52}$$

respectively. Then $q(\cdot)$ is a $(\mathcal{F}_t, P)$-martingale by Proposition 1.5. Thus Girsanov's theorem shows that

$$\hat{W}(t) = W(t) - \int_0^t \Phi(s)\, ds \tag{1.53}$$

is a $(\mathcal{F}_t, \hat{P})$-Wiener process, where $\hat{P} = q \circ P$. By (1.50) and (1.53), $(\xi, \hat{W})$ is a weak solution of (1.49) (see Definition 1.12).

From (1.49) it follows that

$$\sigma(\tilde{W}(s) - \tilde{W}(\theta); s \in [\theta, T]) \subset \sigma(\xi(s); s \in [\theta, T]). \tag{1.54}$$

### *1.2.3 SDEs with Deterministic Borel Coefficients*

In stochastic control problems, we encounter SDEs with Borel measurable coefficients. Here we will overview such equations.

Let $b : [0, \infty) \times \mathbb{R}^d \mapsto \mathbb{R}^d$, and $\alpha : [0, \infty) \times \mathbb{R}^d \mapsto \mathbb{R}^d \otimes \mathbb{R}^m$ be Borel measurable. Let us consider the following SDE, called of Markovian type:

$$dX(t) = b(t, X(t))\, dt + \alpha(t, X(t))\, dW(t), \quad t > \theta, \tag{1.55}$$

with the initial condition

$$X(\theta) = X_\theta \ (\in L^2(\Omega, \mathcal{F}_\theta; \mathbb{R}^d)). \tag{1.56}$$

Firstly we assume $(a_1)$ and $(a_2)$, i.e., we have an SDE with Lipschitz continuous coefficients. By Theorem 1.2, the unique solution $X_{\theta, X_\theta}(t)$ depends only on $(\theta, X_\theta)$ and $(W(s) - W(\theta), s \in [\theta, t])$ (this solution is called a strong solution). Put

$$p(\theta, x; t, A) = P(X_{\theta x}(t) \in A), \ A \in \mathcal{B}(\mathbb{R}^d), \quad \theta < t. \tag{1.57}$$

Since $X_{\theta x}(t) = X_{s, X_{\theta,x}(s)}(t), \ \theta \le s \le t, \ p(\theta, x; t, A)$ satisfies the Chapman–Kolmogorov equation;

$$\begin{aligned} p(\theta, x; t, A) &= E_{\theta x}[\chi_A(X(t))] \\ &= E_{\theta x}[E(\chi_A(X_{s, X(s)}(t))|\mathcal{F}_s)] \\ &= E_{\theta x}[p(s, X(s); t, A)] \\ &= \int_{\mathbb{R}^d} p(\theta, x; s, dy) p(s, y; t, A). \end{aligned} \tag{1.58}$$

Namely, the solution is a Markov process with the transition probability function $p(\cdot)$.

For the transition operator $\mathrm{T}_{\theta t}\phi(x) = E_{\theta x}\phi(X(t))$, Itô's formula gives the expression for the backward evolution operator $\mathcal{A}$. For $\Phi \in C^{12}([0, \infty) \times \mathbb{R}^d)$,

$$\begin{aligned} \mathcal{A}\Phi(\theta, x) &= \lim_{s \to 0} (E_{\theta x}\Phi(\theta + s, X(\theta + s)) - \Phi(\theta, x))/s \\ &= \partial_\theta \Phi(\theta, x) + b(\theta, x) \cdot \partial_x \Phi(\theta, x) \\ &\quad + \frac{1}{2} \mathrm{tr}(a(\theta, x) \partial_{xx} \Phi(\theta, x)), \end{aligned} \tag{1.59}$$

where $a = \alpha\alpha^\top$.

When we drop the continuity condition $(a_1)$ and $(a_2)$, we can hardly expect the existence of strong solutions. This leads us to the definition of the notion of weak solution.

**Definition 1.12 (Weak solution).** Suppose that $\tilde{X}$ and $\tilde{W}$ are given on a filtered probability space $(\tilde{\Omega}, \tilde{\mathcal{F}}, (\tilde{\mathcal{F}}_t), \tilde{P})$. $(\tilde{X}, \tilde{W})$ is called a weak solution of (1.55)–(1.56), if

(a) $\tilde{X}(\theta)$ is a $d$-dimensional random variable, having the same probability distribution as $X_\theta$,
(b) $\tilde{W}$ is an $m$-dimensional $(\tilde{\mathcal{F}}_t, \tilde{P})$-Wiener process,
(c) $\tilde{X}$ is a $d$-dimensional continuous $(\tilde{\mathcal{F}}_t)$-adapted process,
(d) $\tilde{X}(t) = \tilde{X}(\theta) + \int_\theta^t b(s, \tilde{X}(s))\, ds + \int_\theta^t \alpha(s, \tilde{X}(s))\, d\tilde{W}(s), \ \theta \le t, \ \tilde{P}$-a.s.

Sometimes $\tilde{X}$ itself or the probability distribution of $\tilde{X}$ are called the weak solution of (1.55)–(1.56).

By uniqueness of the weak solution we mean uniqueness in law of $\tilde{X}$, namely, if $(\tilde{X}, \tilde{W})$ and $(X^*, W^*)$ are weak solutions, then the probability distributions of $\tilde{X}$ and $X^*$ coincide.

We have already considered a weak solution in Example 1.4, as an application of the Girsanov transformation.

Let us state an existence and uniqueness results for weak solutions.

**Proposition 1.6 ([Kr09], 2–6, Theorem 1).** *Suppose that $d = m$ and the coefficients $\alpha(\cdot)$ and $b(\cdot)$ are bounded Borel functions. Further assume that $\alpha(\cdot)$ is symmetric and uniformly parabolic, i.e., there exists $\lambda_0 > 0$, such that*

$$y^\top \alpha(t,x) y \geq \lambda_0 |y|^2, \quad \forall y \in \mathbb{R}^d, \quad \forall t, x. \tag{1.60}$$

*Then, the* SDE (1.55)–(1.56) *admits a weak solution provided $E|X_\theta|^4 < \infty$.*

**Proposition 1.7 ([SV79], Theorem 7.2.1).** *Let $b : [0,\infty) \times \mathbb{R}^d \mapsto \mathbb{R}^d$ and $\alpha : [0,\infty) \times \mathbb{R}^d \mapsto \mathbb{R}^d \otimes \mathbb{R}^m$ be bounded and Borel measurable. Suppose that $\alpha(t,x)$ is uniformly continuous w.r.t. $x$, uniformly in $t$ and $\alpha\alpha^\top$ is uniformly parabolic. Then the* SDE (1.55)–(1.56) *admits the unique weak solution.*

### *1.2.4 Backward Stochastic Differential Equations*

In this subsection, we will overview the terminal value problems for SDEs, called backward SDE, introduced by Pardoux and Peng [PP90].

Let $W$ be a $d$-dimensional Wiener process defined on $(\Omega, \mathcal{F}, (\mathcal{F}_t^W), P)$. Let $h : [0,T] \times \mathbb{R}^1 \times \mathbb{R}^d \times \Omega \mapsto \mathbb{R}^1$ be measurable and satisfy the conditions
$(a_1)^*$ $h(\cdot, y, z, \cdot) \in L^2([0,T] \times \Omega, (\mathcal{F}_t^W); \mathbb{R}^1)$, $\forall y, z$
and
$(a_2)^*$ $\exists l > 0$ such that $\forall t, \omega$,

$$|h(t, y_1, z_1, \omega) - h(t, y_2, z_2, \omega)| \leq l(|y_1 - y_2| + |z_1 - z_2|)$$
$$\forall y_1, y_2 \in \mathbb{R}^1, \quad z_1, z_2 \in \mathbb{R}^d. \tag{1.61}$$

Consider the real SDE

$$dY(t) = h(t, Y(t), Z(t), \omega)\, dt + Z(t)\, dW(t), \quad t \in [0,T) \tag{1.62}$$

with the lateral boundary condition

$$Y(T) = \xi \; (\in L^2(\Omega, \mathcal{F}_T^W; \mathbb{R}^1)). \tag{1.63}$$

The SDE (1.62)–(1.63) is called a backward SDE on $[0,T]$.

Let $Y$ and $Z$ be $(\mathcal{F}_t^W)$-progressively measurable real and $d$-dimensional processes respectively.

**Definition 1.13.** $(Y, Z)$ is called a solution of (1.62)–(1.63), if

(a) $Y$ is continuous,
(b) $E[\sup_{0\le t\le T} |Y(t)|^2 + \int_0^T |Z(t)|^2\, dt] < \infty$
(c) $Y(t) = \xi - \int_t^T h(s, Y(s), Z(s), \omega)\, ds - \int_t^T Z(s)\, dW(s), \ \forall t \in [0, T], \ P$-a.s.

We identify two solution $(Y_1, Z_1)$ and $(Y_2, Z_2)$ whenever

$$\begin{aligned} &Y_1(t) = Y_2(t), \quad \forall t \in [0, T], \\ &Z_1(t) = Z_2(t) \text{ for almost all } t \in [0, T], \quad P\text{-a.s.} \end{aligned} \tag{1.64}$$

Now we have;

**Theorem 1.4.** *Under the conditions* $(a_1)^*$ *and* $(a_2)^*$*, problem* (1.62)–(1.63) *admits a unique solution.*

(see [ElKPQ97] and [YZ99], Chapter 7 for details).

*Example 1.5 (Backward SDE associated with a quasi-linear parabolic equation).* Let $F : [0, T] \times \mathbb{R}^d \times \mathbb{R}^1 \times \mathbb{R}^d \mapsto \mathbb{R}^1$ be a continuous function satisfying

$$|F(t, x, y_1, z_1) - F(t, x, y_2, z_2)| \le l(|y_1 - y_2| + |z_1 - z_2|), \quad \forall t, x, \tag{1.65}$$

and

$$|F(t, x, 0, 0)| \le K(1 + |x|^p), \quad \forall t, \tag{1.66}$$

with positive constants $l$, $K$ and $p$.

Let $\xi$ be the diffusion process given by the SDE;

$$\begin{cases} d\xi(t) = b(t, \xi(t))\, dt + \alpha(t, \xi(t))\, dW(t), \quad t \in (0, T), \\ \xi(0) = c \ (\in \mathbb{R}^d) \end{cases} \tag{1.67}$$

where coefficients $b$ and $\alpha$ are deterministic and satisfy $(a_1)$ and $(a_2)$ with $d = m$.

Now consider a backward SDE

$$\begin{cases} dY(t) = F(t, \xi(t), Y(t), Z(t))\, dt + Z(t)\, dW(t), \quad t \in [0, T), \\ Y(T) = g(\xi(T)), \end{cases} \tag{1.68}$$

with a continuous function $g$ satisfying a linear growth condition.

The following quasi-linear parabolic equation is related to (1.68):

$$\partial_t v(t, x) + \mathrm{G}_t v(t, x) - F(t, x, v(t, x), \alpha(t, x)^\top \partial_x v(t, x)^\top) = 0 \ \text{ on } [0, T) \times \mathbb{R}^d, \tag{1.69}$$

with the lateral boundary condition

$$v(T,x) = g(x), \quad x \in \mathbb{R}^d, \tag{1.70}$$

where $\mathrm{G}_t$ is the generator of transition semigroup of $\xi$.

We have

**Proposition 1.8.** *If the Cauchy problem* (1.69)–(1.70) *admits a classical solution* $v(\cdot)$*, then*

$$\begin{cases} Y(t) = v(t,\xi(t)), \\ Z(t) = \alpha(t,\xi(t))\partial_x v(t,\xi(t))^\top, \end{cases} \tag{1.71}$$

*whenever* $E[\sup_t |Y(t)|^2 + \int_0^T |Z(t)|^2 dt] < \infty$.

Indeed, applying Itô's formula to $v(t,\xi(t))$ and using (1.69), we obtain (1.71). □

## 1.3 Asset Pricing Problems

Let us consider a contract to sell a contingent claim with payoff $G$ and maturity $T$, at time 0. The problem is to price this contingent claims. Suppose that the agent invests his own money in a bond and stocks whose prices evolve according to SDEs and does not want to run any risk of losing money. The lowest price satisfying this condition is called the selling price (refer [ElKQ95, ElKPQ97]).

In Sect. 1.3.1, we will formulate the problem and, by using the dynamics of bond and stocks, we study the selling price and portfolio strategies in Sects. 1.3.2 and 1.3.3.

### *1.3.1 Formulation*

Let us consider a financial market consisting of one bond (riskless asset) and $d$ stocks (risky assets). We suppose that the price of the bond is given by

$$S^0(t) = e^{rt}, \quad t \geq 0, \tag{1.72}$$

where $r$ is a positive constant, and the price of $i$-th stock evolves according to the SDE

$$\begin{cases} dS^i(t) = S^i(t)(\mu^i(t, S(t))\,dt + \sum_{j=1}^{d} \sigma^i_j(t, S(t))\,dW^j(t)), \quad t > 0, \\ S^i(0) = s^i > 0 \quad (i = 1, \ldots, d) \end{cases} \tag{1.73}$$

where $W(t) = (W^1(t), \ldots, W^d(t))$ is a $d$-dimensional Wiener process, defined on $(\Omega, \mathcal{F}, (\mathcal{F}^W_t), P)$, $\sigma(\cdot) = (\sigma^i_j(\cdot))_{i,j=1,\ldots,d}$ and $\mu(\cdot) = (\mu^1(\cdot), \ldots, \mu^d(\cdot))$ are called the volatility and the mean return rate, respectively.

Let $X(t)$ denote the agent's wealth at time $t$. The agent invests the number of shares $\pi^i(t)$ in the $i$-th stock at $t$ and $\pi^0(t) := \frac{X(t)-\sum_{i=1}^d \pi^i(t)S^i(t)}{S^0(t)}$ in the bond. Now we admit $\pi^i(t) \in \mathbb{R}^1 (i = 1, \ldots, d)$ where $\pi^i(t) < 0$ means selling $i$-th stock and $\pi^0(t) < 0$ means borrowing money. Hence $X$ evolves according to the SDE

$$dX(t) = (X(t) - \pi(t) \cdot S(t))r dt + \pi(t)\, dS(t), \quad t > 0, \tag{1.74}$$

with the initial condition

$$X(0) = x \; (> 0). \tag{1.75}$$

Let us consider the contract between the agent and a quest at time 0, such that the agent pays $G$ $(\in L^2(\Omega, \mathcal{F}^W_t; [0, \infty)))$ to the quest at the maturity $T$. The problem is what is the price of $G$ at 0. $\pi(t); [0, T] \times \Omega \mapsto \mathbb{R}^m$ is called a portfolio strategy, when $\pi(\cdot) \cdot S(\cdot) \in L^2([0, T] \times \Omega, (\mathcal{F}^W_t); \mathbb{R}^1)$. By $X^{\pi(\cdot)}_x$, we denote the solution of (1.74)–(1.75). Since the agent does not want to run any risk of losing money, he chooses a portfolio strategy $\pi(\cdot)$ such that

$$X^{\pi(\cdot)}_x(T) \geq G, \quad P\text{-a.s.} \tag{1.76}$$

Thus the problem is to seek $x^G$, defined by

$$x^G = \inf\{x > 0, \exists \pi(\cdot) \text{ such that (1.76) holds}\}. \tag{1.77}$$

$x^G$ is called the selling price of $G$ at $t = 0$.

### *1.3.2 Backward SDE for the Selling Price*

We are going to study the selling price by using a backward SDE. We assume that the following conditions are satisfied:

(a) $\sigma : [0, T] \times \mathbb{R}^d \mapsto \mathbb{R}^d \otimes \mathbb{R}^d$, $\mu : [0, T] \times \mathbb{R}^d \mapsto \mathbb{R}^d$ are bounded and uniformly Lipschitz continuous w.r.t. $x$,
(b) $\sigma(t, x)$ is invertible and $\sigma(t, x)^{-1}$ is bounded.

Then (1.73) has the unique strong solutions $S(t) = (S^1(t), \dots, S^d(t))$ and

$$S^i(t) = s^i \exp\Big( \int_0^t \sigma^i(\theta, S(\theta))\, dW(\theta) + \int_0^t \Big( \mu^i(\theta, S(\theta)) - \frac{1}{2} |\sigma^i(\theta, S(\theta))|^2 \Big)\, ds \Big) > 0. \tag{1.78}$$

Further (a) and (1.78) yield

$$E\Big[ \sup_{0 \le s \le T} |S(s)|^{2p} \Big] < \infty, \quad \text{for any } p \ge 1,$$

and (b) yields

$$\mathcal{F}_t^W = \mathcal{F}_t^S, \quad \forall t \in [0, T]. \tag{1.79}$$

Assume that $G$ is given by

$$G = g(S(T)), \tag{1.80}$$

with a linearly growing continuous function $g$ on $\mathbb{R}_+^d$.

When $\pi(\cdot)$ is applied, the wealth evolves according to (1.74), namely

$$dX(t) = (rX(t) + \zeta(t) \cdot (\mu(t, S(t)) - r1^d))\, dt + \zeta(t)^\top \sigma(t, S(t))\, dW(t) \tag{1.81}$$

with a given initial value, where $\zeta^i(t) = \pi^i(t) S^i(t)$, $i = 1, \dots, d$ and $1^d = (1, \dots, 1) \in \mathbb{R}^d$.

Since $X_x^{\pi(\cdot)}(T)$ is strictly increasing in $x$, we will seek an initial value $\hat{x}$ such that $X_{\hat{x}}^{\hat{\pi}(\cdot)}(T) = g(S(T))$ with some $\hat{\pi}(\cdot)$. The SDE (1.81) leads to the following backward SDE

$$dX(t) = (rX(t) + \zeta(t) \cdot (\mu(t, S(t)) - r1^d))dt + \zeta(t)^\top \sigma(t, S(t))\, dW(t), \quad t \in [0, T) \tag{1.82}$$

with the lateral boundary condition

$$X(T) = g(S(T)). \tag{1.83}$$

Noticing that the backward SDE

$$
\begin{cases} dY(t) = (rY(t) + Z(t)^\top \sigma(t, S(t))^{-1}(\mu(t, S(t)) - r1^d))dt \\ \qquad\qquad + Z(t)^\top dW(t), \quad t \in [0, T), \\ Y(T) = g(S(T)) \end{cases} \tag{1.84}
$$

has the unique solution $(Y, Z)$ by Theorem 1.4, and comparing (1.82) with (1.84), we conclude that $x^G = Y(0)$ and

$$
\hat{\pi}^i(t) = \sum_{j=1}^d (\sigma(t, S(t))^{-1})_i^j Z^j(t)/S^i(t), \quad i = 1, \ldots, d, \tag{1.85}
$$

provides a claimed portfolio strategy.

We note that $Y(t)$ gives the selling price at time $t$.

### 1.3.3 *Parabolic Equation Associated with* (1.84)

In order to get an explicit formula for $x^G$ and $\hat{\pi}(\cdot)$, we study the Cauchy problem for a parabolic equation related to (1.84).

Using Proposition 1.8 and $S^i(t) > 0, i = 1, \ldots, d$, we have

$$
\begin{cases} \partial_t v(t, x) + \dfrac{1}{2} \displaystyle\sum_{i,j=1}^d a^{ij}(t, x) x^i x^j \partial_{ij} v + r \sum_{i=1}^d x^i \partial_i v - rv = 0, \\ \qquad\qquad\qquad\qquad (t, x) \in [0, T) \times \mathbb{R}_+^d, \\ v(T, x) = g(x), \quad x \in \mathbb{R}_+^d \end{cases} \tag{1.86}
$$

where $a = \sigma\sigma^\top$.

If (1.86) admits a classical solution, then we have

$$
\begin{cases} v(0, S(0)) = x^G, \\ \partial_i v(t, S(t)) = \hat{\pi}^i(t), \quad i = 1, \ldots, d. \end{cases} \tag{1.87}
$$

Since $V(t) := v(t, S(t))$ satisfies (1.82)–(1.83), $V(t)$ provides the selling price at time $t$. Accordingly $V(\cdot)$ is called the selling price process. Applying Itô's formula to $v(\cdot)$, we have

$$
v(t, x) = E_{tx}[e^{r(T-t)} g(\xi(T))],
$$

where $\xi(t) = (\xi^1(t), \ldots, \xi^d(t))$ is the diffusion process given by

$$
d\xi^i(t) = \xi^i(t)(rdt + \sigma^i(t, \xi(t))\, dW(t)). \tag{1.88}
$$

*Example 1.6 (Constant coefficient markets).* Let $\mu(\cdot)$ and $\sigma(\cdot)$ be constants. Then (1.86) reduces to

$$\begin{cases} \partial_t v + \dfrac{1}{2}\displaystyle\sum_{ij=1}^{d} a^{ij} x^j x^j \partial_{ij} v + r \sum_{i=1}^{d} x^i \partial_i v - rv = 0, & \text{on } [0,T) \times \mathbb{R}_+^d, \\ v(T,x) = g(x), \quad x \in \mathbb{R}_+^d. \end{cases} \tag{1.89}$$

Referring to (1.88), we have

$$\xi^i(t) = \xi^i(\theta) \exp\left( \sum_{j=1}^{d} \sigma_j^i (W^i(t) - W^j(\theta)) + \left( r - \frac{1}{2}|\sigma^i|^2 \right)(t-\theta) \right) \tag{1.90}$$

and

$$v(t,x) = e^{-r(T-t)} \int_{\mathbb{R}^d} g(\zeta(T-t,x,\sigma z)) p_d(T-t,z)\, dz, \tag{1.91}$$

where $p_d(s,z) = (2\pi s)^{-\frac{d}{2}} \exp\left(-\frac{|z|^2}{2s}\right)$ and $\zeta = (\zeta^1,\dots,\zeta^d)$ with $\zeta^i(s,x,y) = x^i \exp(y^i + (r - \frac{1}{2}|\sigma^i|^2)s)$, $i = 1,\dots,d$. In particular, for $d = 1$,

$$v(t,x) = \sigma^{-r(T-t)} \int_{-\infty}^{\infty} g\left( x \exp\left( \sigma\sqrt{T-t}z + \left( r - \frac{\sigma^2}{2} \right)(T-t) \right) \right) p_1(1,z)\, dz \tag{1.92}$$

holds. Suppose

$$g(x) = (x-K) \vee 0, \tag{1.93}$$

with a positive constant $K$.

Then we can calculate (1.92) explicitly and obtain

$$\begin{aligned} v(t,x) &= \int_{-\rho_-}^{\infty} \left( x \exp\left( \sigma\sqrt{T-t}z - \frac{1}{2}\sigma^2(T-t) \right) - Ke^{-r(T-t)} \right) p_1(1,z)\, dz \\ &= x \int_{-\infty}^{\rho_+} p_1(1,z)\, dz - Ke^{-r(T-t)} \int_{-\infty}^{\rho_-} p_1(1,z)\, dz, \end{aligned} \tag{1.94}$$

where

$$\rho_\pm = \left\{ \log\left( \frac{x}{K} \right) + \left( r \pm \frac{1}{2}\sigma^2 \right)(T-t) \right\} / \sigma\sqrt{T-t}.$$

Equation (1.94) is the famous Black–Scholes formula [BS73].

# Chapter 2
# Optimal Control for Diffusion Processes

**Abstract** This chapter deals with completely observable stochastic control problems for diffusion processes, described by SDEs. The decision maker chooses an optimal decision among all possible ones to achieve the goal. Namely, for a control process, its response evolves according to a (controlled) SDE and the payoff on a finite time interval is given. The controller wants to minimize (or maximize) the payoff by choosing an appropriate control process from among all possible ones. Here we consider three types of control processes:

1. $(\mathcal{F}_t)$-progressively measurable processes.
2. Brownian-adapted processes.
3. Feedback controls.

In order to analyze the problems, we mainly use the dynamic programming principle (DPP) for the value function.

The reminder of this chapter is organized as follows. Section 2.1 presents the formulation of control problems and basic properties of value functions, as preliminaries for later sections. Section 2.2 focuses on DPP. Although DPP is known as a two stage optimization method, we will formulate DPP by using a semigroup and characterize the value function via the semigroup. In Sect. 2.3, we deal with verification theorems, which give recipes for finding optimal Markovian policies. Section 2.4 considers a class of Merton-type optimal investment models, as an application of previous results.

## 2.1 Introduction

This section is devoted to formulating the time horizon stochastic control and analyze basic notions. We introduce control processes, payoffs and value functions in Sect. 2.1.1, and investigate their properties in Sect. 2.1.2.

Before we formulate the stochastic control problem, we give a typical example, called the linear quadratic control.

*Example 2.1 (Linear quadratic* (LQ) *control).* Consider a $d$-dimensional stochastic system with an external random force $\gamma(\cdot)$,

$$dX(t) = (b(t)X(t) + \gamma(t))\,dt + dW(t),$$

© Springer Japan 2015

M. Nisio, *Stochastic Control Theory*, Probability Theory and Stochastic Modelling 72, DOI 10.1007/978-4-431-55123-2_2

where $W$ is a $d$-dimensional $(\mathcal{F}_t)$-Wiener process, $b : [0,T] \mapsto \mathbb{R}^d \otimes \mathbb{R}^d$ and $\gamma : [0,T] \times \Omega \mapsto \mathbb{R}^d$ are $(\mathcal{F}_t)$-progressively measurable.

Let $S_+^d$ denote the set of $d \times d$ non-negative definite matrices. Suppose that $M, N : [0,T] \mapsto S_+^d$ are given. The problem is to choose $\gamma(\cdot)$ so that the payoff:

$$J(\theta, x; \gamma(\cdot)) = E_{\theta x}\Big[\int_\theta^T (X(t)^\top M(t)X(t) + \gamma(t)^\top N(t)\gamma(t))\, dt + |X(T)|^2\Big]$$

is minimized and calculate the value function;

$$v(\theta, x) = \inf_{\gamma(\cdot)} J(\theta, x; \gamma(\cdot)).$$

In particular, when $d = 1, M = N = b = 0$ and $|\gamma(\cdot)| \le 1$ are assumed, the choice $\hat{\gamma}(t) := -\mathrm{sgn} X(t)$ is optimal (see Examples 2.4 and 2.5).

### 2.1.1 Formulations

We are going to formulate the finite time horizon stochastic control problem. Let $(\Omega, \mathcal{F}, (\mathcal{F}_t), P)$ be a filtered probability space and $W$ an $m$-dimensional $(\mathcal{F}_t)$-Wiener process. We call $(\Omega, \mathcal{F}, (\mathcal{F}_t), P, W)$ a reference probability system. Let $T > 0$ be given. A $\sigma$-compact and convex subset $\Gamma$ of $\mathbb{R}^q$ is called a control region, where $q$ is a positive integer.

**Definition 2.1.** The $\Gamma$-valued $(\mathcal{F}_t)$-progressively measurable process $(\gamma(t), t \in [0,T])$ is called a control process, if $\gamma(\cdot) \in L^\infty([0,T] \times \Omega, (\mathcal{F}_t); \Gamma)$, namely, there is a compact set $\Gamma_{\gamma(\cdot)} (\subset \Gamma)$, such that $\gamma(t,\omega) \in \Gamma_{\gamma(\cdot)}$, for almost all $(t,\omega)$.

The 6-tuple $A = (\Omega, \mathcal{F}, (\mathcal{F}_t), P, W, \gamma(\cdot))$ is called an admissible control. We denote by $\mathbb{A}$ the set of all admissible controls.

Let $b : [0,T] \times \mathbb{R}^d \times \Gamma \mapsto \mathbb{R}^d$ and $\alpha : [0,T] \times \mathbb{R}^d \times \Gamma \mapsto \mathbb{R}^d \otimes \mathbb{R}^m$ be given and satisfy the following condition:

$(b_1)$

$$\begin{aligned} &|b(t_1, x_1, \gamma_1) - b(t_2, x_2, \gamma_2)| + |\alpha(t_1, x_1, \gamma_1) - \alpha(t_2, x_2, \gamma_2)| \\ &\le l|x_1 - x_2| + \bar{m}(|t_1 - t_2| + |\gamma_1 - \gamma_2|) \end{aligned} \tag{2.1}$$

and

$$|b(t, 0, \gamma)| + |\alpha(t, 0, \gamma)| \le K, \quad \forall (t, \gamma) \in [0,T] \times \Gamma, \tag{2.2}$$

where $l$ and $K$ are positive constants and $\bar{m}(\cdot)$ is a bounded modulus function, say $\bar{m}(\cdot) \le \bar{M}$.

When $A(\in \mathbb{A})$ is applied, the stochastic system evolves according to the SDE (more precisely, controlled SDE)

$$dX(t) = b(t, X(t), \gamma(t))\, dt + \alpha(t, X(t), \gamma(t))\, dW(t), \quad (0 \le)\theta < t \le T, \tag{2.3}$$

with the initial condition

$$X(\theta) = x(\in \mathbb{R}^d). \tag{2.4}$$

By Theorem 1.2, there exists a unique solution of SDE (2.3)–(2.4), denoted by $X^A_{\theta x}$. This solution is sometimes called the response for $A$. $X^A_{\theta x}(t)$ is clearly $\sigma(\gamma(s), W(s') - W(\theta), s, s' \in [\theta, t])$-measurable.

We omit the indices $\theta$, $x$ and $A$, when no confusion occurs.

Let us introduce three functions $f, \phi$ and $\kappa$ with the conditions $(b_2)$–$(b_5)$.

$(b_2)$ $f : [0, T] \times \mathbb{R}^d \times \Gamma \mapsto \mathbb{R}^1$ is continuous and $f(t, x, \gamma)$ is continuous w.r.t. $x$, uniformly in $(t, \gamma)$,

$(b_3)$ $\phi : \mathbb{R}^d \mapsto \mathbb{R}^1$ is continuous,

$(b_4)$ $|f(t, x, \gamma)| + |\phi(x)| \le \hat{k}(1 + |x|^2), \forall t, x, \gamma$ with a constant $\hat{k}$,

$(b_5)$ $\kappa : [0, T] \times \mathbb{R}^d \times \Gamma \mapsto [0, c_0]$, satisfies $(b_2)$.

$f$ and $\phi$ are called the running cost and the terminal cost, respectively, and $\kappa$ is the discount rate.

For the response $X = X^A_{\theta x}$, the cost function on time interval $[s, t]$ $(\subset [\theta, T])$ is given by

$$\begin{aligned} &C(t, s, A; \phi) \\ &= \int_s^t \exp\{-\int_s^\lambda \kappa(h, X(h), \gamma(h))\, dh\} f(\lambda, X(\lambda), \gamma(\lambda))\, d\lambda \\ &\quad + \exp\{-\int_s^t \kappa(h, X(h), \gamma(h))\, dh\}\phi(X(t)). \end{aligned} \tag{2.5}$$

When the response $X$ stops at time $t$, we define the payoff (or criterion) by

$$J(t, \theta, x, A; \phi) = E_{\theta x} C(t, \theta, A; \phi), \tag{2.6}$$

where $(\theta, x)$ refer to the initial condition of $X$. Clearly $J(t, \theta, x, A; \phi)$ depends on the joint probability of $(\gamma(s), W(s') - W(\theta); s, s' \in [\theta, t])$, but not on $A$ itself.

We want to minimize (or maximize) the payoff, by choosing an appropriate admissible control.

**Definition 2.2.** $v(\cdot)$, defined by

$$v(t, \theta, x; \phi) = \inf_{A \in \mathbb{A}} J(t, \theta, x, A; \phi) \quad (\text{or } \sup_{A \in \mathbb{A}} J(t, \theta, x, A; \phi)) \tag{2.7}$$

is called the value function. If $A \in \mathbb{A}$ gives the infimum (or supremum) of the RHS, $A$ is called an optimal control.

Thus we are concerned with the characterization of the value function and an optimal control.

Let us introduce different classes of admissible controls: Brownian adapted controls and feedback controls.

1. Let $A = (\Omega, \mathcal{F}, (\mathcal{F}_t), P, W, \gamma(\cdot))$ be an admissible control. $A$ is called Brownian adapted, if $\gamma(\cdot)$ is $(\mathcal{F}_t^W)$-progressively measurable. $\mathbb{A}^W$ denotes the set of all Brownian adapted controls. For a fixed reference probability system $(\Omega, \mathcal{F}, (\mathcal{F}_t), P, W)$, $\boldsymbol{\Gamma}^W$ denotes the set of all $(\mathcal{F}_t^W)$-progressively measurable control processes. Since the payoff is calculated in terms of the joint probability distribution of $(W(\cdot), \gamma(\cdot))$, for any given a reference probability system $(\Omega, \mathcal{F}, (\mathcal{F}_t), P, W)$, it holds that

$$\inf_{A \in \mathbb{A}^W} J(t, \theta, x, A; \phi) = \inf_{\gamma(\cdot) \in \boldsymbol{\Gamma}^W} J(t, \theta, x, A; \phi)$$

2. We control a system by using the data on the system, in the customary manners. A $\Gamma$-valued Borel function $\hat{\gamma}(\cdot)$, defined on $[0, T] \times \mathbb{R}^d$, is called a Markovian policy, if it is bounded. When we apply $\hat{\gamma}(\cdot)$, the system evolves according to the SDE

$$dX(t) = b(t, X(t), \hat{\gamma}(t, X(t)))\, dt + \alpha(t, X(t), \hat{\gamma}(t, X(t)))\, dW(t). \tag{2.8}$$

We can hardly expect that a strong solution will exist for (2.8), but a weak solution does exist under mild conditions (see Sect. 1.3.3). Hence there exist $W^*$ and $X^*$, on an appropriate filtered probability space $(\Omega^*, \mathcal{F}^*, (\mathcal{F}_t^*), P^*)$, such that (2.8) holds. Putting $\gamma^*(s) = \hat{\gamma}(s, X^*(s))$, we have $A^* = (\Omega^*, \mathcal{F}^*, (\mathcal{F}_t^*), P^*, W^*, \gamma^*(\cdot)) \in \mathbb{A}$, but not in $\mathbb{A}^W$. We call $A^*$ an admissible control associated with the Markovian policy $\hat{\gamma}(\cdot)$.

### 2.1.2 Value Functions: Basic Properties

Recalling Theorem 1.2, we first list some basic properties of responses.

**Proposition 2.1.** *Let $p \geq 1$ be given. Then there exists a constant $K_p(> 0)$, such that the following estimates hold:*

(i) *For any $(\theta, x) \in [0, T] \times \mathbb{R}^d$,*

$$E_{\theta x}\Big[\sup_{\theta \leq t \leq T} |X^A(t)|^{2p}\Big] \leq K_p(1 + |x|^{2p}), \quad \forall A \in \mathbb{A}. \tag{2.9}$$

(ii) *For any $(\theta, x) \in [0, T] \times \mathbb{R}^d$ and $\theta \leq t_1 < t_2 \leq T$,*

$$E_{\theta x}\Big[\sup_{t_1\le s\le t_2}|X^A(s)-X^A(t_1)|^{2p}\Big]\le K_p(1+|x|^{2p})(t_2-t_1)^p,\quad \forall A\in\mathbb{A}. \tag{2.10}$$

(iii) *For any* $x, y \in \mathbb{R}^d$ *and* $\theta \in [0, T]$,

$$E\Big[\sup_{\theta\le s\le T}|X^A_{\theta x}(s)-X^A_{\theta y}(s)|^2\Big]\le K_1|x-y|^2,\quad \forall A\in\mathbb{A}. \tag{2.11}$$

(iv) *For* $0 \le \theta_1 < \theta_2 \le T$ *and* $x \in \mathbb{R}^d$,

$$E\Big[\sup_{\theta_2\le s\le T}|X^A_{\theta_1 x}(s)-X^A_{\theta_2 x}(s)|^2\Big]\le K_1(1+|x|^2)(\theta_2-\theta_1),\quad \forall A\in\mathbb{A}. \tag{2.12}$$

Regarding the dependence on the control process, we have;

**Proposition 2.2.** *There exists a constant* $K_0$ *such that, for any controls* $A = (\Omega, \mathcal{F}, (\mathcal{F}_t), P, W, \gamma(\cdot))$ *and* $\hat{A} = (\Omega, \mathcal{F}, (\mathcal{F}_t), P, W, \hat{\gamma}(\cdot))$,

$$\begin{aligned}E_{\theta x}\Big[\sup_{\theta\le s\le t}|X^A(s)-X^{\hat{A}}(s)|^2\Big] &\le K_0\int_\theta^t E[\bar{m}(|\gamma(s)-\hat{\gamma}(s)|)^2]\,ds\\ &\le K_0\Big\{a^2(t-\theta)+\bar{M}^2\frac{E\mathbf{1}\gamma(s)-\hat{\gamma}(s)\mathbf{1}^2_{\theta t}}{(\bar{m}^{-1}(a))^2}\Big\},\\ &\forall a>0,\quad \forall 0\le\theta\le t\le T,\quad x\in\mathbb{R}^d\end{aligned} \tag{2.13}$$

*where* $\bar{M}$ *is a bound of* $\bar{m}(\cdot)$, *and* $\mathbf{1}\gamma(\cdot)-\hat{\gamma}(\cdot)\mathbf{1}_{\theta t} = \|\gamma(\cdot)-\hat{\gamma}(\cdot)\|_{L^2[\theta,t]}$. *We put* $\mathbf{1}*\mathbf{1} = \mathbf{1}*\mathbf{1}_{\theta T}$.

*Proof.* Set

$$Z(t)=X^A(t)-X^{\hat{A}}(t),\quad \Delta b(t)=b(t,X^A(t),\gamma(t))-b(t,X^{\hat{A}}(t),\hat{\gamma}(t)),$$

and similarly for $\alpha(\cdot)$. Then

$$\sup_{\theta\le s\le t}|Z(s)|\le\int_\theta^t|\Delta b(s)|\,ds+\sup_{\theta\le s\le t}\Big|\int_\theta^s\Delta\alpha(h)\,dW(h)\Big|$$

holds by (2.3). Therefore, the Burkholder–Davis–Gundy inequality yields

$$\rho(t)\le C\Big[\int_\theta^t\rho(s)\,ds+\int_\theta^t E[\bar{m}(|\gamma(s)-\hat{\gamma}(s)|)^2]\,ds\Big] \tag{2.14}$$

with a constant $C > 0$, where

$$\rho(t) = E\Big[\sup_{\theta \le s \le t} |Z(s)|^2\Big].$$

Now Gronwall's inequality and (2.14) lead to the left inequality in (2.13).

Now using the estimation

$$\begin{aligned} E\bar{m}(|\xi|)^2 &\le a^2 + E[\bar{m}(|\xi|)^2; |\xi| > \bar{m}^{-1}(a)] \\ &\le a^2 + \bar{M}^2 P(|\xi| > \bar{m}^{-1}(a)) \\ &\le a^2 + \bar{M}^2 \frac{E|\xi|^2}{(\bar{m}^{-1}(a))^2}, \end{aligned}$$

we obtain the right inequality in (2.13). This completes the proof. □

Hence, when $A, A_n, n = 1, 2, \ldots$ have the same reference probability system, we obtain

**Corollary 2.1.** *The response depends on its control process continuously in $L^2([0,T] \times \Omega)$. Further, if*

$$\lim_{n \to \infty} E \mathbf{1}\gamma_n(\cdot) - \gamma(\cdot)\mathbf{1}^2 = 0,$$

*then we can choose a subsequence $n'$ such that, for any $(\theta, x) \in [0,T] \times \mathbb{R}^d$,*

$$\lim_{n' \to \infty} \sup_{\theta \le s \le T} |X_{\theta x}^{A_{n'}}(s) - X_{\theta x}^{A}(s)| = 0 \quad P\text{-a.s.}$$

*and*

$$\lim_{n' \to \infty} E_{\theta x}\Big[\sup_{\theta \le s \le T} |X^{A_{n'}}(s) - X^{A}(s)|^2\Big] = 0.$$

Next we discuss continuity properties of the payoff and the value function, replying on Propositions 2.1 and 2.2.

**Theorem 2.1.** *Suppose $(b_1)$–$(b_5)$ hold. Then $J(t,\theta,x,A;\phi)$ and $v(t,\theta,x;\phi)$ have the following properties:*

(i) *There is a constant $\bar{K}$ such that, for any $0 \le \theta < t \le T$ and $x \in \mathbb{R}^d$,*

$$\begin{aligned} &|J(t,\theta,x,A;\phi)| \le \bar{K}(1 + |x|^2), \quad \forall A \in \mathbb{A}, \\ &|v(t,\theta,x;\phi)| \le \bar{K}(1 + |x|^2). \end{aligned} \tag{2.15}$$

(ii) ***Continuous dependence on initial conditions***
*Let $R$ and $\varepsilon(>0)$ be given. Then there is a constant $\delta_{\varepsilon R}>0$ such that, for any $x_1, x_2 \in S_R$ and $0 \le \theta_1 < \theta_2 < t \le T$,*

$$|J(t,\theta_1,x_1,A;\phi) - J(t,\theta_2,x_2,A;\phi)| < \varepsilon, \quad \forall A \in \mathbb{A}, \tag{2.16}$$

*whenever $|x_1 - x_2| + |\theta_1 - \theta_2| < \delta_{\varepsilon R}$, and*

$$|v(t,\theta_1,x_1;\phi) - v(t,\theta_2,x_2;\phi)| < \varepsilon$$

*whenever $|x_1 - x_2| + |t_1 - t_2| < \delta_{\varepsilon R}$.*

(iii) ***Continuous dependence on the terminal time***
*Let $R$ and $\varepsilon(>0)$ be given. Then there is a constant $\Delta_{\varepsilon R}>0$ such that, for any $x \in S_R$ and $0 \le \theta \le t_1 < t_2 \le T$,*

$$|J(t_1,\theta,x,A;\phi) - J(t_2,\theta,x,A;\phi)| < \varepsilon, \quad \forall A \in \mathbb{A} \tag{2.17}$$

*and*

$$|v(t_1,\theta,x;\phi) - v(t_2,\theta,x;\phi)| < \varepsilon,$$

*whenever $|t_1 - t_2| < \Delta_{\varepsilon R}$.*

*Proof.* We assume $\kappa = 0$, because the proof in the case of general $\kappa$ is similar.

(i) is immediate from $(b_4)$ and (2.9).

(ii) Fix $\theta$ and $A$ and put $X(t,x) = X^A_{\theta x}(t)$ for simplicity. For given $\varepsilon > 0$ and $R > 0$, set

$$\gamma_{\varepsilon R} = (K_1(1+(R+1)^2)\varepsilon^{-1})^{\frac{1}{2}},$$

with $K_1$ of (2.9). Then, for any $x \in S_{R+1}$,

$$P\Big(\sup_{\theta \le t \le T} |X(t,x)| > \gamma_{\varepsilon R}\Big) \le E\Big[\sup_{\theta \le t \le T} |X(t,x)|^2\Big]\gamma_{\varepsilon R}^{-2} < \varepsilon. \tag{2.18}$$

Next choose a small $\delta_0 = \delta_0(\varepsilon, R) > 0$ such that, for any $x, y \in S_R$,

$$|\phi(x) - \phi(y)| + \sup_{\gamma,t} |f(t,x,\gamma) - f(t,y,\gamma)| < \varepsilon,$$
$$\text{whenever } |x-y| < \delta_0, \tag{2.19}$$

which is possible thanks to $(b_2)$ and $(b_3)$.

Let us fix $y \in S_R$ arbitrarily. Then

$$P\Big(\sup_{\theta \le t \le T} |X(t,x) - X(t,y)| > \delta_0\Big)$$
$$< E\Big[\sup_{\theta \le t \le T} |X(t,x) - X(t,y)|^2\Big]\delta_0^{-2} < K_1|x-y|^2\delta_0^{-2} \tag{2.20}$$

follows from (2.11).

Now using (2.18) and (2.20), we introduce two sets $\Omega_x$ and $\tilde{\Omega}_x$ by

$$\Omega_x = \Big\{\omega \in \Omega;\ \sup_{\theta \le t \le T} |X(t,x;\omega)| \le \gamma_{\varepsilon R}\Big\} \tag{2.21}$$

for $x$ with $|x-y| \le 1$, and

$$\tilde{\Omega}_x = \Big\{\omega \in \Omega;\ \sup_{\theta \le t \le T} |X(t,x;\omega) - X(t,y;\omega)| < \delta_0\Big\}. \tag{2.22}$$

Then (2.18) and (2.20) imply

$$P(\Omega_x) > 1 - \varepsilon, \qquad P(\Omega_y) > 1 - \varepsilon, \tag{2.23}$$

and

$$P(\tilde{\Omega}_x) > 1 - \varepsilon \tag{2.24}$$

whenever $|x-y|^2 < \varepsilon\delta_0^2 K_1^{-1}$.

Let us estimate the quantity

$$I_1(x,y) := E\Big[\int_\theta^t |f(s, X(s,x), \gamma(s)) - f(s, X(s,y), \gamma(s))|\, ds\Big].$$

We have

$$\begin{aligned} &I_1(x,y)\\ &\le E\Big[\int_\theta^t |f(s, X(s,x), \gamma(s)) - f(s, X(s,y), \gamma(s))|\, ds; \Omega_x \cap \Omega_y \cap \tilde{\Omega}_x\Big]\\ &+ \hat{K}E\Big[\int_\theta^t (2 + |X(s,x)|^2 + |X(s,y)|^2)\, ds; \Omega_x^c \cup \Omega_y^c \cup \tilde{\Omega}_x^c\Big] \end{aligned} \tag{2.25}$$

thanks to $(b_4)$. By (2.9) and (2.22)–(2.25), we can find a constant $c_1 > 0$, independent of $\theta, t, A, \varepsilon$, and $R$, such that

$$I_1(x,y) < c_1\sqrt{\varepsilon}(1 + R^2) \tag{2.26}$$

whenever $x, y \in S_R$ and $|x-y| < \sqrt{\frac{\varepsilon}{K_1}}\delta_0$.

Indeed, (2.19) and (2.22) imply that

$$\text{1st term in the RHS of (2.25)} < \varepsilon T. \tag{2.27}$$

For the 2nd term, (2.9), (2.23), and (2.24) yield

$$\begin{aligned} &E\Big[\int_\theta^t |X(s,x)|^2\,ds;\ \Omega_x^c \cup \Omega_y^c \cup \tilde{\Omega}_x^c\Big] \\ \le &\Big(E\big(\int_\theta^t |X(s,x)|^2\,ds\big)^2\Big)^{\frac{1}{2}}\sqrt{3\varepsilon} \\ \le &TK_2(1+|x|^2)\sqrt{3\varepsilon} \le TK_2(1+(R+1)^2)\sqrt{3\varepsilon} \end{aligned} \tag{2.28}$$

whenever $x, y \in S_R$ and $|x-y| < \sqrt{\frac{\varepsilon}{K_1}}\delta_0$.

Since for $X(s, y)$ one has the same estimate as (2.28), (2.27) and (2.28) together with (2.25) establish (2.26).

Applying the same arguments to the terminal cost and using (2.26), we can find $\delta_1 := \delta_1(\varepsilon, R) > 0$, such that

$$\begin{aligned} &|J(t,\theta,x,A;\phi) - J(t,\theta,y,A;\phi)| < \varepsilon, \\ &\qquad\qquad \forall A \in \mathbb{A}, \quad 0 \le \theta < t \le T, \end{aligned} \tag{2.29}$$

whenever $x, y \in S_R$ and $|x-y| < \delta_1$.

Finally, we consider the initial time. Let $\theta_1 < \theta_2 < t$ and $x \in S_R$. Set $X(s,\theta) = X_{\theta x}^A(s)$, and with the same $\gamma_{\varepsilon R}$ and $\delta_0$, define $\Omega_i$ and $\tilde{\Omega}$ by

$$\Omega_i = \Big\{\omega \in \Omega;\ \sup_{\theta_i \le s \le T} |X(s,\theta_i)| < \gamma_{\varepsilon R}\Big\}, \quad i = 1, 2,$$

and

$$\tilde{\Omega} = \Big\{\omega \in \Omega;\ \sup_{\theta_2 \le s \le T} |X(s,\theta_1) - X(s,\theta_2)| < \delta_0\Big\}.$$

By arguing as in the previous estimations, we can take $\delta_2 = \delta_2(\varepsilon, R) > 0$, so that

$$\begin{aligned} &|J(t,\theta_1,x,A;\phi) - J(t,\theta_2,x,A;\phi)| < \varepsilon, \\ &\qquad\qquad \forall t \ge \theta_2, \quad \forall x \in S_R, \quad \forall A \in \mathbb{A}, \end{aligned} \tag{2.30}$$

whenever $\theta_2 - \theta_1 < \delta_2$.

Now (ii) the proof of follows from (2.29) and (2.30).

Since (iii) the proof of mimics of (ii), the result is established. □

Theorem 2.1 asserts that, if the terminal cost function $\phi$ has quadratic growth, then the value function also has quadratic growth. Put

$$\tilde{C} = \left\{ \psi \in C(\mathbb{R}^d);\ \tilde{\psi}(x) := \frac{\psi(x)}{1+|x|^2} \in C_b(\mathbb{R}^d) \right\}. \tag{2.31}$$

We introduce the norm $\|\psi\|_{\tilde{C}}$ by

$$\|\psi\|_{\tilde{C}} = \|\tilde{\psi}\|_{C(\mathbb{R}^d)} \tag{2.32}$$

and the order $\leq$ by

$$\varphi \leq \psi \Longleftrightarrow \varphi(x) \leq \psi(x), \quad \forall x \in \mathbb{R}^d. \tag{2.33}$$

Then $\tilde{C}$ becomes a Banach lattice.

Recalling $(b_3)$ and $(b_4)$, we define a two-parameter operator $V_{\theta t}; \tilde{C} \mapsto \tilde{C}$ $(0 \leq \theta \leq t \leq T)$ by

$$V_{\theta t}\phi(x) = v(t, \theta, x; \phi). \tag{2.34}$$

**Proposition 2.3.** *$V_{\theta t}$ has the following properties:*

(i) *$V_{\theta\theta}$ is the identity map on $\tilde{C}$.*
(ii) *Monotonicity property: $\phi \leq \psi \Rightarrow V_{\theta t}\phi \leq V_{\theta t}\psi$.*
(iii) *There is a constant $\tilde{K} > 0$, such that*

$$\|V_{\theta t}\phi - V_{\theta t}\psi\|_{\tilde{C}} \leq \tilde{K}\|\phi - \psi\|_{\tilde{C}}, \quad \forall \theta \leq t, \quad \forall \phi, \psi \in \tilde{C}, \tag{2.35}$$

*and*

$$\|V_{\theta t}0\|_{\tilde{C}} \leq \tilde{K}, \quad \forall \theta \leq t. \tag{2.36}$$

*Proof.* (i) and (ii) are clear from the definition of payoff.

(iii) From the inequality $|\phi(x) - \psi(x)| \leq \|\phi - \psi\|_{\tilde{C}}(1+|x|^2)$ it follows that

$$|J(t, \theta, x, A; \phi) - J(t, \theta, x, A; \psi)| \leq \|\phi - \psi\|_{\tilde{C}} E_{\theta x}\Big(1 + \sup_{\theta \leq t \leq T} |X(t)|^2\Big).$$

Now (2.9) yields (iii). □

## 2.2 Dynamic Programming Principle (DPP)

The dynamic programming principle (DPP), introduced by R. Bellman [Be52, Be57], gives a powerful tool for stochastic control problems and is known as a two-stages optimization method. Here we will formulate DPP as a two parameter semigroup on a suitable Banach space, constructed by using time discretization in Sects. 2.1.1–2.2.3 [N76, N81]. We characterize the semigroup as the envelope of Markovian transition semigroups and show that its generator is related to HJB equation in Sect. 2.2.5. We have two kinds of control processes on a fixed reference probability system $(\Omega, \mathcal{F}, (\mathcal{F}_t), P, W)$: one is an $(\mathcal{F}_t)$-progressively measurable control process and the other is an $(\mathcal{F}_t^W)$-progressively measurable one. But we note that these two kind controls give the same value function (see Sect. 2.2.4).

In this section, we always assume the conditions $(b_1)$–$(b_5)$.

### 2.2.1 Discrete-Time Dynamic Programming Principle

Let $\mathcal{D}, 0 = t_0 < t_1 < \cdots < t_p < t_{p+1} = T$ be a division of $[0, T]$. We put $\mathcal{D} = (t_1, \ldots, t_p)$, $\mathscr{P}_{\mathcal{D}} =$ the set of division points of $\mathcal{D}(= \{t_1, \ldots, t_p\})$ and $|\mathcal{D}| = \max_{i=0,\ldots,p} |t_i - t_{i+1}|$. For divisions $\mathcal{D}$ and $\tilde{\mathcal{D}}$, we say that $\mathcal{D} \subset \tilde{\mathcal{D}}$, if $\mathscr{P}_{\mathcal{D}} \subset \mathscr{P}_{\tilde{\mathcal{D}}}$.

Let $\mathbb{A}^{\mathcal{D}}$ denote the set of all admissible controls $(\Omega, \mathcal{F}, (\mathcal{F}_t), P, W, \gamma(\cdot))$ with

$$\gamma(t, \omega) = \gamma(t_i, \omega) \quad \text{for } t \in [t_i, t_{i+1}), \quad i = 0, \ldots, p. \tag{2.37}$$

$\gamma(\cdot)$ is called a switching control at $\mathcal{D}$, or a $\mathcal{D}$-admissible control. When we restrict $\mathbb{A}$ to $\mathbb{A}^{\mathcal{D}}$, the value function $v^{\mathcal{D}}$ is given by

$$v^{\mathcal{D}}(t, \theta, x; \phi) = \inf_{A \in \mathbb{A}^{\mathcal{D}}} J(t, \theta, x, A; \phi). \tag{2.38}$$

Since $v^{\mathcal{D}}$ clearly satisfies (i)–(iii) in Theorem 2.1, $V_{\theta t}^{\mathcal{D}}$ defined by

$$V_{\theta t}^{\mathcal{D}} \phi(x) = v^{\mathcal{D}}(t, \theta, x; \phi) \tag{2.39}$$

is a mapping from $\tilde{C}$ into $\tilde{C}$.

We will show the following subsidiary theorem;

**Theorem 2.2 (Discrete-time *DPP*).** *Let $\Gamma$ be convex and compact. Then for $l < m < n$,*

$$V_{t_l t_m}^{\mathcal{D}}(V_{t_m t_n}^{\mathcal{D}} \phi) = V_{t_l t_n}^{\mathcal{D}} \phi, \tag{2.40}$$

*that is,*

$$\inf_{A \in \mathbb{A}^{\mathcal{D}}} J(t_m, t_l, x, A; v^{\mathcal{D}}(t_n, t_m, \cdot\,; \phi)) = v^{\mathcal{D}}(t_n, t_l, x; \phi). \tag{2.41}$$

Before we embark upon the proof, let us consider constant controls. Set $A_\gamma = (\Omega, \mathcal{F}, (\mathcal{F}_t), P, W, \gamma)$, where $\gamma \in \Gamma$ stands for a constant control process: $\gamma(t, \omega) = \gamma$, for all $t$ and $\omega$. Then its response $X_{\theta y}^{A_\gamma}$ is the strong solution of (2.3), measurable w.r.t. $\sigma(W(\cdot) - W(\theta), y, \gamma)$, and the payoff $J(t, \theta, y, A_\gamma; \phi)$ depends only on $t, \theta, y, \gamma$ and $\phi$. Denote the payoff by $J(t, \theta, y, \gamma; \phi)$. We will seek an optimal $\gamma(\in \Gamma)$ by using the measurable selection theorem (see [SV79], Lemma 12.1.7 and Theorem 12.1.10).

Since $J(t, \theta, y, \gamma; \phi)$ is continuous in $(y, \gamma)$ by (2.11) and (2.13), the compactness of $\Gamma$ implies that, for $y \in \mathbb{R}^d$, the set

$$\Gamma_y = \{\gamma \in \Gamma; J(t, \theta, y, \gamma; \phi) = \inf_{\gamma \in \Gamma} J(t, \theta, y, \gamma; \phi)\} \tag{2.42}$$

is non-empty and compact. Hence there is a minimum selector (Borel function) $\gamma_{\theta t}(\cdot; \phi); \mathbb{R}^d \mapsto \Gamma$, such that

$$\gamma_{\theta t}(y; \phi) \in \Gamma_y, \quad \forall y \in \mathbb{R}^d \tag{2.43}$$

(see [FR75], Lemma B in Appendix).

Consequently

$$J(t, \theta, y, \gamma_{\theta t}(y; \phi); \phi) = \inf_{\gamma \in \Gamma} J(t, \theta, y, \gamma; \phi) =: \nu_{\theta t}\phi(y). \tag{2.44}$$

Further, by (2.9) and $(b_4)$, there is a constant $c > 0$, such that

$$\begin{aligned} |f(s, x, \gamma)| + |\nu_{\theta t}\phi(x)| &\le c(1 + |x|^2), \\ \forall x \in \mathbb{R}^d, \quad \forall \theta \le s &\le t, \quad \forall \gamma \in \Gamma. \end{aligned} \tag{2.45}$$

Let us prove Theorem 2.2 by using the minimum selector.

*Proof.* For the proof, we assume $\kappa = 0$, because, for the general $\kappa$, the proof is entirely similar. Putting

$$\Big(\prod_{k=i}^{j} \nu_{t_k t_{k+1}}\Big)\phi = \nu_{t_i t_{i+1}} \nu_{t_{i+1} t_{i+2}} \cdots \nu_{t_j t_{j+1}} \phi, \tag{2.46}$$

we will show the inequality,

$$V^{\mathcal{D}}_{t_i t_{j+1}} \phi(x) \ge \Big(\prod_{k=i}^{j} \nu_{t_k t_{k+1}}\Big)\phi(x) \tag{2.47}$$

and its opposite,

$$V^{\mathcal{D}}_{t_i t_{j+1}} \phi(x) \le \Big(\prod_{k=i}^{j} \nu_{t_k t_{k+1}}\Big)\phi(x). \tag{2.48}$$

Put $X^A(\cdot) = X^A_{t_i x}(\cdot)$. Since $X^A(t_j)$ and $\gamma(t_j)$ are $\mathcal{F}_{t_j}$-measurable and $W(\cdot + t_j) - W(t_j)$ is independent of $\mathcal{F}_{t_j}$,

$$E(C(t_{j+1}, t_j, A; \phi)|\mathcal{F}_{t_j}) = J(t_{j+1}, t_j, X^A(t_j), \gamma(t_j); \phi) \\ \geq \nu_{t_j t_{j+1}} \phi(X^A(t_j)) \quad P\text{-a.s.} \tag{2.49}$$

holds for any $A \in \mathbb{A}^{\mathcal{D}}$. Hence, we have

$$J(t_{j+1}, t_i, x, A; \phi) \geq J(t_j, t_i, x, A; \nu_{t_j t_{j+1}} \phi) \geq \Big(\prod_{k=i}^{j} \nu_{t_k t_{k+1}}\Big) \phi(x). \tag{2.50}$$

Taking the infimum of LHS over $A \in \mathbb{A}^{\mathcal{D}}$, we obtain (2.47).

For (2.48), we construct a control process $\gamma^*(\cdot)$ for which the RHS of (2.47) is attained. This is done by using the minimum selector $\gamma_{\theta t}(\cdot)$ of (2.43). On a reference probability system $(\Omega, \mathcal{F}, (\mathcal{F}_t), P, W)$, we first define $\gamma^*(t_i)$ by

$$\gamma^*(t_i) = \gamma_{t_i t_{i+1}}\Big(x; \Big(\prod_{k=i+1}^{j} \nu_{t_k t_{k+1}}\Big)\phi\Big), \tag{2.51}$$

and, consider the SDE

$$\begin{cases} dX(t) = b(t, X(t), \gamma^*(t_i))\, dt + \alpha(t, X(t), \gamma^*(t_i))\, dW(t), \quad t \in (t_i, t_{i+1}], \\ X(t_i) = x. \end{cases} \tag{2.52}$$

Then we define $\gamma^*(t_{i+1})$ by

$$\gamma^*(t_{i+1}) = \gamma_{t_{i+1} t_{i+2}}\Big(X(t_{i+1}); \Big(\prod_{k=i+2}^{j} \nu_{t_k t_{k+1}}\Big)\phi\Big). \tag{2.53}$$

Repeating this procedure, we obtain an $(\mathcal{F}^W_t)$-progressively measurable $\gamma^*(\cdot)$, such that $A^* := (\Omega, \mathcal{F}, (\mathcal{F}_t), P, W, \gamma^*(\cdot))$ is in $\mathbb{A}^{\mathcal{D}}$ and

$$J(t_{j+1}, t_i, x, A^*; \phi) = \Big(\prod_{k=i}^{j} \nu_{t_k t_{k+1}}\Big)\phi(x), \tag{2.54}$$

which yields (2.48).

Now from (2.47) and (2.48) it follows that

$$\begin{aligned} V^{\mathcal{D}}_{t_l t_n}\phi(x) &= \Big(\prod_{k=l}^{m-1} v_{t_k t_{k-1}}\Big)(V^{\mathcal{D}}_{t_m t_n}\phi)(x) \\ &= V^{\mathcal{D}}_{t_l t_n}(V^{\mathcal{D}}_{t_m t_n}\phi)(x). \end{aligned} \tag{2.55}$$

This completes the proof. □

Put $\boldsymbol{\Gamma}^W = L^\infty([0,T]\times\Omega,(\mathcal{F}^W_t);\boldsymbol{\Gamma})$ and $\boldsymbol{\Gamma}^{W,\mathcal{D}} = \{\gamma(\cdot)\in\boldsymbol{\Gamma}^W$; switching at $\mathcal{D}\}$. Then referring to the proof above and to the minimum selector, we have

*Remark 2.1.* For $\theta, t \in \mathscr{P}_{\mathcal{D}}\cup\{0,T\}$ and $(\Omega,\mathcal{F},(\mathcal{F}_t),P,W)$, there is $\gamma^*(\cdot)\in\boldsymbol{\Gamma}^{W,\mathcal{D}}$, such that

$$V^{\mathcal{D}}_{\theta t}\phi(x) = J(t,\theta,x,A^*;\phi),$$

where $A^* = (\Omega,\mathcal{F},(\mathcal{F}_t),P,W,\gamma^*(\cdot))$.

Next we study the discrete-time DP property.

**Proposition 2.4.** *Let $\mathcal{D} = (t_1,\dots,t_p)$ be given.*

(i) *Let $\gamma^*(\cdot)$ be an optimal control process in $\boldsymbol{\Gamma}^{W,\mathcal{D}}$, given by the minimum selector. Then, for any $(\mathcal{F}^W_t)$-stopping time $\tau$ taking values in $\mathscr{P}_{\mathcal{D}}\cup\{0,T\}$,*

$$\begin{aligned} V^{\mathcal{D}}_{0T}\phi(x) =& E_{0x}\Big[\int_0^\tau \exp\{-\int_0^s \kappa(\lambda,X^*(\lambda),\gamma^*(\lambda))\,d\lambda\} f(s,X^*(s),\gamma^*(s))\,ds \\ &+ \exp\{-\int_0^\tau \kappa(\lambda,X^*(\lambda),\gamma^*(\lambda))\,d\lambda\} V^{\mathcal{D}}_{\tau T}\phi(X^*(\tau))\Big], \end{aligned} \tag{2.56}$$

*where $X^*$ is the response for $\gamma^*(\cdot)$, with $X^*(0) = x$,*

(ii) *Let $\gamma(\cdot)\in\boldsymbol{\Gamma}^{W,\mathcal{D}}$ and $X$ its response with $X(0) = x$. Then, for any $(\mathcal{F}^W_t)$-stopping time $\tau$ taking values in $\mathscr{P}_{\mathcal{D}}\cup\{0,T\}$,*

$$\begin{aligned} V^{\mathcal{D}}_{0T}\phi(x) \le& E_{0x}\Big[\int_0^\tau \exp\{-\int_0^s \kappa(\lambda,X(\lambda),\gamma(\lambda))\,d\lambda\} f(s,X(s),\gamma(s))\,ds \\ &+ \exp\{-\int_0^\tau \kappa(\lambda,X(\lambda),\gamma(\lambda))\,d\lambda\} V^{\mathcal{D}}_{\tau T}\phi(X(\tau))\Big]. \end{aligned} \tag{2.57}$$

*Proof.* $P(\tau = 0) = 1$ or 0, because $(\tau = 0)\in\mathcal{F}^W_0$. When $P(\tau = 0) = 1$, (2.56) and (2.57) are trivial. Hence we may assume in the proof that the set of values of $\tau$ is $\{t_1,\dots,t_p,T\}$.

We again assume $\kappa = 0$.

(i) We have

$$C(T,0,\gamma^*(\cdot);\phi)$$
$$=\sum_{k=1}^{p+1}\chi(\tau = t_k)\Big(\int_0^{t_k} f(s,X^*(s),\gamma^*(s))\,ds + C(T,t_k,\gamma^*(\cdot);\phi)\Big), \tag{2.58}$$

where $t_{p+1} = T$.

From the definition of $\gamma^*(\cdot)$, it follows that

$$E(C(T,t_p,\gamma^*(\cdot);\phi)|\mathcal{F}^W_{t_p}) = \nu_{t_pT}\phi(X^*(t_p)),$$
$$E(C(T,t_{p-1},\gamma^*(\cdot);\phi)|\mathcal{F}^W_{t_{p-1}})$$
$$= E\Big(\big(\int_{t_{p-1}}^{t_p} f(s,X^*(s),\gamma^*(s))\,ds + E(C(T,t_p,\gamma^*(\cdot);\phi)|\mathcal{F}^W_{t_p})\big)|\mathcal{F}^W_{t_{p-1}}\Big)$$
$$= E(C(t_p,t_{p-1},\gamma^*(\cdot);\nu_{t_pT}\phi)|\mathcal{F}^W_{t_{p-1}})$$
$$= \nu_{t_{p-1},t_p}(\nu_{t_pT}\phi)(X^*(t_{p-1}))$$
$$= V^{\mathcal{D}}_{t_{p-1}T}\phi(X^*(t_{p-1}))$$

and

$$E(C(T,t_k,\gamma^*(\cdot);\phi)|\mathcal{F}^W_{t_k}) = V^{\mathcal{D}}_{t_kT}\phi(X^*(t_k)). \tag{2.59}$$

In view of (2.58) and (2.59), we obtain

$$V^{\mathcal{D}}_{0T}\phi(x) = E_{0x}C(T,0,\gamma^*(\cdot);\phi)$$
$$= \sum_{k=1}^{p+1} E_{0x}\Big[\chi(\tau{=}t_k)\Big\{\int_0^{t_k} f(s,X^*(s),\gamma^*(s))\,ds + E(C(T,t_k,\gamma^*(\cdot);\phi)|\mathcal{F}^W_{t_p})\Big\}\Big]$$
$$= E_{0x}\Big[\int_0^{\tau} f(s,X^*(s),\gamma^*(s))\,ds + V^{\mathcal{D}}_{\tau T}\phi(X^*(\tau))\Big].$$

(ii) Let us define $\tilde{\gamma}(t)$ by

$$\tilde{\gamma}(t) = \sum_{k=1}^{p+1}\chi(\tau = t_k)(\gamma(t)\chi(t < t_k) + \gamma^*(t)\chi(t \geq t_k)).$$

Then $\tilde{\gamma}(\cdot) \in \boldsymbol{\Gamma}^{W,\mathcal{D}}$. Further (2.59) yields

$$\begin{aligned} V_{0T}^{\mathcal{D}}\phi(x) &\le J(T,0,x,\tilde{\gamma}(\cdot);\phi) \\ &= \sum_{k=1}^{p+1} E_{0x}\Big[\chi(\tau = t_k)\Big(\int_0^{t_k} f(s,X(s),\gamma(s))\,ds + E(C(T,t_k,\gamma^*(\cdot);\phi)|\mathcal{F}_{t_k}^W)\Big)\Big] \\ &= \text{RHS of (2.57)}, \end{aligned}$$

which concludes the proof of (ii). □

## 2.2.2 Approximation Theorem

Before we prove the dynamic programming principle, we establish the following approximation result.

**Theorem 2.3.** *Let $\Gamma$ be convex and compact and $\mathcal{D}_n = (t_{n,1},\ldots,t_{n,j(n)})$, $n = 1,2,\ldots$.*

*Suppose that*

$$\mathcal{D}_n \subset \mathcal{D}_{n+1}, \quad n = 1,2,\ldots \quad \textit{and} \quad \lim_{n\to\infty} |\mathcal{D}_n| = 0. \tag{2.60}$$

*Let $A = (\Omega,\mathcal{F},(\mathcal{F}_t),P,W,\gamma(\cdot))$ $(\in \mathbb{A})$ be given. Then there exists $A_n = (\Omega,\mathcal{F},(\mathcal{F}_t),P,W,\gamma_n(\cdot))$ of $\mathbb{A}^{\mathcal{D}_n}$ such that*

$$\lim_{n\to\infty} E\mathbf{1}\gamma_n(\cdot) - \gamma(\cdot)\mathbf{1}^2 = 0 \tag{2.61}$$

*and*

$$\lim_{n\to\infty} |\gamma_n(s) - \gamma(s)| = 0 \quad a.e. \quad \text{on } [0,T]\times\Omega. \tag{2.62}$$

*Proof.* Since $\Gamma$ is convex and compact, we can apply a routine.

First fix $\gamma_0 \in \Gamma$ arbitrarily and set

$$\gamma(s) = \gamma_0 \quad \text{for } s < 0.$$

With this convention, we define $\bar{\gamma}_l(\cdot)$ by

$$\bar{\gamma}_l(t) = 2^l \int_{t-2^{-l}}^{t} \gamma(s)\,ds, \quad t \ge 0.$$

Then $\bar{\gamma}_l(\cdot)$ is a continuous control process, satisfying

$$\lim_{l\to\infty} E\mathbf{1}\bar{\gamma}_l(\cdot) - \gamma(\cdot)\mathbf{1}^2 = 0. \tag{2.63}$$

Second, we define $\bar{\gamma}_{l,m}(\cdot)$ by

$$\bar{\gamma}_{l,m}(t) = \bar{\gamma}_l(t_{m,j}) \quad \text{for } t \in [t_{m,j}, t_{m,j+1}), \quad j = 0, 1 \ldots, j(m) - 1. \tag{2.64}$$

Then $\bar{\gamma}_{l,m}(\cdot)$ provides a switching control at $\mathcal{D}_m$, and, as $m \to \infty$, $\bar{\gamma}_{l,m}(\cdot)$ approaches $\bar{\gamma}_l(\cdot)$ uniformly on $[0, T]$, $P$-a.s.

Third, we can choose $(l_n, m_n)$ so that $m_n \geq n$ and

$$E[\mathbf{1}\gamma_{l_n,m_n}(\cdot) - \gamma(\cdot)\mathbf{1}^2] < 2^{-n}. \tag{2.65}$$

Finally we define $\gamma_p(\cdot)$ as follows:

$$\gamma_1(t) = \cdots = \gamma_{m_1-1}(t) = \gamma_0, \quad \forall t \in [0, T],$$
$$\gamma_{m_n}(t) = \cdots = \gamma_{m_{n+1}-1}(t) = \gamma_{l_n,m_n}(t), \quad \forall t \in [0, T], \quad n = 1, 2, \ldots.$$

Then $\gamma_p(\cdot)$, $p = m_n, \ldots, m_{n+1} - 1$ become switching controls at $\mathcal{D}_{m_n}$, and hence at $\mathcal{D}_p$, by (2.60). By (2.65), $\gamma_p(\cdot)$ satisfies (2.61).

Appealing to the Borel–Cantelli Lemma, we can choose a subsequence $\gamma_{p_j}(\cdot)$, so that (2.62) holds. This completes the proof. □

Regarding the responses $X^A$ and $X^{A_n}$, (2.13) and (2.61) lead to

$$\lim_{n\to\infty} E_{\theta x}\Big[\sup_{\theta\leq t\leq T} |X^{A_n}(t) - X^A(t)|^2\Big] = 0, \quad \forall(\theta, x). \tag{2.66}$$

Set

$$\mathbb{A}_s = \bigcup_{\mathcal{D}} \mathbb{A}^{\mathcal{D}} = \text{set of all switching controls.}$$

Using (2.66), we get

**Corollary 2.2.** *Let $\Gamma$ is convex and compact. Then*

$$\begin{aligned} V_{\theta t}\phi(x) &= \inf_{A\in\mathbb{A}_s} J(t, \theta, x, A; \phi) \\ &= \lim_{n\to\infty} \inf_{A\in\mathbb{A}^{\mathcal{D}_n}} J(t, \theta, x, A; \phi), \quad \forall\phi \in \tilde{C} \end{aligned} \tag{2.67}$$

*whenever $\mathcal{D}_n, n = 1, 2, \ldots$ satisfy* (2.60).

### 2.2.3 Dynamic Programming Principle

Now we are ready to prove DPP

**Theorem 2.4.** *Let $\Gamma$ be convex and $\sigma$-compact. We assume $(b_1)$–$(b_5)$ in Sect.* 2.1.1. *Then, for any $\theta < \theta_1 < \theta_2 \leq T$,*

$$V_{\theta\theta_2}\phi = V_{\theta\theta_1}(V_{\theta_1\theta_2}\phi) \quad \textit{for } \phi \in \tilde{C}. \tag{2.68}$$

*Proof.* We divide the proof into two steps.

**Step 1.** Let $\Gamma$ be convex and compact. Take a sequence $\mathcal{D}_n, n = 1, 2, \ldots$ such that $\theta, \theta_1, \theta_2 \in \mathscr{P}_{\mathcal{D}_1} \cup \{0, T\}$ and (2.60) holds.
Then Theorem 2.2 shows that

$$V_{\theta\theta_2}^{\mathcal{D}_n}\phi = V_{\theta\theta_1}^{\mathcal{D}_n} V_{\theta_1\theta_2}^{\mathcal{D}_n}\phi. \tag{2.69}$$

Since $V_{\theta_1\theta_2}^{\mathcal{D}_n}\phi \geq V_{\theta_1\theta_2}\phi$, the monotonicity property of $V_{\theta\theta_1}^{\mathcal{D}_n}$ leads to

$$V_{\theta\theta_1}^{\mathcal{D}_n}(V_{\theta_1\theta_2}^{\mathcal{D}_n}\phi) \geq V_{\theta\theta_1}^{\mathcal{D}_n}(V_{\theta_1\theta_2}\phi). \tag{2.70}$$

Hence from (2.69) and (2.70), it follows that

$$V_{\theta\theta_2}^{\mathcal{D}_n}\phi \geq V_{\theta\theta_1}^{\mathcal{D}_n}(V_{\theta_1\theta_2}\phi) \tag{2.71}$$

Letting $n \to \infty$, and using Corollary 2.2 we get

$$V_{\theta\theta_2}\phi \geq V_{\theta\theta_1}(V_{\theta_1\theta_2}\phi) \tag{2.72}$$

Next we shall show the converse of inequality (2.72). Since $V_{\theta\theta_2}^{\mathcal{D}_{n+m}}\phi(x) = V_{\theta\theta_1}^{\mathcal{D}_{n+m}}(V_{\theta_1\theta_2}^{\mathcal{D}_{n+m}}\phi) \leq V_{\theta\theta_1}^{\mathcal{D}_{n+m}}(V_{\theta_1\theta_2}^{\mathcal{D}_m}\phi)(x)$, we obtain, letting $n \to \infty$,

$$\begin{aligned} V_{\theta\theta_2}\phi(x) &\leq V_{\theta\theta_1}(V_{\theta_1\theta_2}^{\mathcal{D}_m}\phi)(x) \quad (m = 1, 2, \ldots) \\ &\leq J(\theta_1, \theta, x, A; V_{\theta_1\theta_2}^{\mathcal{D}_m}\phi), \quad \forall A \in \mathbb{A}. \end{aligned} \tag{2.73}$$

Using (2.15) and (2.9), we have

$$\begin{aligned} E_{\theta x}|V_{\theta_1\theta_2}^{\mathcal{D}_m}\phi(X^A(\theta_1))|^2 &\leq c_1 E_{\theta x}(1 + |X^A(\theta_1)|^2)^2 \\ &\leq c_2(1 + |x|^4), \end{aligned} \tag{2.74}$$

with constants $c_1$ and $c_2$ independent of $A$ and $\mathcal{D}_m$.
Since Corollary 2.2 implies

$$\lim_{m\to\infty} V_{\theta_1\theta_2}^{\mathcal{D}_m}\phi(X(\theta_1)) = V_{\theta_1\theta_2}\phi(X(\theta_1)) \quad P\text{-a.s.}, \tag{2.75}$$

(2.74) and the convergence theorem lead to

$$\lim_{m\to\infty} J(\theta_1, \theta, x, A; V_{\theta_1\theta_2}^{\mathcal{D}_m}\phi) = J(\theta_1, \theta, x, A; V_{\theta_1\theta_2}\phi). \tag{2.76}$$

Therefore the inequality

$$V_{\theta\theta_2}\phi(x) \le J(\theta_1, \theta, x, A; V_{\theta_1\theta_2}\phi), \quad \forall A \in \mathbb{A} \tag{2.77}$$

follows from (2.73) and (2.76). Taking the infimum of the RHS of (2.77) over $A \in \mathbb{A}$, we obtain the converse of inequality (2.72).

This completes the proof of (2.68) when $\Gamma$ is convex and compact.

**Step 2.** Let $\Gamma$ be convex and $\sigma$-compact.

We can take a sequence of convex and compact sets $\Gamma_n, n = 1, 2, \cdots$ such that

$$\Gamma_n \subset \Gamma_{n+1}, \quad n = 1, 2, \ldots \quad \text{and} \quad \bigcup_n \Gamma_n = \Gamma. \tag{2.78}$$

We denote by $\mathbb{A}^n$ the set of all admissible controls with control region $\Gamma_n$. Then

$$\mathbb{A}^n \subset \mathbb{A}^{n+1} \quad \text{and} \quad \bigcup_n \mathbb{A}^n = \mathbb{A}, \tag{2.79}$$

by Definition 2.1. Observing that (2.79) implies

$$J(t, \theta, x, A; \phi) \ge \lim_{n\to\infty} V^n_{\theta t}\phi(x), \quad \forall A \in \mathbb{A}, \tag{2.80}$$

where

$$V^n_{\theta t}\phi(x) = \inf_{A\in\mathbb{A}^n} J(t, \theta, x, A; \phi), \tag{2.81}$$

and taking the infimum of (2.80) w.r.t. $A$ over $\mathbb{A}$, we have

$$V_{\theta t}\phi(x) = \lim_{n\to\infty} V^n_{\theta t}\phi(x), \quad \forall \theta \in t, \quad x \in \mathbb{R}^d, \tag{2.82}$$

because $V_{\theta t}\phi(x) \le V^n_{\theta t}\phi(x)$, $\forall n$.

Since $V^n_{\theta t}$ satisfies DPP,

$$\begin{aligned} V_{\theta\theta_2}\phi(x) &= \lim_{n\to\infty} V^n_{\theta\theta_2}\phi(x) \\ &= \lim_{n\to\infty} V^n_{\theta\theta_1}(V^n_{\theta_1\theta_2}\phi)(x) \\ &\ge \lim_{n\to\infty} V^n_{\theta\theta_1}(V_{\theta_1\theta_2}\phi)(x) \quad (\text{by } V^n_{\theta_1\theta_2}\phi \ge V_{\theta_1\theta_2}\phi) \\ &= V_{\theta\theta_1}(V_{\theta_1\theta_2}\phi)(x). \end{aligned} \tag{2.83}$$

For the converse inequality, we have

$$\begin{aligned} V^{n+m}_{\theta\theta_2}\phi(x) &= V^{n+m}_{\theta\theta_1}(V^{n+m}_{\theta_1\theta_2}\phi)(x) \\ &\le V^{n+m}_{\theta\theta_1}(V^{m}_{\theta_1\theta_2}\phi)(x), \quad m, n = 1, 2, \ldots. \end{aligned} \tag{2.84}$$

Letting $n \to \infty$, we get

$$\begin{aligned} V_{\theta\theta_2}\phi(x) &\le V_{\theta\theta_1}(V^m_{\theta_1\theta_2}\phi)(x), \quad m = 1, 2, \ldots \\ &\le J(\theta_1, \theta, x, A; V^m_{\theta_1\theta_2}\phi), \quad \forall A \in \mathbb{A}, \quad m = 1, 2, \ldots. \end{aligned} \tag{2.85}$$

Again (2.74) and the convergence theorem yield

$$\lim_{m\to\infty} J(\theta_1, \theta, x, A; V^m_{\theta_1\theta_2}\phi) = J(\theta_1, \theta, x, A; V_{\theta_1\theta_2}\phi). \tag{2.86}$$

Thus, the converse of inequality (2.83) follows from (2.85) and (2.86).

This completes the proof of Theorem 2.4. □

### 2.2.4 Brownian Adapted Controls

In this section, we will comment on the value function in the case where one restricts the control processes to the class of Brownian adapted ones.

We denote the set of all admissible controls, $A = (\Omega, \mathcal{F}, (\mathcal{F}_t), P, W, \gamma(\cdot))$ where $\gamma(\cdot)$ is $(\mathcal{F}^W_t)$-progressively measurable, by $\mathbb{A}^W$. Put

$$V^W_{\theta t}\phi(x) = \inf_{A\in\mathbb{A}^W} J(t, \theta, x, A; \phi). \tag{2.87}$$

Then we have:

**Proposition 2.5.**

$$V^W_{\theta t}\phi = V_{\theta t}\phi, \quad \forall\theta < t. \tag{2.88}$$

*Proof.* Since we can easily see that (2.82) is also valid for $V^W_{\theta t}$, we may assume for the proof that $\Gamma$ is convex and compact. Now the proposition follows from Remark 2.1 and Corollary 2.2. □

For $A \in \mathbb{A}^W$, $J(t, \theta, x, A; \phi)$ can be calculated by means of the joint probability distribution $(W, \gamma(\cdot))$. So, we fix a reference probability system $(\Omega, \mathcal{F}, (\mathcal{F}_t), P, W)$ and identify $A$ with its control process $\gamma(\cdot)$.

For a stopping time $\tau$, we again denote the payoff by $J(\cdot)$:

$$\begin{aligned} J(\tau, \theta, x, \gamma(\cdot); \psi) =& E_{\theta x}\Big[\int_\theta^\tau \exp\{-\int_\theta^s \kappa(\lambda, X(\lambda), \gamma(\lambda))\,d\lambda\} f(s, X(s), \gamma(s))\,ds \\ &+ \exp\{-\int_\theta^\tau \kappa(\lambda, X(\lambda), \gamma(\lambda))\,d\lambda\}\psi(X(\tau))\Big], \end{aligned} \tag{2.89}$$

where $X = X^{\gamma(\cdot)}$.

**Proposition 2.6 (DP Property).**
*Let $(\Omega, \mathcal{F}, (\mathcal{F}_t), P, W)$ and $(\theta, x) \in [0, T) \times \mathbb{R}^d$ be given.*

(i) *For a $[\theta, T]$-valued $(\mathcal{F}_t^W)$-stopping time $\tau$ and $\gamma(\cdot) \in \boldsymbol{\Gamma}^W$,*

$$V_{\theta T}\phi(x) \leq J(\tau, \theta, x, \gamma(\cdot); V_{\tau T}\phi). \tag{2.90}$$

(ii) *For $\varepsilon > 0$, there is $\gamma^{\varepsilon}(\cdot) \in \boldsymbol{\Gamma}^W$, such that*

$$V_{\theta t}\phi(x) + \varepsilon \geq J(\tau, \theta, x, \gamma^{\varepsilon}(\cdot); V_{\tau T}\phi) \tag{2.91}$$

*for any $[\theta, T]$-valued $(\mathcal{F}_t^W)$-stopping time $\tau$.*

*Proof.* Although the DP property was given in [FS06], Theorem 7.1, we will sketch the proof by using Proposition 2.4. Assume that $\theta = 0, \kappa = 0$, and $\Gamma$ is convex and compact.

Put $\tau_n = 2^{-n}[1+2^n\tau] \wedge T$. For $\gamma(\cdot) \in \boldsymbol{\Gamma}^W$, we are mainly interested in estimating $J(\tau_n, 0, x, \gamma(\cdot); V_{\tau_n T}\phi) - J(\tau, 0, x, \gamma(\cdot); V_{\tau T}\phi)$. We have

$$\begin{aligned} I_n &:= E_{0x} \int_{\tau}^{\tau_n} |f(s, X(s), \gamma(s))|\, ds \\ &\leq \sqrt{T} 2^{-\frac{n}{2}} \Big( E_{0x} \int_0^t K(1 + |X(s)|^4)\, ds \Big)^{\frac{1}{2}} \quad \text{(use (b}_4\text{))} \\ &\leq 2^{-\frac{n}{2}} c_1 (1 + |x|^2) \end{aligned} \tag{2.92}$$

with a constant $c_1$ independent of $n$ and $\gamma(\cdot)$.

Since $V_{tT}\phi(y)$ is continuous w.r.t. $(t, y)$,

$$|V_{\tau_n T}\phi(X(\tau_n))| \leq \bar{K}\Big(1 + \sup_{0 \leq t \leq T} |X(t)|^2\Big) \tag{2.93}$$

by (2.15), and the RHS of (2.93) is integrable, the dominated convergence theorem yields

$$\lim_{n \to \infty} E_{0x} |V_{\tau_n T}\phi(X(\tau_n)) - V_{\tau T}\phi(X(\tau))| = 0. \tag{2.94}$$

Let us fix $\tau_n$ and take a switching controls, $\gamma_l(\cdot) \in \boldsymbol{\Gamma}^W, l = 1, 2, \ldots,$ approaching $\gamma(\cdot)$, so that

$$\lim_{l \to \infty} E_{0x} \Big[ \sup_t |X^{\gamma_l(\cdot)}(t) - X(t)|^2 \Big] = 0 \tag{2.95}$$

and

$$\lim_{l \to \infty} \sup_t |X^{\gamma_l(\cdot)}(t) - X(t)| = 0 \quad P\text{-a.s.} \tag{2.96}$$

We take a sequence of divisions $\mathcal{D}_l, l = 1, 2, \ldots$, such that (2.60) holds and $\mathcal{D}_l$ contains all values of $\tau_n$ and switching times of $\gamma_l(\cdot)$. Then Corollary 2.2 yields

$$\lim_{m\to\infty} V_{\tau_n T}^{\mathcal{D}_{l+m}}\phi(X^{\gamma_l(\cdot)}(\tau_n)) = V_{\tau_n T}\phi(X^{\gamma_l(\cdot)}(\tau_n)) \quad P\text{-a.s. and in } L^1(\Omega). \tag{2.97}$$

On the other hand, Proposition 2.4 (ii) implies

$$V_{0T}^{\mathcal{D}_{l+m}}\phi(x) \le J(\tau_n, 0, x, \gamma_l(\cdot); V_{\tau_n T}^{\mathcal{D}_{l+m}}\phi). \tag{2.98}$$

Letting $m \to \infty$ and using (2.97), we obtain

$$V_{0T}\phi(x) \le E_{0x}\Big[\int_0^{\tau_n} f(s, X^{\gamma_l(\cdot)}(s), \gamma_l(s))\,ds + V_{\tau_n T}\phi(X^{\gamma_l(\cdot)}(\tau_n))\Big]. \tag{2.99}$$

Now taking $l \to \infty$ and $n \to \infty$ one obtains (i).

(ii) Let $\varepsilon > 0$ be given. By Corollary 2.2 and Remark 2.1, we can choose a switching control $\gamma^\varepsilon(\cdot) \in \boldsymbol{\Gamma}^W$, such that

$$V_{0T}\phi(x) + \varepsilon > J(T, 0, x, \gamma^\varepsilon(\cdot); \phi). \tag{2.100}$$

Suppose that $\tau$ is a finitely-many valued $(\mathcal{F}^W)$-stopping time, say $(t_k, k = 1, \ldots, p)$. Since $\gamma^\varepsilon(t)$ is a functional of $(W(\theta), \theta \le t_k)$ and $(W(s) - W(t_k), s \in [t_k, t])$, under $P(\cdot|\mathcal{F}_{t_k}^W)$, $\gamma^\varepsilon(\cdot)$ becomes a control process by freezing $(W(\theta), \theta \le t_k)$, and similarly for its response $X^\varepsilon(t)$. Hence

$$E\Big(\int_{t_k}^T f(s, X^\varepsilon(s), \gamma^\varepsilon(s))\,ds + \phi(X^\varepsilon(T))|\mathcal{F}_{t_k}^W\Big) \ge V_{t_k T}\phi(X^\varepsilon(t_k)) \quad P\text{-a.s.}. \tag{2.101}$$

Considering

$$\begin{aligned} J(T, 0, x, \gamma^\varepsilon(\cdot); \phi) = E_{0x}\sum_{k=1}^p \chi(\tau = t_k)\Big\{&\int_0^{t_k} f(s, X^\varepsilon(s), \gamma^\varepsilon(s))\,ds \\ &+ \int_{t_k}^T f(s, X^\varepsilon(s), \gamma^\varepsilon(s))\,ds + \phi(X^\varepsilon(T))\Big\} \end{aligned} \tag{2.102}$$

and applying (2.101) to (2.102), we have

$$\begin{aligned} &J(T, 0, x, \gamma^\varepsilon(\cdot); \phi) \\ &\ge E_{0x}\sum_{k=1}^p \chi(\tau = t_k)\Big\{\int_0^{t_k} f(s, X^\varepsilon(s), \gamma^\varepsilon(s))\,ds + V_{t_k T}\phi(X^\varepsilon(T))\Big\} \\ &= J(\tau, 0, x, \gamma^\varepsilon(\cdot); V_{\tau T}\phi). \end{aligned} \tag{2.103}$$

Since any stopping time can be approached by finitely-many valued ones, (2.100) and (2.103) complete the proof. □

*Example 2.2* (DPP *with hitting time*). Assume $f(\cdot) = 0$ and $\kappa(\cdot) = 0$. Let $\mathcal{O}$ be an open set. For $\gamma(\cdot) \in \boldsymbol{\Gamma}^W$, $X^{\gamma(\cdot)}$ and $\tau^{\gamma(\cdot)}$ denote its response and the hitting time of $\mathcal{O}$ by $X^{\gamma(\cdot)}$ respectively. Then DP property yields that the value function $v(\theta, x) = V_{\theta T}\phi(x)$ satisfies

$$v(\theta, x) = \inf_{\gamma(\cdot)\in\boldsymbol{\Gamma}^W} E_{\theta x} v(\tau^{\gamma(\cdot)},\ X^{\gamma(\cdot)}(\tau^{\gamma(\cdot)})) \quad \text{for } \theta \in [0, T), \quad x \notin \mathcal{O}. \tag{2.104}$$

### 2.2.5 Characterization of the Semigroup ($V_{\theta t}, \theta \le t$)

Referring to Theorems 2.1 and 2.4 we summarize the basic properties of the semigroup $(V_{\theta t}, \theta \le t), \theta, t \in [0, T]$.

**Proposition 2.7.** *$V_{\theta t} : \tilde{C} \to \tilde{C}$ satisfies the followings:*

(i) *$V_{\theta\theta} =$ identity, $\forall \theta \in [0, T]$,*

(ii) ***Semigroup property***

$$V_{\theta t} = V_{\theta s} V_{st} \quad \textit{for } \theta \le s \le t,$$

(iii) ***Monotonicity***

$$\phi \le \psi \Longrightarrow V_{\theta t}\phi \le V_{\theta t}\psi,$$

(iv) *$\exists \tilde{k} > 0$ such that $\forall \phi, \psi \in \tilde{C}$ and $\theta \le t$,*

$$\begin{cases} \| V_{\theta t}\phi - V_{\theta t}\psi \|_{\tilde{C}} \le \tilde{k} \|\phi - \psi\|_{\tilde{C}}, \\ \| V_{\theta t} 0 \|_{\tilde{C}} \le \tilde{k}, \end{cases} \tag{2.105}$$

(v) ***Continuity w.r.t. the time parameters***
*Let $R > 0$ be given. For any $\hat{t}$ and $\hat{\theta}$, we have*

$$\begin{cases} \lim_{t\to\hat{t}} \sup_{|x|\le R} |V_{\theta t}\phi(x) - V_{\theta\hat{t}}\phi(x)| = 0, \ \ \forall \theta < \hat{t}, \\ \lim_{\theta\to\hat{\theta}} \sup_{|x|\le R} |V_{\theta t}\phi(x) - V_{\hat{\theta} t}\phi(x)| = 0, \ \forall t > \hat{\theta}. \end{cases} \tag{2.106}$$

Next we will characterize $(V_{\theta t}, \theta \le t)$ from the point of view of semigroups.

For a constant control $\gamma(t) = \gamma$, $\forall t \in [0, T]$, its response is a diffusion governed by the SDE

$$dX(t) = b(t, X(t), \gamma)\, dt + \alpha(t, X(t), \gamma)\, dW(t).$$

Let us define $\mathrm{H}^{\gamma}_{\theta t} : \tilde{C} \mapsto \tilde{C}$ $(\theta \le t)$, by

$$\mathrm{H}^{\gamma}_{\theta t}\phi(x) = J(t,\theta,x,\gamma;\phi), \quad \forall x \in \mathbb{R}^d. \tag{2.107}$$

Then $(\mathrm{H}^{\gamma}_{\theta t};\theta \le t)$ is a semigroup satisfying properties (i)–(v) of Proposition 2.7 and

$$V_{\theta t}\phi \le \mathrm{H}^{\gamma}_{\theta t}\phi, \quad \forall \phi \in \tilde{C}. \tag{2.108}$$

**Theorem 2.5.** *Let* $\mathrm{U}_{\theta t}; \tilde{C} \mapsto \tilde{C}, \theta \le t$, *be a semigroup with the monotonicity property. If*

$$\mathrm{U}_{\theta t}\phi \le \mathrm{H}^{\gamma}_{\theta t}\phi, \quad \forall \gamma \in \Gamma, \quad \forall \phi \in \tilde{C}, \quad \forall \theta \le t, \tag{2.109}$$

*then*

$$\mathrm{U}_{\theta t}\phi \le V_{\theta t}\phi, \quad \forall \phi \in \tilde{C}, \quad \forall \theta \le t. \tag{2.110}$$

*In other words,* $(V_{\theta t},\theta \le t)$ *is the maximal element in the set of monotone semigroups satisfying* (2.109).

$(V_{\theta t},\theta \le t)$ is called the envelope of $\{(\mathrm{H}^{\gamma}_{\theta t},\theta \le t);\gamma \in \Gamma\}$.

The proof is easy, using (2.47), (2.48), and Corollary 2.2.

Finally, using the properties of the value function, we compute the generator of $(V_{\theta t},\theta \le t)$. Put

$$\mathcal{D} = \{\psi \in \tilde{C} \cap C^2_p(\mathbb{R}^d); \partial_x\psi \text{ and } \partial_{xx}\psi \text{ satisfy the polynomial growth condition}\}.$$

Then

$$\mathrm{G}^{\gamma}_{\theta} = \frac{1}{2}\mathrm{tr}(a(\theta,x,\gamma)\partial_{xx}) + b(\theta,x,\gamma)\cdot\partial x$$

is the generator of $X^{\gamma}$, with domain $\mathcal{D}$, and

$$\mathrm{G}^{\gamma}_{\theta}\psi(x) - \kappa(\theta,x,\gamma)\psi(x) + f(\theta,x,\gamma)$$

is the generator of $(\mathrm{H}^{\gamma}_{\theta t},\theta \le t)$. Put

$$\mathrm{G}_{\theta}\psi(x) := \inf_{\gamma\in\Gamma}(\mathrm{G}^{\gamma}_{\theta}\psi(x) - \kappa(\theta,x,\gamma)\psi(x) + f(\theta,x,\gamma)).$$

**Proposition 2.8.** *For any* $R_0 > 0$ *and* $\psi \in \mathcal{D}$,

$$\lim_{t\to\theta}\sup_{|x|\le R_0}\left|\frac{1}{t-\theta}(V_{\theta t}\psi(x) - \psi(x)) - \mathrm{G}_{\theta}\psi(x)\right| = 0. \tag{2.111}$$

Straightforward computations together with Proposition 2.1 give the following lemma.

**Lemma 2.1.** *Suppose that the Borel function $h$ defined on $[0,T] \times \mathbb{R}^d \times \Gamma$ is continuous w.r.t. $(t,x)$ uniformly on $\Gamma$ and satisfies*

$$|h(t,x,\gamma)| \leq K(1+|x|^{2p}), \quad \forall t,x,\gamma, \tag{2.112}$$

*with constants $K>0$ and $p \geq 1$. Then, for any positive $\varepsilon$ and $R_0$, there is $\delta_{\varepsilon R_0} > 0$ such that for any $\gamma(\cdot) \in \boldsymbol{\Gamma}^W$,*

$$\sup_{|x|\leq R_0} E_{\theta x}\Big[\sup_{\theta\leq s\leq t} |h(s,X^{\gamma(\cdot)}(s),\gamma(s)) - h(\theta,x,\gamma(s))|\Big] < \varepsilon \tag{2.113}$$

*whenever $|t-\theta| < \delta_{\varepsilon R_0}$.*

*Proof (Proposition 2.8).* Put

$$\begin{aligned} h(t,x,\gamma) =& \frac{1}{2}\mathrm{tr}(a(t,x,\gamma)\partial_{xx}\psi(x)) + b(t,x,\gamma)\cdot\partial_x\psi(x) \\ &- \kappa(t,x,\gamma)\psi(x) + f(t,x,\gamma), \quad \psi \in \mathcal{D}. \end{aligned}$$

Since $h(\cdot)$ satisfies the conditions of Lemma 2.1,

$$\begin{aligned} J_1 :=& \Big| \inf_{\gamma(\cdot)\in\boldsymbol{\Gamma}^W} E_{\theta x}\Big[\int_\theta^t h(s,X^{\gamma(\cdot)}(s),\gamma(s))\,ds\Big] \\ &- \inf_{\gamma(\cdot)\in\boldsymbol{\Gamma}^W} E_{\theta x}\Big[\int_\theta^t h(\theta,x,\gamma(s))\,ds\Big]\Big| \\ &< \varepsilon(t-\theta), \quad \forall |x| \leq R_0 \end{aligned} \tag{2.114}$$

whenever $|t-\theta| < \delta_{\varepsilon R_0}$.

On the other hand, (2.112) and $(b_5)$ yield

$$\begin{aligned} J_2 :=& E_{\theta x}\Big[\int_\theta^t |e^{-\int_\theta^s \kappa(\lambda,X^{\gamma(\cdot)}(\lambda),\gamma(\lambda))\,d\lambda} - 1||h(s,X^{\gamma(\cdot)}(s),\gamma(s))|\,ds\Big] \\ &< c_3 E_{\theta x}\int_\theta^t (s-\theta)(1+|X^{\gamma(\cdot)}(s)|^{2p})\,ds \\ &\leq \frac{c_3}{2}(t-\theta)^2 E_{\theta x}\Big[\sup_{\theta\leq s\leq t}(1+|X^{\gamma(\cdot)}(s)|^{2p})\Big] \\ &\leq c_4(t-\theta)^2(1+|x|^{2p}), \quad \forall \gamma(\cdot) \in \boldsymbol{\Gamma}^W, \end{aligned} \tag{2.115}$$

with constant $c_3$ and $c_4$ independent of $\gamma(\cdot), t, \theta$, and $x$.

Thus, Itô's formula together with (2.114) and (2.115) yield

$$\begin{aligned}&\left|V_{\theta t}\psi(x) - \psi(x) - \inf_{\gamma(\cdot)\in\boldsymbol{\Gamma}^W}\left[E_{\theta x}\int_\theta^t h(\theta,x,\gamma(s))\,ds\right]\right|\\ &\le J_1 + J_2 < (t-\theta)(c_4(1+R_0^{2p})(t-\theta)+\varepsilon), \quad \forall |x|\le R_0,\end{aligned} \tag{2.116}$$

whenever $|t-\theta| < \delta_{\varepsilon R_0}$.

Using the inequalities

$$\begin{aligned}\inf_{\gamma(\cdot)\in\boldsymbol{\Gamma}^W} E_{\theta x}\int_\theta^t h(\theta,x,\gamma(s))\,ds &\ge (t-\theta)\inf_{\gamma\in\boldsymbol{\Gamma}} h(\theta,x,\gamma)\\ &= \inf_{\gamma\in\Gamma}\int_\theta^t h(\theta,x,\gamma)\,ds\\ &\ge \inf_{\gamma(\cdot)\in\boldsymbol{\Gamma}^W} E_{\theta x}\int_\theta^t h(\theta,x,\gamma(s))\,ds,\end{aligned}$$

we have

$$\inf_{\gamma(\cdot)\in\boldsymbol{\Gamma}^W} E_{\theta x}\int_\theta^t h(\theta,x,\gamma(s))\,ds = (t-\theta)\inf_{\gamma\in\Gamma} h(\theta,x,\gamma). \tag{2.117}$$

Inserting (2.117) into (2.114) and referring to Proposition 2.5, we complete the proof. □

Proposition 2.8 asserts that if the value function is smooth, then it satisfies the Cauchy problem for the (nonlinear) parabolic equation;

$$-\frac{\partial V}{\partial t}(t,x) + \mathrm{H}(t,x,\partial_{xx}V(t,x),\partial_x V(t,x),V(t,x)) = 0, \quad \forall (t,x)\in[0,T)\times\mathbb{R}^d \tag{2.118}$$

with the lateral boundary condition

$$V(T,x) = \phi(x), \quad x\in\mathbb{R}^d, \tag{2.119}$$

where $\mathrm{H} : [0,T]\times\mathbb{R}^d\times S^d\times\mathbb{R}^d\times\mathbb{R}^1 \mapsto \mathbb{R}^1$, is given by

$$\begin{aligned}&\mathrm{H}(t,x,A,P,v)\\ &= \sup_{\gamma\in\Gamma}\Big(-\frac{1}{2}\mathrm{tr}(a(t,x,\gamma)A) - b(t,x,\gamma)\cdot p + \kappa(t,x,\gamma)v - f(t,x,\gamma)\Big).\end{aligned} \tag{2.120}$$

The parabolic equation (2.118) is called Hamilton–Jacobi–Bellman equation (HJB equation in short), or the dynamic programming equation.

Viscosity solutions of HJB equations will be considered in Chap. 3.

*Example 2.3 (Time-homogeneous case).* Suppose that all the coefficients $\alpha, b, \kappa$ and $f$ are independent of the time variable. Here we use Brownian adapted control processes. In this case, we have

$$\nu_{\theta t} = \nu_{0(t-\theta)}. \tag{2.121}$$

Indeed, for $\gamma(\cdot) \in \boldsymbol{\Gamma}^W$, its response is the solution of the SDE

$$\begin{cases} dX(t) = b(X(t), \gamma(t))\,dt + \alpha(X(t), \gamma(t))\,dW(t), \quad t > \theta, \\ X(\theta) = x. \end{cases} \tag{2.122}$$

If $\gamma(\cdot + \theta)$ is adapted to $W^{\theta}(\cdot) := W(\cdot + \theta) - W(\theta)$, then $\hat{\gamma}(t) := \gamma(t + \theta)$ can be regarded as an element of $\boldsymbol{\Gamma}^{W^{\theta}}$ and the solution $\hat{X}$ of the SDE

$$\begin{cases} d\hat{X}(t) = b(\hat{X}(t), \hat{\gamma}(t))\,dt + \alpha(\hat{X}(t), \hat{\gamma}(t))\,dW^{\theta}(t), \quad t > 0, \\ \hat{X}(0) = x, \end{cases} \tag{2.123}$$

is nothing but $X(\cdot + \theta)$. Consequently,

$$\begin{aligned} V_{\theta t}\phi(x) &\le \inf_{\gamma(\cdot) \in \boldsymbol{\Gamma}^{W^{\theta}}} J(t, \theta, x, \gamma(\cdot); \phi) \\ &= V_{0(t-\theta)}\phi(x). \end{aligned} \tag{2.124}$$

However, we have seen in (2.101) that for any $\gamma(\cdot) \in \boldsymbol{\Gamma}^W$,

$$\begin{aligned} E_{\theta x} C(t, \theta, \gamma(\cdot); \phi) &= E_{\theta x}\left[E(C(t, \theta, \gamma(\cdot); \phi) | \mathcal{F}^W_{\theta})\right] \\ &\ge \inf_{\gamma(\cdot) \in \boldsymbol{\Gamma}^{W^{\theta}}} J(t, \theta, x, \gamma(\cdot); \phi). \end{aligned} \tag{2.125}$$

Now (2.124) and (2.125) yield (2.121).

Putting $V_t := V_{0t}$, we have the semigroup $(V_t, t \in [0, T])$ with the generator G given by

$$\begin{aligned} \mathrm{G}\psi(x) &= \lim_{t \to 0} \frac{1}{t}(V_t\psi(x) - \psi(x)) \\ &= \inf_{\gamma \in \Gamma}\Big(\frac{1}{2}\mathrm{tr}(a(x, \gamma)\partial_{xx}\psi) + b(x, \gamma) \cdot \partial_x \psi - \kappa(x, \gamma)\psi + f(x, \gamma)\Big) \end{aligned} \tag{2.126}$$

for $\psi \in \mathcal{D}$.

## 2.3 Verification Theorems and Optimal Controls

In this section we show how to construct optimal controls or optimal Markovian policies using the HJB equations. In the previous section, we have shown, using DPP and Itô's formula, that under appropriate regularity assumption the value function satisfies the HJB equation. However, it is rather difficult to verify the regularity of value function. But, if a classical solution of HJB equation exists, then we can take an optimal Markovian policy by using the minimum selector and obtain the value function. This assertion is called a verification theorem.

In Sect. 2.3.1, we state verification theorems, which are useful in seeking optimal controls in practical problems. In Sect. 2.3.2, we give three simple examples.

### *2.3.1 Verification Theorems*

Here we assume the following conditions $(b_1)'$ and $(b_2)'$ weaker than $(b_1)$–$(b_5)$:

$(b_1)'$

$$b : [0,T] \times \mathbb{R}^d \times \Gamma \longmapsto \mathbb{R}^d, \quad \alpha : [0,T] \times \mathbb{R}^d \times \Gamma \longmapsto \mathbb{R}^d \otimes \mathbb{R}^m$$

are uniformly continuous and satisfy

$$\begin{aligned}&|b(t_1,x_1,\gamma_1) - b(t_2,x_2,\gamma_2)| + |\alpha(t_1,x_1,\gamma_1) - \alpha(t_2,x_2,\gamma_2)|\\ \le& l|x_1 - x_2| + m(|t_1 - t_2| + |\gamma_1 - \gamma_2|),\end{aligned} \tag{2.127}$$

with a constant $l$ and $a$ modulus function $m(\cdot)$, and

$$|b(t,x,\gamma)| + |\alpha(t,x,\gamma)| \le K(1 + |x| + |\gamma|), \tag{2.128}$$

with a constant $K$.

$(b_2)'$

$$\kappa : [0,T] \times \mathbb{R}^d \times \Gamma \mapsto [0,c_0], \quad f : [0,T] \times \mathbb{R}^d \times \Gamma \mapsto \mathbb{R}^1, \quad \phi : \mathbb{R}^d \mapsto \mathbb{R}^1$$

are continuous and

$$|f(t,x,\gamma)| + |\phi(x)| \le \hat{K}(1 + |x|^2 + |\gamma|^2), \tag{2.129}$$

with a constant $\hat{K}$.

Here we admit a slightly bigger class of control processes. Specifically let $(\Omega, \mathcal{F}, (\mathcal{F}_t), P, W)$ be a reference probability system. We take $(\gamma_*(t), t \in [0,T])$ to be a $\Gamma$-valued $(\mathcal{F}_t)$-progressively measurable process, satisfying

$$E\Big[\int_0^T |\gamma_*(t)|^n\,ds\Big] < \infty \quad \text{for } n = 1, 2, \ldots. \tag{2.130}$$

This condition is sometimes convenient for establishing the existence of optimal controls (see Example 2.4).

For a control $A_* = (\Omega, \mathcal{F}, (\mathcal{F}_t), P, W, \gamma_*(\cdot))$, its response evolves according to the SDE

$$dX(t) = b(t, X(t), \gamma_*(t))\,dt + \alpha(t, X(t), \gamma_*(t))\,dW(t), \quad t \in [\theta, T], \tag{2.131}$$

with the initial condition

$$X(\theta) = x. \tag{2.132}$$

We clearly have a unique solution of (2.3)–(2.4), satisfying

$$E_{\theta x}\Big[\sup_{\theta\le s\le t} |X(s)|^{2p}\Big] \le c_p\Big(1 + |x|^{2p} + E\int_\theta^t |\gamma_*(s)|^{2p}\,ds\Big), \quad \forall \theta, t, x, A_*, \tag{2.133}$$

with a constant $c_p$.

By (2.131) and $(b_2)'$, we can define the payoff $J(T, \theta, x, A_*; \phi)$ by (2.6) and the value function $v_*(\cdot)$ by

$$v_*(\theta, x) = \inf_{A_*} J(T, \theta, x, A_*; \phi), \tag{2.134}$$

respectively.

*Remark 2.2.*

$$v_*(\theta, x) = v(\theta, x).$$

*Proof.* That $v_*(\theta, x) \le v(\theta, x)$ is clear. For the opposite inequality, we fix $A_* = (\Omega, \mathcal{F}, (\mathcal{F}_t), P, W, \gamma_*(\cdot))$ arbitrarily and claim that

$$J(T, \theta, x, A_*; \phi) \ge v(\theta, x). \tag{2.135}$$

Put $\gamma_N(t) = \gamma_*(t)\chi(|\gamma_*(t)| \le N)$ and $A_N = (\Omega, \mathcal{F}, (\mathcal{F}_t), P, \gamma_N(\cdot))$. Then $A_N \in \mathbb{A}$ and

$$\lim_{N\to\infty} E\Big[\int_0^T |\gamma_*(s)-\gamma_N(s)|^n\,ds\Big] = \lim_{N\to\infty} E\Big[\int_0^T |\gamma_*(s)|^n \chi(|\gamma_*(s)| > N)\,ds\Big] = 0. \tag{2.136}$$

$X$ and $X_N$ denote the responses for $\gamma_*(\cdot)$ and $\gamma_N(\cdot)$, respectively.

Let $\rho_N(t) := E_{\theta x}[\sup_{\theta \le s \le t} |X(s) - X_N(s)|^2]$. From (2.127) and the Burkholder–Davis–Gundy inequality, we easily deduce that

$$\rho_N(t) \le cE\Big(\int_\theta^t \{\rho_N(s) + (1 + |\gamma_*(s)|^2)\chi(|\gamma_*(s)| \ge N)\}\, ds\Big), \tag{2.137}$$

with a constant $c$ independent of $\theta, x, t, N$, and $\gamma_*(\cdot)$.

Hence, (2.136) and (2.137) yield

$$\lim_{N\to\infty} E_{\theta x}\Big[\sup_{\theta \le s \le T} |X(s) - X_N(s)|^2\Big] = 0.$$

Choosing a subsequence $N_j$ so that

$$\lim_{N_j\to\infty} \sup_{\theta \le s \le T} |X(s) - X_{N_j}(s)| = 0 \quad P\text{-a.s.}$$

and using (2.133), we conclude that

$$\lim_{N_j\to\infty} J(T, \theta, x, A_{N_j}; \phi) = J(T, \theta, x, A_*; \phi),$$

which in turn yields (2.135). □

Let us consider the HJB equation;

$$-\frac{\partial V}{\partial t}(t,x) + \mathrm{H}(t, x, \partial_{xx}V(t,x), \partial_x V(t,x), V(t,x)) = 0, \quad \forall (t,x) \in [0,T) \times \mathbb{R}^d \tag{2.118$'$}$$

with the lateral boundary condition

$$v(T, x) = \phi(x), \quad x \in \mathbb{R}^d. \tag{2.119$'$}$$

The following result holds true.

**Theorem 2.6 (([FS06], p. 159) *Verification theorem*).**

*Let $\omega \in C^{12}([0,T) \times \mathbb{R}^d) \cap C([0,T] \times \mathbb{R}^d)$ be a solution of (2.118′)–(2.119′), satisfying*

$$|\omega(t,x)| \le K_1(1 + |x|^2), \quad \forall t, x, \tag{2.138}$$

*with a constant $K_1 > 0$.*

*Then*

(i) *For any $A_* = (\Omega, \mathcal{F}, (\mathcal{F}_t), P, W, \gamma_*(\cdot))$,*

$$\omega(t,x) \le J(T, t, x, A_*; \phi), \quad \forall t, x. \tag{2.139}$$

(ii) *Let* $(t, x)$ *be given. Suppose that there exists* $\hat{A}_* = (\hat{\Omega}, \hat{\mathcal{F}}, (\hat{\mathcal{F}}_t), \hat{P}, \hat{W}, \hat{\gamma}_*(\cdot))$ *such that*

$$\begin{aligned}&\min_{\gamma\in\Gamma}(\mathrm{G}_s^{\gamma}\omega(s,\hat{X}(s)) - \kappa(s,\hat{X}(s),\gamma)\omega(s,\hat{X}(s)) + f(s,\hat{X}(s),\gamma))\\ =&\mathrm{G}_s^{\hat{\gamma}_*(s)}\omega(s,\hat{X}(s)) - \kappa(s,\hat{X}(s),\hat{\gamma}_*(s))\omega(s,\hat{X}(s)) + f(s,\hat{X}(s),\hat{\gamma}_*(s)),\\ &\forall s\in[t,T],\quad \hat{P}\text{-a.s.},\end{aligned} \tag{2.140}$$

*where* $\hat{X}$ *is the response for* $\hat{A}_*$ *with* $\hat{X}(t) = x$.

*Then*

$$\omega(t,x) = J(T,t,x,\hat{A}_*;\phi). \tag{2.141}$$

(iii) $\hat{A}_*$ *is optimal for* $(t, x)$, *i.e.*,

$$v_*(t,x) = \omega(t,x) = J(T,t,x,\hat{A}_*;\phi). \tag{2.142}$$

For the case of optimal Markovian policies, we deduce from Theorem 2.6 and Proposition 1.6 the following

**Corollary 2.3.** *In addition to* $(b_1)'$ *and* $(b_2)'$, *we assume that*

(a) $\Gamma$ *is convex and compact,*
(b) $d = m$,
(c) $b(\cdot)$ *is bounded,*
(d) $\alpha(\cdot)$ *is bounded, symmetric and uniformly parabolic.*

*Suppose that the HJB equation (2.118′)–(2.119′) has a classical solution, satisfying the growth condition* (2.138). *Then the maximum selector of* $\mathrm{H}(\cdot)$ *provides an optimal Markovian policy.*

Regarding the classical solution of HJB equation, we recall the following result.

**Theorem 2.7 ([FS06], p. 163).** *Let* $\kappa = 0$ *and assume;*

(a) $\Gamma$ *is compact and* $d = m$,
(b) $\alpha(\cdot)$ *is uniformly parabolic and independent of* $\gamma$. *Moreover,* $\alpha(\cdot) \in C^{12}([0,T]\times \mathbb{R}^d)$ *and* $\alpha(\cdot), \alpha(\cdot)^{-1}$, *and* $\partial_x\alpha(\cdot)$ *are bounded,*
(c) $b(t,x,\gamma) = \tilde{b}(t,x) + \alpha(t,x)\Theta(t,x,\gamma)$, *where* $\tilde{b}(\cdot) \in C^{12}([0,T]\times\mathbb{R}^d)$ *and* $\partial_x\tilde{b}(\cdot)$ *is bounded, and* $\Theta(\cdot)$ *and* $\partial_x\Theta(\cdot)$ *are in* $C_b([0,T]\times\mathbb{R}^d\times\Gamma)$,
(d) $f(\cdot)$ *and* $\partial_x f(\cdot)$ *are in* $C_p([0,T]\times\mathbb{R}^d\times\Gamma)$,
(e) $\phi(\cdot) \in C^3(\mathbb{R}^d)\cap C_p^1(\mathbb{R}^d)$.

*Then the* HJB *equation* (2.118′)–(2.119′) *has a unique classical solution.*

### 2.3.2 Examples of Optimal Control

We are concerned with explicit formulations of optimal controls for three simple models. The first two examples are related to Gaussian diffusions effected by linear control processes. The third one is a stochastic control with state constraint.

*Example 2.4 (Linear Gaussian quadratic regulator).* Let $\Gamma = \mathbb{R}^q$ and suppose that matrix-valued continuous functions of suitable sizes $A, B, \sigma, M, N$ and $D$ are given on $[0, T]$. For an admissible control $A_* = (\Omega, \mathcal{F}, (\mathcal{F}_t), P, W, \gamma_*(\cdot))$, we have the $d$-dimensional SDE

$$dX(t) = (A(t)X(t) + B(t)\gamma_*(t))\,dt + \sigma(t)\,dW(t), \quad t \in [\theta, T], \tag{2.143}$$

with the initial condition

$$X(\theta) = x (\in \mathbb{R}^d) \tag{2.144}$$

and the payoff

$$\begin{aligned} &J(T, \theta, x, A_*) \\ &= E_{\theta x}\Big[\int_\theta^T (X(s)^\top M(s)X(s) + \gamma_*(s)^\top N(s)\gamma(s))\,ds + X(T)^\top DX(T)\Big]. \end{aligned} \tag{2.145}$$

We assume that for any $t \in [0, T]$,

(a) $D, M(t)$ are non-negative definite symmetric $d \times d$ matrices;
(b) $N(t)$ is a positive definite symmetric $q \times q$ matrix.

Since the corresponding HJB equation reads

$$\begin{aligned} 0 =& \frac{\partial \omega}{\partial t}(t, x) + \frac{1}{2}\mathrm{tr}(a(t)\partial_{xx}\omega(t, x)) + (A(t)x) \cdot \partial_x\omega(t, x) \\ &+ x^\top M(t)x + \inf_{\gamma \in \Gamma}((B(t)\gamma) \cdot \partial_x\omega(t, x) + \gamma^\top N(t)\gamma), \\ & \qquad t \in [0, T), \quad x \in \mathbb{R}^d, \end{aligned} \tag{2.146}$$

with the lateral boundary condition

$$\omega(T, x) = x^\top Dx, \tag{2.147}$$

we are looking for the solution of (2.146)–(2.147).

Observing that the expression whose inf is taken in (2.146) is

$$\left|\frac{1}{2}N(t)^{-\frac{1}{2}}B(t)^\top\partial_x\omega(t, x) + N(t)^{\frac{1}{2}}\gamma\right|^2 - \frac{1}{4}|N(t)^{-\frac{1}{2}}B(t)^\top\partial_x\omega(t, x)|^2 \tag{2.148}$$

and assuming that $\omega(\cdot)$ is quadratic, we have

$$\omega(t,x) = x^\top P(t)x + \int_t^T \operatorname{tr}(a(s)P(s))\,ds \tag{2.149}$$

where $P(\cdot)$ is the solution of the following Riccati equation:

$$\begin{aligned} 0 =& \frac{dP}{dt}(t) + M(t) + A(t)^\top P(t) + P(t)A(t) \\ &- P(t)B(t)N(t)^{-1}B(t)^\top P(t), \quad t \in [0,T) \end{aligned} \tag{2.150}$$

with

$$P(T) = D. \tag{2.151}$$

Hence, from (2.148) and (2.149) it follows that an optimal Markovian policy is given by $-N(t)^{-1}B(t)^\top P(t)x$.

More details for LQ problems are given in [YZ99], Chapter 6.

*Example 2.5 (1-dimensional bang-bang control).* Let $\Gamma = [-1,1]$. Suppose that $g \in C_p^1(\mathbb{R}^1)$ is even, $g(0) = 0$ and, $g'(x) \geq 0$, on $[0,\infty)$.

Let us consider the following time-homogeneous simple model. For $A = (\Omega, \mathcal{F}, (\mathcal{F}_t), P, W, \gamma(\cdot))$, its response $X^A$ evolves according to the SDE

$$dX(t) = \gamma(t)\,dt + dW(t), \quad t \in (0,T] \tag{2.152}$$

with the initial condition

$$X(0) = x \ (\in \mathbb{R}^1), \tag{2.153}$$

and the payoff $j(\cdot)$ is given by

$$j(t,x,A) = E_x g(X^A(t)). \tag{2.154}$$

The (formal) HJB equation for the value function $v(t,x) = \inf_{A\in\mathbb{A}} j(t,x,A)$ reads

$$\begin{cases} \dfrac{\partial v}{\partial t}(t,x) &= \dfrac{1}{2}\partial_{xx}v(t,x) + \inf\limits_{|\gamma|\leq 1}(\gamma\partial_x v(t,x)) \\ &= \dfrac{1}{2}\partial_{xx}v(t,x) - |\partial_x v(t,x)|, \quad t > 0, \quad x \in \mathbb{R}^1, \\ v(0,x) &= g(x), \quad x \in \mathbb{R}^1. \end{cases} \tag{2.155}$$

Now we look for the explicit solution of (2.155).

**Proposition 2.9 (Communicated by F. Asakura).**

$$
\begin{aligned}
v(t,x) &= \frac{1}{\sqrt{2\pi t}} e^{-\frac{t}{2}} \int_{-\infty}^{\infty} g(y) \exp\Big(-\frac{|x|^2 - y}{2t} + |x| - |y|\Big)\, dy \\
&- \frac{1}{2\sqrt{2\pi}} \int_{-\infty}^{\infty} g(y) e^{|x|-|y|} \Big( \int_t^{\infty} s^{-\frac{3}{2}} (|x|+|y| + s) \exp\Big(-\frac{|x| + |y| + s^2}{2s}\Big)\, ds\Big)\, dy \\
&+ \int_{-\infty}^{\infty} e^{-2|y|} g(y)\, dy
\end{aligned}
\tag{2.156}
$$

*and*

$$
\operatorname{sgn}(\partial_x v(t,x)) = \operatorname{sgn} x = \begin{cases} 1, & x > 0, \\ 0, & x = 0, \\ -1, & x < 0. \end{cases} \tag{2.157}
$$

*Outline of proof.* Consider the auxiliary equation

$$
\frac{\partial \omega}{\partial t}(t,x) = \partial_{xx}\omega(t,x), \quad t > 0, \quad x > 0 \tag{2.158}
$$

with the boundary and initial conditions

$$
\begin{cases} \partial_x \omega(t,0) + \dfrac{1}{2}\omega(t,0) = 0, & t > 0, \\ \omega(0,x) = e^{-\frac{x}{2}} g\Big(\dfrac{x}{2}\Big), & x > 0. \end{cases} \tag{2.159}
$$

We solve (2.158)–(2.159) using the Laplace transformation. Putting $u(t,x) = \omega(t,x)e^{\frac{2x-t}{4}}$, we have

$$
\partial_x u(t,x) > 0 \quad \text{for } x > 0 \tag{2.160}
$$

and

$$
\lim_{t\to 0} u(t,x) = g\Big(\frac{x}{2}\Big). \tag{2.161}
$$

Further, $v(t,x) := u(2t, 2|x|)$, $t > 0$, $x \in \mathbb{R}^1$ is given by expression (2.156) and satisfies (2.155). From (2.156), it follows that

$$
\operatorname{sgn}(\partial_x v(t,x)) = \operatorname{sgn} x. \tag{2.162}
$$

This completes the proof. □

By Theorem 2.6, $v(\cdot)$ is the value function and the minimum selector $\hat{\gamma}(t,x) = -\mathrm{sgn}(\partial_x v(t,x)) = -\mathrm{sgn}\, x$ gives an optimal Markovian policy. Its response $X$ evolves according to SDE

$$dX(t) = -\mathrm{sgn}\, X(t) + dW(t). \tag{2.163}$$

Since (2.163) admits a unique strong solution, $X$ is the diffusion with generator $\frac{1}{2}\frac{d^2}{dx^2} - (\mathrm{sgn}\, x)\frac{d}{dx}$ and $e^{-2|x|}$ gives the density of the corresponding invariant probability measure. From (2.156) it follows that for any $R > 0$, there is a constant $C_R > 0$ such that

$$\sup_{|x|\le R} \left| v(t,x) - \int_{-\infty}^{\infty} e^{-2|y|} g(y)\, dy \right| \le C_R e^{-\frac{t}{2}}, \quad \forall t > 1. \tag{2.164}$$

For related topics consult [Be75, BSW80] and [IW81], Chap. 6.

Finally we will consider stochastic control with state constraint. We assume that the coefficients $\alpha, b$, and $f$ are independent of the time variable, and satisfy $(b_1)'$ and $(b_2)'$.

Let $\mathcal{O}$ be a bounded and open set of $\mathbb{R}^d$ with smooth boundary $\partial\mathcal{O}$. For an admissible control $A = (\Omega, \mathcal{F}, (\mathcal{F}_t), P, W, \gamma(\cdot))$, the response $X^A$ evolves according to the SDE

$$\begin{cases} dX(t) = b(x(t), \gamma(t))\, dt + \alpha(X(t), \gamma(t))\, dW(t), & t > 0, \\ X(0) = x. \end{cases} \tag{2.165}$$

By $\tau^A$ we denote the exit time of $X^A$ from $\mathcal{O}$, that is

$$\tau^A = \begin{cases} \inf\{s > 0; X^A(s) \notin \mathcal{O}\}, \\ \infty, \quad \text{if } \{\cdots\} \text{ is empty.} \end{cases} \tag{2.166}$$

For a given continuous function $g$ on $\mathcal{O}$ and a positive constant $\kappa$, we define the payoff $J(\cdot)$ by

$$J(x, A) = E_x\Big[\int_0^{\tau^A} e^{-\kappa s} f(X^A(s), \gamma(s))\, ds + e^{-\kappa\tau^A} g(X^A(\tau^A))\Big], \quad x \in \mathcal{O}; \tag{2.167}$$

here, when $\tau^A = \infty$, $e^{-\kappa\tau^A} g(X^A(\tau^A))$ stands for 0.

When $\gamma(\cdot)$ is constant $\gamma \in \Gamma$, its response $X^A$ is the diffusion with the generator $\mathrm{G}^\gamma$ given by

$$\mathrm{G}^\gamma u = \frac{1}{2}\mathrm{tr}(a(x,\gamma)\partial_{xx} u) + b(x,\gamma) \cdot \partial_x u \tag{2.168}$$

and $J(\cdot, A)$ satisfies the elliptic PDE with boundary value $g$,

$$\begin{cases} \mathrm{G}^{\gamma} J(x, A) - \kappa J(x, A) + f(x, \gamma) = 0, & \forall x \in \mathcal{O}, \\ \lim\limits_{y \to x} J(y, A) = g(x), & x \in \partial \mathcal{O}, \end{cases} \tag{2.169}$$

provided $J(\cdot, A)$ is smooth up to boundary. $v(x) := \inf_{A \in \mathbb{A}} J(x, A)$ is called the value function.

Let us consider the following Dirichlet problem for the HJB equation:

$$\begin{cases} \inf\limits_{\gamma \in \Gamma} (\mathrm{G}^{\gamma} u(x) - \kappa u(x) + f(x, \gamma)) = 0, & x \in \mathcal{O}, \\ u(x) = g(x), & x \in \partial \mathcal{O}. \end{cases} \tag{2.170}$$

Then, the following result holds true.

**Theorem 2.8 (Verification theorem).** *Let $u(\cdot)$ be a classical solution of* (2.170). *Then, it is valid that*

(i) $u(x) \leq J(x, A), \ \forall x \in \mathcal{O}, \forall A \in \mathbb{A}$,
(ii) *Let $x_0 \in \mathcal{O}$ be given. If there exists $A^* = (\Omega^*, \mathcal{F}^*, (\mathcal{F}_t^*), P^*, W^*, \gamma^*(\cdot))$ and its response $X^*$ with $X^*(0) = x_0$, such that*

$$\begin{aligned} 0 = & \min_{\gamma \in \Gamma} (\mathrm{G}^{\gamma} u(X^*(t)) - \kappa u(X^*(t)) + f(X^*(t), \gamma)) \\ = & \mathrm{G}^{\gamma^*(t)} u(X^*(t)) - \kappa u(X^*(t)) + f(X^*(t), \gamma^*(t)), \\ & \forall t < \tau^{A^*} \quad P\text{-a.s.}, \end{aligned} \tag{2.171}$$

*then*

$$u(x_0) = J(x_0, A^*). \tag{2.172}$$

*From* (i) *and* (2.172), *it follows that $u(x_0) = v(x_0)$ and $A^*$ gives an optimal control for the initial state $x_0$.*

Regarding the classical solution of (2.170), we can refer to [E83], [E98], [Mo10], Sect. 5.3.

*Example 2.6 (Production planning problem).* Consider the production planning for one commodity, in the presence of random demand, fluctuating according to the SDE

$$dZ(t) = b\, dt - \sigma\, dW, \tag{2.173}$$

where the positive constant $b$ represents the expected demand rate and $\sigma$ and $W$ denote a positive constant and a real Wiener process. The firm adjusts its production rate $\gamma(\cdot)$ and the finished products are stored in a buffer of size $\bar{K}$. Since the running cost $f(\cdot)$ and the terminal cost $g(\cdot)$ are needed, the firm's objective is to minimize its expected total cost by choosing a suitable $\gamma(\cdot)$.

Let us formulate the problem in precise terms. Let $K$ be the production capacity and put $\Gamma = [0, K]$. Then, for $A = (\Omega, \mathcal{F}, (\mathcal{F}_t), P, W, \gamma(\cdot))$, the inventory level $X(t)$ evolves according to the SDE

$$\begin{cases} dX(t) = \gamma(t)\,dt - dZ(t) = (\gamma(t) - b)\,dt + \sigma\,dW(t), & t > 0, \\ X(0) = x. \end{cases}$$

$X(t) > 0$ (resp. $< 0$) means a surplus (resp. backlog) of product. Since the inventory state cannot exceed the buffer size $\overline{K}$ and the firm fixes the backlog size $\underline{K}(< 0)$, because a new policy is needed for a large amount of backlog, we give the payoff by

$$J(x, A) = E_x\Big[\int_0^{\tau_A} e^{-\kappa s} f(X^A(s), \gamma(s))\,ds + e^{-\kappa\tau_A} g(X^A(\tau^A))\Big], \tag{2.174}$$

where $\tau^A$ = exit time of $X^A$ from $(\underline{K}, \overline{K}) := \mathcal{O}$.

We suppose that $f(\cdot)$ is given by

$$f(x, \gamma) = p(x \vee 0) + \gamma^2, \tag{2.175}$$

with a positive constant $p$.

The corresponding HJB equation reads

$$\frac{1}{2}\sigma^2 u''(x) - bu'(x) - \kappa u(x) + \inf_{\gamma\in\Gamma}(\gamma u'(x) + \gamma^2) + p(x \vee 0) = 0, \quad \forall x \in (\underline{K}, \overline{K}), \tag{2.176}$$

with the boundary conditions

$$u(\underline{K}) = g(\underline{K}), \quad u(\overline{K}) = g(\overline{K}).$$

Since there exists a unique classical solution of (2.176), the verification theorem asserts that the solution is equal to the value function and an optimal Markovian policy is given by the minimum selector;

$$\hat{\gamma}(x) = \begin{cases} 0 & \text{on } \{x \in \mathcal{O}; u'(x) \geq 0\}, \\ -\dfrac{u'(x)}{2} & \text{on } \{x \in \mathcal{O}; u'(x) \in (-2K, 0)\}, \\ K & \text{on } \{x \in \mathcal{O}; u'(x) \leq -2K\}. \end{cases}$$

Refer to [Mo10], Chapter 6 for related topics.

## 2.4 Optimal Investment Models

This section is devoted to finite time horizon optimal investment problems. We consider a market with one bond and $m(\geq 1)$ risky assets, where $d$ factor processes, $X^1, \ldots, X^d$, determine the performance of the market. This model, called a factor model, was introduced by Merton [Me71] (refer to [Me73, HP81, Na03]).

Suppose that the factor process $X = (X^1, \ldots, X^d)$ evolves according to the SDE

$$dX = b(t, X(t))\, dt + d W(t). \tag{2.177}$$

The prices of the bond and of the $i$-th asset are given by

$$\begin{cases} dS^0(t) = S^0(t) r(t, X(t))\, dt, \quad t > 0, \\ S^0(0) = s^0 > 0, \end{cases} \tag{2.178}$$

and by the SDE

$$\begin{cases} dS^i(t) = S^i(t)\Big(g^i(t, X(t))\, dt + \sum_{j=1}^{d+m} \sigma^i_j(t, X(t))\, dW^j\Big), \quad t > 0, \\ S^i(0) = s^i > 0, \quad i = 1, \ldots, m, \end{cases} \tag{2.179}$$

respectively, where $W = (W^1, \ldots, W^d)$ is the Wiener process of (2.177).

Let us consider an agent who invests at any $t \in (0, T)$ a proportion $\pi^i(t)$ of his/her wealth in the $i$-th risky asset ($i = 1, \ldots, m$) and $\pi^0(t) = 1 - \sum_{i=l}^m \pi^i(t)$ in the bond. The agent wants to maximize the expected utility from the terminal wealth by choosing a suitable $(\pi^i(\cdot), i = 1, \ldots, m)$.

As an application of our previous results, we study the problem by using a DPP argument. We note that the martingale approach is also powerful (refer to [HP81] and [HP83]).

### 2.4.1 *Formulations*

Let

$$\begin{aligned} \sigma &: [0, T] \times \mathbb{R}^d \longmapsto \mathbb{R}^m \otimes \mathbb{R}^{d+m}, \\ b &: [0, T] \times \mathbb{R}^d \longmapsto \mathbb{R}^d, \\ g &: [0, T] \times \mathbb{R}^d \longmapsto \mathbb{R}^m, \\ r &: [0, T] \times \mathbb{R}^d \longmapsto [0, K_0] \end{aligned}$$

be continuous and Lipschitz continuous in the variable $x \in \mathbb{R}^d$, uniformly on $[0, T]$. We assume that

(a) $a(t,x) := \sigma(t,x)\sigma(t,x)^\top$ is uniformly parabolic, i.e.,

$$y^\top a(t,x)y \geq \lambda_0|y|^2, \quad \forall y \in \mathbb{R}^m, \quad \forall t,x,$$

for a positive constant $\lambda_0$,

(b) $\sigma(t,x)^\top a(t,x)^{-1}\sigma(t,x)$ is bounded,

(c) $g(t,x)^\top a(t,x)^{-1}g(t,x)$ and $\sigma(t,x)^\top a(t,x)^{-1}g(t,x)$ satisfy linear growth condition in $x$ uniformly on $[0,T]$.

Let $\tilde{W} = (W^1,\ldots,W^d,W^{d+1},\ldots,W^{d+m})$ be a $(d+m)$-dimensional Wiener process, defined on $(\Omega,\mathcal{F},P)$, and put $W = (W^1,\ldots,W^d)$ (= first $d$-component of $\tilde{W}$).

Hence, the strong solution of SDE (2.177) is a diffusion process with the generator $\mathrm{G}_t$:

$$\mathrm{G}_t\psi = \frac{1}{2}\Delta\psi + b(t,x)\cdot\partial_x\psi, \tag{2.180}$$

where $\Delta$ is the $d$-dimensional Laplacian operator.

Further,

$$\mathcal{F}_t^W = \mathcal{F}_t^X, \quad \forall t \in [0,T],$$

if $X(0)$ =constant.

Since $S^0$ and $S^i$ evolve according to (2.178) and (2.179), respectively, we have

$$S^0(t) = s^0\exp\Big(\int_0^t r(s,X(s))\,ds\Big) \tag{2.181}$$

and

$$S^i(t) = s^i\exp\Big\{\int_0^t\Big(g(s,X(s))^i - \frac{1}{2}|\sigma(s,X(s))^i|^2\Big)\,ds + \int_0^t \sigma(s,X(s))^i\,d\tilde{W}(s)\Big\}. \tag{2.182}$$

$g(t,X(t))$ and $\sigma(t,X(t))$ are called the mean return process and the volatility process, respectively.

We assume

$$g(t,x)^i > r(t,x), \quad \forall t,x, \quad i = 1,\ldots,m. \tag{2.183}$$

By using the data of $X$ and $S^i, i = 1,\ldots,m$, the agent invests at any time $t \in (0,T)$ a proportion $\pi^i(t)$ of its wealth in the $i$-th asset, $i = 1,\ldots,m$, and $\pi^0(t) = 1 - \sum_{i=1}^m \pi^i(t)$ in the bond. ($\pi^i(t) > 1$) and ($\pi^i(t) < 0$) stand for borrowing money and selling, respectively. We call $\pi(\cdot) = (\pi^1(\cdot),\ldots,\pi^m(\cdot))$ an investment strategy or an admissible strategy, if

$$\pi(\cdot)^\top\sigma(\cdot, X(\cdot)) \in L^\infty([0,T]\times\Omega, (\mathcal{F}_t^{\tilde{W}}); \mathbb{R}^1 \otimes \mathbb{R}^{d+m}). \tag{2.184}$$

Let $\mathscr{A}$ denote the set of all admissible strategies.

For $\pi(\cdot) \in \mathscr{A}$, the agent's wealth $Z^{\pi(\cdot)}$ evolves according to the SDE

$$\begin{aligned}&\frac{dZ^{\pi(\cdot)}(t)}{Z^{\pi(\cdot)}(t)}\\ &=\sum_{i=1}^m \pi^i(t)(g(t,x(t))^i\,dt + \sigma(t,X(t))^i\,d\tilde{W}(t)) + \pi^0(t)r(t,X(t))\,dt\\ &=(r(t,X(t)) + \pi(t)\cdot(g(t,X(t)) - r(t,X(t))1^m))\,dt + \pi(t)^\top\sigma(t,X(t))\,d\tilde{W}(t)\end{aligned} \tag{2.185}$$

by (2.178) and (2.179). The quantity

$$\mu(t,x) := g(t,x) - r(t,x)1^m \tag{2.186}$$

is the excess rate of return from risky assets, where $1^m = (1,1,\ldots,1) \in \mathbb{R}^m$.

From (2.185), we deduce that

$$\begin{aligned}Z^{\pi(\cdot)}(t) =& z\exp\Big\{\int_0^t (r(s,X(s)) + \pi(s)\cdot\mu(s,X(s)))\,ds\\ &+\int_0^t \pi(s)^\top\sigma(s,X(s))\,d\tilde{W}(s) - \frac{1}{2}\int_0^t |\pi(s)^\top\sigma(s,X(s))|^2\,ds\Big\},\end{aligned} \tag{2.187}$$

where $z =$ the initial wealth $> 0$.

$U \in C^2((0,\infty))$ is called a utility function, if $U' > 0$ (increasing) and $U'' < 0$ (concave).

We are interested in HARA (hyperbolic absolute risk aversion) utility functions

$$\text{power utility function: } U(x) = \frac{x^\delta}{\delta}, \quad x > 0 \quad \text{with } \delta < 1, \neq 0, \tag{2.188}$$

and

$$\text{logarithmic utility function: } U(x) = \log x, \quad x > 0. \tag{2.189}$$

The agent's objective is to maximize the expected utility from the terminal wealth, $E[U(Z^{\pi(\cdot)}(T))]$, by choosing an appropriate investment strategy.

### 2.4.2 Investment Problems for Power Utility Function

By (2.187) and (2.188), we have

$$U(Z^{\pi(\cdot)}(t)) = \frac{z^\delta}{\delta} M^{\pi(\cdot)}(t) \exp\Big(\delta \int_0^t \eta(s, X(t), \pi(s))\, ds\Big) \tag{2.190}$$

where

$$M^{\pi(\cdot)}(t) = \exp\Big(\int_0^t \delta\pi(s)^\top \sigma(s, X(s))\, d\tilde{W}(s) - \frac{1}{2}\int_0^t |\delta\pi(s)^\top \sigma(s, X(s))|^2\, ds\Big) \tag{2.191}$$

and

$$\eta(s, x, \pi(s)) = -\frac{1-\delta}{2}|\pi(s)^\top \sigma(s, x)|^2 + \pi(s)\cdot \mu(s, x) + r(s, x). \tag{2.192}$$

$M^{\pi(\cdot)}$ is an exponential $(\mathcal{F}_t^{\tilde{W}})$-martingale, because of the boundedness of $\pi(\cdot)^\top \sigma(\cdot)$.
For the computation of $EU(Z^{\pi(\cdot)}(t))$, we apply Girsanov's transformation;

$$P^{\pi(\cdot)} = M^{\pi(\cdot)}(T) \circ P \quad \text{on } \mathcal{F}_T^{\tilde{W}}. \tag{2.193}$$

The following (i)–(iii) are valid with respect to $P^{\pi(\cdot)}$

(i) $\tilde{W}^{\pi(\cdot)}$ given by (2.194) is an $(\mathcal{F}_t^{\tilde{W}})$-Wiener process, i.e.,

$$\tilde{W}^{\pi(\cdot)}(t) = \tilde{W}(t) - \delta \int_0^t \sigma(s, X(s))^\top \pi(s)\, ds, \quad t \in [0, T]. \tag{2.194}$$

(ii) The factor process $X$ is described by SDE.

$$dX(t) = \big(b(t, X(t)) + \delta\bar{\sigma}(t, X(t))^\top \pi(t)\big)\, dt + dW^{\pi(\cdot)}(t), \tag{2.195}$$

where $W^{\pi(\cdot)}$ is the first $d$-component of $\tilde{W}^{\pi(\cdot)}$ and $\bar{\sigma}(\cdot)$ is the $m \times d$-matrix consisting of the first $d$ columns of $\sigma(\cdot)$.

(iii) Since $\eta(\cdot)$ is independent of $z$,

$$E_{zx}U(Z^{\pi(\cdot)}(t)) = z^\delta E_x^{\pi(\cdot)}\Big[\frac{1}{\delta}\exp\Big(\delta \int_0^t \eta(s, X(s), \pi(s))\, ds\Big)\Big], \tag{2.196}$$

where $E^{\pi(\cdot)}$ denotes the expectation w.r.t. $P^{\pi(\cdot)}$.

Referring to (2.195), we consider the following control problem. Let $\Gamma = \mathbb{R}^m$ and fix a reference probability system $(\Omega, \mathcal{F}, (\mathcal{F}_t), P, \tilde{B})$ with a $(d+m)$-dimensional Wiener process $\tilde{B}$. For a control process $\pi(\cdot) \in \boldsymbol{\Gamma}^{\tilde{B}}$, its response $Y^{\pi(\cdot)}$ evolves according to the SDE

$$dY(s) = \big((b(s, Y(s)) + \delta\bar{\sigma}(s, Y(s))^\top \pi(s)\big)\, ds + dB(s), \quad s \in (t, T], \tag{2.197}$$

with the initial condition

$$Y(t) = x \ (\in \mathbb{R}^d), \tag{2.198}$$

where $B$ is the first $d$-dimensional component of $\tilde{B}$. We define the payoff $j(\cdot)$ and the value function $u(\cdot)$ by

$$j(t, x, \pi(\cdot)) = E_{tx}\Big[\frac{1}{\delta}\exp\Big(\delta \int_t^T \eta(s, Y(s), \pi(s))\, ds\Big)\Big] \tag{2.199}$$

and

$$u(t, x) = \sup_{\pi(\cdot)} j(t, x, \pi(\cdot)), \tag{2.200}$$

respectively. Now let us apply the logarithmic transformation,

$$v(t, x) = \frac{1}{\delta}\log(\delta u(t, x)). \tag{2.201}$$

Calculating formally, we obtain

$$\begin{aligned} 0 = & \frac{\partial v}{\partial t} + \frac{1}{2}(\Delta v + \delta|\partial_x v|^2 + 2b(t, x)\cdot\partial_x v + 2r(t, x)) \\ & + \sup_{\pi\in\mathbb{R}^m}\Big\{-\frac{1-\delta}{2}|\pi^\top\bar{\sigma}(t, x)|^2 + \pi^\top(\mu(t, x) + \delta\bar{\sigma}(t, x)\partial_x v)\Big\}. \end{aligned} \tag{2.202}$$

Since the supremum is attained for $\pi = \hat{\Pi}$, where

$$\hat{\Pi}(t, x) = \frac{1}{1-\delta}\{a(t, x)^{-1}(\mu(t, x) + \delta\bar{\sigma}(t, x)\partial_x v(t, x))\}, \tag{2.203}$$

we obtain

$$0 = \frac{\partial v}{\partial t} + \frac{1}{2}\Delta v + \mathrm{H}(t, x, \partial_x v(t, x)) \quad \text{on } [0, T) \times \mathbb{R}^d, \tag{2.204}$$

with the lateral boundary condition

$$v(T, x) = 0, \quad x \in \mathbb{R}^d, \tag{2.205}$$

where

$$\begin{aligned}
\mathrm{H}(t,x,p) =& \frac{\delta}{2}\left\{|p|^2 + \frac{\delta}{1-\delta}p(\bar{\sigma}^\top a^{-1}\bar{\sigma})(t,x)p^\top\right\} \\
&+ b(t,x)\cdot p + \frac{\delta}{2(1-\delta)}\{(\mu^\top a^{-1}\bar{\sigma})(t,x)p^\top + p(\bar{\sigma}^\top a^{-1}\mu)(t,x)\} \\
&+ r(t,x) + \frac{1}{2(1-\delta)}(\mu^\top a^{-1}\mu)(t,x). \qquad (2.206)
\end{aligned}$$

### 2.4.3 Optimal Investment Strategy

For the existence of optimal strategy, we need to take a broader class than $\mathscr{A}$. Put

$$\mathscr{A}_2 = \{\pi(\cdot); \pi(\cdot)^\top\sigma(\cdot, X(\cdot)) \in L^2([0,T]\times\Omega, (\mathcal{F}_t^{\tilde{W}}); \mathbb{R}^1 \otimes \mathbb{R}^m\}. \qquad (2.207)$$

By the condition (a),

$$\pi(\cdot) \in L^2([0,T]\times\Omega, (\mathcal{F}_t^{\tilde{W}}); \mathbb{R}^m) \quad \text{if } \pi(\cdot) \in \mathscr{A}_2 \qquad (2.208)$$

and

$$\pi(\cdot) \in L^\infty([0,T]\times\Omega, (\mathcal{F}_t^{\tilde{W}}); \mathbb{R}^m) \quad \text{if } \pi(\cdot) \in \mathscr{A}. \qquad (2.209)$$

First we prepare Lemma, by using (2.208).

**Lemma 2.2 (Approximation).**

(i) *For $\pi(\cdot) \in \mathscr{A}_2$, there is a sequence $\pi_n(\cdot), n = 1,2,\ldots$, in $\mathscr{A}$, such that*

$$\lim_{n\to\infty} Z^{\pi_n(\cdot)}(T) = Z^{\pi(\cdot)}(T) \quad P\text{-a.s.} \qquad (2.210)$$

(ii)

$$\sup_{\mathscr{A}_2} E_{zx}[U(Z^{\pi(\cdot)}(T))] = \sup_{\mathscr{A}} E_{zx}[U(Z^{\pi(\cdot)}(T))]. \qquad (2.211)$$

*Proof.* We divide the proof for (i) into three steps.

**Step 1.** For $\pi(\cdot) \in \mathscr{A}_2$, we can take a bounded $\pi_l(\cdot)$, such that

$$E\mathbf{1}\pi_l(\cdot) - \pi(\cdot)\mathbf{1}^2 < 2^{-l} \quad (l = 1,2,\ldots). \qquad (2.212)$$

On the other hand, for $\varepsilon > 0$, there is a positive constant $k = k_\varepsilon$, such that

$$P_x\left(\sup_{0\le s\le T}|X(s)| \ge k\right) < \varepsilon. \qquad (2.213)$$

Put

$$\pi_{l\varepsilon}(t) = \pi_l(t)\chi(\tau_k > t), \tag{2.214}$$

where

$$\tau_k = \begin{cases} \inf\{t < T; |X(t)| \geq k\}, \\ T \quad \text{if } \{\cdots\} = \text{empty}. \end{cases}$$

Then $\pi_{l.\varepsilon}(\cdot)$ is in $\mathscr{A}$.

**Step 2.** We evaluate the stochastic integral term

$$\xi := \int_0^T \pi(s)^\top \sigma(s, X(s))\, d\tilde{W}(s). \tag{2.215}$$

Replacing $\pi(\cdot)$ by $\pi_l(\cdot)$ and $\pi_{l\varepsilon}(\cdot)$, we define $\xi_l$ and $\xi_{l\varepsilon}$, respectively. Compute

$$P_x(|\xi_{l\varepsilon} - \xi| > \varepsilon) = P_x(|\xi_{l\varepsilon} - \xi| > \varepsilon; \tau_k = T) + P_x(|\xi_{l\varepsilon} - \xi| > \varepsilon; \tau_k < T). \tag{2.216}$$

Since $\pi_l(s) = \pi_{l\varepsilon}(s)$, for all $s(< T)$, for $\tau_k = T$, one has that,

$$\begin{aligned} \text{1st term in RHS} &= P_x(|\xi_l - \xi| > \varepsilon;\ \tau_k = T) \\ &\leq P_x\left(\left|\int_0^T (\pi_l(s) - \pi(s))^\top \sigma(s, X(s))\, d\tilde{W}(s)\right| > \varepsilon\right) \\ &\leq \frac{1}{\varepsilon^2} E[\mathbf{1}\pi_l(\cdot) - \pi(\cdot)\mathbf{1}^2] c_1^2 (1 + k^2), \end{aligned} \tag{2.217}$$

because $|\sigma(s,x)| \leq c_1(1 + |x|)$.
Combining (2.212), (2.213), (2.216), and (2.217), we can take a large integer $l_\varepsilon$ such that

$$P_x(|\xi_{l_\varepsilon} - \xi| > \varepsilon) < 2\varepsilon \quad \text{whenever } l > l_\varepsilon. \tag{2.218}$$

**Step 3.** By (2.212), (2.214), and (2.218), we take a sequence $\pi_n(\cdot), n = 1, 2, \ldots$ in $\mathscr{A}$, such that

$$\lim_{n\to\infty} \mathbf{1}\pi_n(\cdot) - \pi(\cdot)\mathbf{1} = 0 \quad P\text{-a.s.} \tag{2.219}$$

and

$$\lim_{n\to\infty} \int_0^T \pi_n(s)^\top \sigma(s, X(s))\, d\tilde{W}(s) = \int_0^T \pi(s)^\top \sigma(s, X(s))\, d\tilde{W}(s) \quad P\text{-a.s.} \tag{2.220}$$

From (2.187), (2.219), and (2.220), we obtain (2.210).

(ii) (2.211) is an easy consequence of (2.210) together and Fatou's lemma. This completes the proof. □

Now we are going to seek an optimal strategy.

**Theorem 2.9.** *Assume* (a)–(c). *Suppose that* (2.204)–(2.205) *has a solution* $\tilde{v} \in C^{12}([0,T] \times \mathbb{R}^d)$*, satisfying*

$$|\partial_x \tilde{v}(t,x)| \le K(1+|x|), \quad \forall t, x, \tag{2.221}$$

*with a positive constant $K$. Then*

$$\sup_{\mathscr{A}_2} E_{zx}[U(Z^{\pi(\cdot)}(T))] = z^{\delta} \tilde{u}(0,x), \tag{2.222}$$

*where*

$$\tilde{u}(t,x) = \frac{1}{\delta} \exp(\delta \tilde{v}(t,x)). \tag{2.223}$$

*Further, $\hat{\Pi}(\cdot)$ given by* (2.203) *provides an optimal strategy.*

*Proof.* We divide the proof into two steps.

**Step 1.** We show that $\hat{\pi}(t) := \hat{\Pi}(t, X(t))$ is in $\mathscr{A}_2$ and $M^{\hat{\pi}(\cdot)}$ is an exponential $(\mathcal{F}_t^{\tilde{W}})$-martingale.
Indeed, since $\hat{\Pi}(\cdot)$ is the maximum selector of (2.202), (b), (c), and (2.221) show that $\hat{\Pi}(t,x)^{\top}\sigma(t,x)$ is linearly growing w.r.t. $x$ uniformly in $t$. Hence, $\hat{\pi}(\cdot)$ is in $\mathscr{A}_2$. Thus, Proposition 1.5 concludes step 1.

**Step 2.** We have

$$E_{zx}[U(Z^{\hat{\pi}(\cdot)}(T))] \ge \sup_{\mathscr{A}} E_{zx}[U(Z^{\pi(\cdot)}(T))]. \tag{2.224}$$

Indeed, for $\pi(\cdot) \in \mathscr{A}$, $M^{\pi(\cdot)}$ is an exponential $(\mathcal{F}_t^{\tilde{W}})$-martingale. Put

$$\zeta^{\pi(\cdot)}(t) = \tilde{u}(t, X(t)) \exp\Big(\delta \int_0^t \eta(s, X(s), \pi(s))\, ds\Big). \tag{2.225}$$

Then Itô's formula and (2.201) yield

$$\begin{aligned} E_{zx}[U(Z^{\pi(\cdot)}(T))] &= z^{\delta} E_x^{\pi(\cdot)}\Big[\frac{1}{\delta}\exp\Big(\int_0^T \delta\eta(s, X(s), \pi(s))\, ds\Big)\Big] \\ &= z^{\delta} E_x^{\pi(\cdot)}[\zeta^{\pi(\cdot)}(T)] \quad \Big(\text{by } \tilde{u}(T,\cdot) = \frac{1}{\delta}\Big). \end{aligned} \tag{2.226}$$

Since (2.202) and (2.203) imply

$$E_x^{\pi(\cdot)}[\zeta^{\pi(\cdot)}(T)] \le E_x^{\pi(\cdot)}[\zeta^{\pi(\cdot)}(0)] = \tilde{u}(0,x) \tag{2.227}$$

and

$$E_x^{\hat{\pi}(\cdot)}[\zeta^{\hat{\pi}(\cdot)}(T)] = E_x^{\hat{\pi}(\cdot)}[\zeta^{\hat{\pi}(\cdot)}(0)] = \tilde{u}(0,x), \tag{2.228}$$

we have (2.224).

Now, using Lemma 2.2 we conclude that

$$E_{zx}[U(Z^{\hat{\pi}(\cdot)}(T))] = \sup_{\mathscr{A}_2} E_{zx}[U(Z^{\pi(\cdot)}(T))],$$

which completes the proof. □

Refer to [Na03, FSh00, FSh02] for details and related topics.

*Example 2.7 (Linear Gaussian model).* Suppose that

$$b(t,x) = b_0 + b_1 x, \quad \sigma(t,x) = \sigma,$$
$$g(t,x) = g_0 + g_1 x, \quad r(t,x) = r,$$

where $b_0, b_1, \sigma, g_0, g_1$ and r are constant and $a := \sigma\sigma^\top$ is a regular matrix. Hence the $d$-dimensional factor process $X$ and the asset prices $S^i, i = 1, \dots, m$, evolve according to the SDEs

$$\begin{cases} dX(t) = (b_0 + b_1 X(t))\,dt + dW(t), & t \in (0,T], \\ X(0) = x \ (\in \mathbb{R}^d) \end{cases} \tag{2.229}$$

and

$$\begin{cases} dS^i(t) = S^i(t)((g_0^i + g_1^i X(t))\,dt + \sigma^i\,d\tilde{W}(t)), & t \in (0,T], \\ S^i(0) = s^i > 0 \quad (i = l, \dots, m), \end{cases} \tag{2.230}$$

respectively.

The bond price $S^0(t)$ is given by

$$S^0(t) = s^0 e^{rt}. \tag{2.231}$$

Following Theorem 2.9, we will seek an optimal investment strategy. In order to solve the semilinear parabolic equation (2.204)–(2.205) for the Gauss model, we assume the quadratic from of $v(\cdot)$:

$$v(t,x) = \frac{1}{2} x^\top Q(t) x + R(t) \cdot x + S(t), \tag{2.232}$$

with $d \times d$ symmetric matrix $Q(t)$, $R(t) \in \mathbb{R}^d$ and $S(t) \in \mathbb{R}^1$. Then straightforward computations lead to the following equations:

$$\begin{cases} 0 = \dfrac{dQ}{dt}(t) + Q(t)K_0Q(t) \\ \quad + K_1^\top Q(t) + Q(t)K_1 + \dfrac{1}{2(1-\delta)} g_1^\top a^{-1} g_1, \quad t \in [0,T), \\ Q(T) = 0, \end{cases} \tag{2.233}$$

where

$$K_0 = \delta I_d + \frac{\delta}{1-\delta}\bar{\sigma}^\top a^{-1}\bar{\sigma}, \quad K_1 = b_1 + \frac{\delta}{1-\delta}\bar{\sigma}^\top a^{-1} g_1;$$

$$\begin{cases} 0 = \dfrac{dR}{dt}(t) + (K_1 + K_0 Q(t))^\top R(t) + Q(t)K_2 + K_3, \quad t \in [0,T), \\ R(T) = 0, \end{cases} \tag{2.234}$$

where

$$K_2 = b_0 + \frac{\delta}{1-\delta}\bar{\sigma}^\top a^{-1}(g_0 - r1^m), \quad K_3 = \frac{1}{1-\delta} g_1^\top a^{-1}(g_0 - r1^m);$$

and

$$\begin{cases} 0 = \dfrac{dS}{dt}(t) + \dfrac{1}{2}\mathrm{tr}Q(t) + \dfrac{1}{2}R(t)^\top K_0 R(t) \\ \quad + K_2 \cdot R(t) + r + \dfrac{1}{1-\delta}(g_0 - r1^m)^\top a^{-1}(g_0 - r1^m), \quad t \in [0,T), \\ S(T) = 0. \end{cases} \tag{2.235}$$

If the Riccati equation (2.233) has a solution, then we can solve (2.234) and (2.235), and $v(\cdot)$ given by (2.232) becomes a solution of (2.204)–(2.205) with linearly growing $\partial_x v(\cdot)$.

Hence

$$\hat{\pi}(t) := a^{-1}\Big(g_0 - r1^m + g_1 X(t) + \delta\bar{\sigma}\partial_x v(t, X(t))\Big) \tag{2.236}$$

gives an optimal investment strategy (refer to [KN02, W71, Wo68] for Riccati equations).

*Example 2.8 (Uniformly elliptic volatility model).* We consider a random volatility model presented in [Ph02].

Let $d = m = 1$. Suppose that $\rho \in (-1, 1)$ and $\varepsilon > 0$ are given. We assume the following forms:

$$b(t,x) = b_0 + b_1 x,\ \sigma(t,x) = (\rho\sqrt{c^2x^2+\varepsilon}, \sqrt{1-\rho^2}\sqrt{c^2x^2+\varepsilon}),$$
$$g(t,x) = g_0 + g_1 x,\ r(t,x) = r > 0$$

with constants $b_0, b_1, c, g_0, g_1$ and $r$.

$\rho$ is called the correlation between asset and the factor process. The factor process $X$ and the asset price $S$ evolve according to the SDEs (2.237) and (2.238) respectively,

$$dX(t) = (b_0 + b_1 X(t))\,dt + dW(t), \tag{2.237}$$

and

$$\begin{aligned} dS(t) =& S(t)\Big((g_0 + g_1 X(t))\,dt + \rho\sqrt{c^2X(t)^2+\varepsilon}\,dW(t) \\ &+ \sqrt{1-\rho^2}\sqrt{c^2X(t)^2+\varepsilon}\,dB(t)\Big), \end{aligned} \tag{2.238}$$

where $W$ and $B$ are mutually independent real Wiener processes.

Let $v(\cdot)$ be the logarithmic transformation of the value function. Then,

$$\begin{cases} 0 = \dfrac{\partial v}{\partial t} + \dfrac{1}{2}\dfrac{\partial^2 v}{\partial x^2} + \dfrac{\delta}{2}\Big(1 + \dfrac{\rho^2}{1-\delta}\Big)\Big|\dfrac{\partial v}{\partial x}\Big|^2 \\ \quad + \Big(b_0 + b_1 x + \dfrac{\delta\rho}{1-\delta}\dfrac{g_0 - r + g_1 x}{\sqrt{c^2x^2+\varepsilon}}\Big)\dfrac{\partial v}{\partial x} \\ \quad + \dfrac{1}{2(1-\delta)}\dfrac{|g_0 - r + g_1 x|^2}{c^2x^2+\varepsilon} + r, \quad t \in [0,T), \quad x \in \mathbb{R}^1, \\ v(T,x) = 0, \quad x \in \mathbb{R}^1 \end{cases} \tag{2.239}$$

holds by (2.204). Since [Ph02], Theorem 4.1 provides a solution $v(\cdot) \in C^{12}([0,T) \times \mathbb{R}^1) \cap C([0,T] \times \mathbb{R}^1)$ with $\frac{\partial v}{\partial x}$ satisfying the linear growth condition, we conclude that

$$\hat{\pi}(t) = \frac{1}{\sqrt{c^2X(t)^2+\varepsilon}}\Big(\frac{g_0 - r + g_1 X(t)}{\sqrt{c^2X(t)^2+\varepsilon}} + \delta\rho\frac{\partial v}{\partial x}(t, X(t))\Big)$$

is an optimal investment strategy.

# Chapter 3
# Viscosity Solutions for HJB Equations

**Abstract** The theory of viscosity solutions was originated by M.G. Crandall and P.L. Lions in the early 80s for the Hamilton–Jacobi equations and later P.L. Lions developed it for the HJB equations (Lions, J Commun PDE 8:1101–1134, 1983; Acta Math 16:243–278, 1988; Viscosity solutions of fully nonlinear second-order equations and optimal stochastic control in infinite dimensions. Part II. Optimal control of Zakai equation. In: Da Prato, Tubaro (eds) Stochastic partial differential equations and applications II. Lecture notes in mathematics, vol 1390. Springer, Berlin/Heidelberg, 1989, pp 147–170, 1989; J Funct Anal 86:1–18, 1989). In Chap. 2, we have seen the relation between the value function and the HJB equations. If the value function is smooth, then it provides the classical solution of the HJB equations. Unfortunately, when the diffusion coefficient is degenerate, smoothness does not necessarily hold, even for a simple case, and the HJB equations may in general have no classical equation, either. However, the theory of viscosity solutions gives a powerful tools for studing stochastic control problems. Regarding the viscosity solutions for the HJB equations, we claim only continuity for a solution, not necessarily differentiability. Thus, it has been shown that under mild conditions the value function is the unique viscosity solution of the HJB equation. We will revisit this fact in terms of semigroups in Sect. 3.1.3.

This chapter is organized as follows. In Sects. 3.1 and 3.2, we recall some basic results on viscosity solutions for (nonlinear) parabolic equations for later use. In Sect. 3.3 we consider stochastic optimal control-stopping problems in a framework similar to that of finite time horizon controls.

## 3.1 Formulations

Let $F : [0,T] \times \mathbb{R}^d \times S^d \times \mathbb{R}^d \times \mathbb{R}^1 \mapsto \mathbb{R}^1$ be continuous and satisfy the following two conditions ($d_1$) and ($d_2$);

($d_1$) ***Monotonicity***

$$F(t,x,A,p,z) \le F(t,x,A,p,\hat{z}) \quad \text{if } z \le \hat{z}.$$

($d_2$) ***Ellipticity***

$$F(t,x,A,p,z) \le F(t,x,\hat{A},p,z) \quad \text{if } A - \hat{A} \in S^d_+ \text{ (i.e., } \hat{A} \le A).$$

© Springer Japan 2015

M. Nisio, *Stochastic Control Theory*, Probability Theory and Stochastic Modelling 72, DOI 10.1007/978-4-431-55123-2_3

Here we are concerned with the Cauchy problem for the (non-linear) parabolic equation

$$-\partial_t U(t,x) + F(t,x,\partial_{xx}U(t,x),\partial_x U(t,x),U(t,x)) = 0, \quad (t,x) \in [0,T) \times \mathbb{R}^d, \tag{3.1}$$

with the lateral boundary condition

$$U(T,x) = \phi(x), \quad x \in \mathbb{R}^d, \tag{3.2}$$

where $\phi \in C(\mathbb{R}^d)$.

We first give the definition of viscosity solutions based on parabolic differentials in Sects. 3.1.1. In 3.1.2 we provide some convenient equivalent definitions. Then we deal with the HJB equations by using the semigroups related to DPP.

### 3.1.1 Definition of Viscosity Solution Based on Parabolic Differentials

**Definition 3.1.** (a) ***Parabolic differentials***

(i) For $W \in USC([0,T] \times \mathbb{R}^d)$ and $(t,x) \in [0,T) \times \mathbb{R}^d$, we set

$$\mathcal{P}^{2+}W(t,x) = \Big\{(q,p,A) \in \mathbb{R}^1 \times \mathbb{R}^d \times S^d,$$

$$\limsup_{\substack{(s,y)\to(t,x)\\ s\neq 0}} \frac{W(s,y)-W(t,x)-q(s-t)-p\cdot(y-x)-\frac{1}{2}(y-x)^\top A(y-x)}{|s-t|+|y-x|^2} \leq 0\Big\}. \tag{3.3}$$

$\mathcal{P}^{2+}W(t,x)$ is called the parabolic superdifferential of $W(\cdot)$ at $(t,x)$.

(ii) Similarly, for $V \in LSC([0,T]\times\mathbb{R}^d)$ and $(t,x) \in [0,T)\times\mathbb{R}^d$, the parabolic subdifferential of $V(\cdot)$ at $(t,x)$ is given by

$$\mathcal{P}^{2-}V(t,x) = \Big\{(q,p,A) \in \mathbb{R}^1 \times \mathbb{R}^d \times S^d;$$

$$\liminf_{\substack{(s,y)\to(t,x)\\ s\neq 0}} \frac{V(s,y)-V(t,x)-q(s-t)-p\cdot(y-x)-\frac{1}{2}(y-x)^\top A(y-x)}{|s-t|+|y-x|^2} \geq 0\Big\}. \tag{3.4}$$

(b) ***Closure of parabolic differentials***

(i) For $W \in USC([0,T] \times \mathbb{R}^d)$, we put

$$\bar{\mathcal{P}}^{2+}W(t,x) = \{(q,p,A) \in \mathbb{R}^1 \times \mathbb{R}^d \times S^d;$$

$$\exists (t_n,x_n) \in (0,T) \times \mathbb{R}^d \text{ and } (q_n,p_n,A_n) \in \mathcal{P}^{2+}W(t_n,x_n)$$

$$\text{such that } \lim_{n\to\infty}(t_n,x_n,W(t_n,x_n),q_n,p_n,A_n) = (t,x,W(t,x),q,p,A)\}. \tag{3.5}$$

(ii) For $V \in LSC([0,T] \times \mathbb{R}^d)$, $\bar{\mathcal{P}}^{2-}V(t,x)$ is defined in the same fashion as (3.5) with $\mathcal{P}^{2+}$ replaced by $\mathcal{P}^{2-}$.

The following properties are clear:

$$\begin{cases} \mathcal{P}^{2-}W(t,x) = -\mathcal{P}^{2+}(-W)(t,x), \\ \bar{\mathcal{P}}^{2-}W(t,x) = -\bar{\mathcal{P}}^{2+}(-W)(t,x), \end{cases} \tag{3.6}$$

for $f \in C^{12}((0,T) \times \mathbb{R}^d) \cap C([0,T] \times \mathbb{R}^d)$ and $(t,x) \in (0,T) \times \mathbb{R}^d$,

$$\begin{aligned} &(q,p,A) \in \mathcal{P}^{2\pm}W(t,x) \\ \Longleftrightarrow\ &(q + \partial_t f(t,x), p + \partial_x f(t,x), A + \partial_{xx} f(t,x)) \in \mathcal{P}^{2\pm}(W+f)(t,x). \end{aligned} \tag{3.7}$$

**Definition 3.2.** (i) A function $U \in USC([0,T] \times \mathbb{R}^d)$ is called a viscosity subsolution of the problem (equations) (3.1)–(3.2), if

$$U(T,x) \le \varphi(x), \quad \forall x \in \mathbb{R}^d \tag{3.8}$$

and, for $(t,x) \in [0,T) \times \mathbb{R}^d$,

$$-q + F(t,x,A,p,U(t,x)) \le 0 \tag{3.9}$$

whenever $(q,p,A) \in \mathcal{P}^{2+}U(t,x)$.

(ii) A function $V \in LSC([0,T] \times \mathbb{R}^d)$ is called a viscosity supersolution of the problem (equations) (3.1)–(3.2), if

$$V(T,x) \ge \varphi(x), \quad \forall x \in \mathbb{R}^d \tag{3.8$'$}$$

and, for $(t,x) \in [0,T) \times \mathbb{R}^d$,

$$-q + F(t,x,A,p,V(t,x)) \ge 0 \tag{3.9$'$}$$

whenever $(q,p,A) \in \mathcal{P}^{2-}U(t,x)$.

(iii) $U(\cdot)$ is called a viscosity solution of Eqs. (3.1) and (3.2), if it is both a viscosity subsolution and a supersolution.

*Convention.* Let $U \in USC([0,T] \times \mathbb{R}^d)$. If $U(t_0,x_0) = -\infty$, then we do not define $\mathcal{P}^{2+}U(t_0,x_0)$ and we suppose that (3.9) is valid at $(t_0,x_0)$.

We make the same convention for (3.9′).

From the definitions of $\bar{\mathcal{P}}^{2+}U(t,x)$ and $\bar{\mathcal{P}}^{2-}U(t,x)$, we can easily deduce

**Proposition 3.1 (Equivalent conditions).**

(i) $U \in USC([0,T] \times \mathbb{R}^d)$ *is a viscosity subsolution, if and only if* (3.8) and (3.9) *holds for* $\bar{\mathcal{P}}^{2+}$.

(ii) $V \in LSC([0,T] \times \mathbb{R}^d)$ *is a viscosity supersolution, if and only if* (3.8′) and (3.9′) *holds for* $\bar{\mathcal{P}}^{2-}$.

Regarding the relation between the notions of classical solutions and viscosity solutions, the following proposition says that they are consistent.

**Proposition 3.2.** *For* $U \in C^{12}([0,T) \times \mathbb{R}^d) \cap C([0,T] \times \mathbb{R}^d)$, $U(\cdot)$ *is a classical solution of* (3.1)–(3.2), *if and only if it is a viscosity solution of* (3.1)–(3.2).

*Proof.* For $U \in C^{12}([0,T) \times \mathbb{R}^d)$, it holds that

$$\begin{aligned}
&\mathcal{P}^{2+}U(t,x) \ni (q,p,A) \\
\iff &\begin{cases} q = \partial_t U(t,x) + c, & \text{where } c = 0 \text{ for } t \neq 0,\ c \geq 0 \text{ for } t = 0, \\ p = \partial_x U(t,x), \\ A = \partial_{xx} U(t,x) + B, & \text{with } \ B \in S^d_+. \end{cases}
\end{aligned}$$

Suppose that $U(\cdot)$ is a classical solution. Then the ellipticity condition ($d_2$) implies

$$\begin{aligned}
&-q + F(t,x,A,p,U(t,x)) \\
\leq &-\partial_t U(t,x) - c + F(t,x,\partial_{xx}U(t,x),\partial_x U(t,x),U(t,x)) \\
= &-c \leq 0.
\end{aligned}$$

Hence, $U(\cdot)$ is a viscosity subsolution. Similarly $U(\cdot)$ is a viscosity supersolution. So, $U(\cdot)$ is a viscosity solution.

Conversely, suppose that $U(\cdot)$ is a viscosity solution. Then

$$-\partial_t U(t,x) + F(t,x,\partial_{xx}U(t,x),\partial_x U(t,x),U(t,x)) = 0,$$

because $(\partial_t U(t,x), \partial_x U(t,x), \partial_{xx} U(t,x)) \in \mathcal{P}^{2+}U(t,x) \cap \mathcal{P}^{2-}U(t,x)$.

This completes the proof. □

### 3.1.2 Equivalent Definitions

We will introduce an equivalent definition of viscosity solution, which is suitable for stochastic control problems. Firstly we recall;

**Proposition 3.3 (See [FS06], p. 211).** *Let* $W \in C([0,T] \times \mathbb{R}^d)$ *and* $(t,x) \in [0,T) \times \mathbb{R}^d$ *be given. The following two statements are equivalent:*

(a) $(q,p,A) \in \mathcal{P}^{2+}W(t,x)$;
(b) *There exists* $u \in C^{12}([0,T) \times \mathbb{R}^d) \cap C([0,T] \times \mathbb{R}^d)$ *such that*

$$\begin{cases} u(t,x) = W(t,x), \ \partial_t u(t,x) = q, \\ \partial_x u(t,x) = p, \qquad \partial_{xx} u(t,x) = A, \end{cases} \tag{3.10}$$

*and* $W(\cdot) - u(\cdot)$ *attains its global maximum at* $(t,x)$.

"Global maximum" in (b) can be replaced by "global strict maximum". Indeed, put $\tilde{u}(s,y) = u(s,y) + (s-t)^2 + (y-x)^4$. Then $\tilde{u}(\cdot)$ satisfies (3.10) and $W(\cdot) - \tilde{u}(\cdot)$ attains its global strict maximum at $(t,x)$.

The following remark is useful.

*Remark 3.1.* Suppose that $W(\cdot)$ satisfies the polynomial growth condition

$$|W(s,y)| \le K(1+|y|^{2p}) \quad \text{on } [0,T] \times \mathbb{R}^d \tag{3.11}$$

with $K > 0$ and a non-negative integer $p$. Let $u(\cdot)$ be the function figuring in Proposition 3.3 (b). Then for any $0 < l < m < 1$, we can take $\tilde{u} \in C^{12}([0,T) \times \mathbb{R}^d) \cap C([0,T] \times \mathbb{R}^d)$, such that

$$\tilde{u}(\cdot) = u(\cdot) \quad \text{on } B_l := \{(s,y) \in [0,T] \times \mathbb{R}^d; |s-t|^2 + |y-x|^4 \le l^2\}, \tag{3.12}$$

$$\tilde{u}(s,y) = K(1+|y|^{2p}) \quad \text{on } [0,T] \times \mathbb{R}^d - B_m, \tag{3.13}$$

and

$$\tilde{u}(\cdot) \ge W(\cdot) \quad \text{on } [0,T] \times \mathbb{R}^d. \tag{3.14}$$

Moreover, $\tilde{u}(\cdot)$ satisfies (3.10) and $W(\cdot) - \tilde{u}(\cdot)$ attains its global maximum at $(t,x)$.

*Proof.* Let $\underline{\chi}, \overline{\chi} : [0,\infty) \mapsto [0,1]$ be smooth functions, decreasing and increasing, respectively, and satisfying

$$\underline{\chi}(\rho) = \begin{cases} 1, & \rho \in \left[0, \dfrac{l+m}{2}\right], \\ 0, & \rho \in [m,\infty), \end{cases} \tag{3.15}$$

and

$$\overline{\chi}(\rho) = \begin{cases} 0, & \rho \in [0,l], \\ 1, & \rho \in \left[\dfrac{l+m}{2}, \infty\right). \end{cases} \tag{3.16}$$

We define $\tilde{u}(\cdot)$ by

$$\begin{aligned}\tilde{u}(s,y) =&\underline{\chi}(|s-t|^2+|y-x|^4)u(s,y)\\&+\overline{\chi}(|s-t|^2+|y-x|^4)K(1+|y|^{2p}).\end{aligned}\tag{3.17}$$

Then $\tilde{u}\in C^{12}([0,T)\times\mathbb{R}^d)\cap C([0,T]\times\mathbb{R}^d)$ satisfies all claimed conditions. □

We note that

$$\tilde{u}(s,y) = K(1+|y|^{2p}) \quad \text{for } |y|\geq |x|+1. \tag{3.18}$$

The following theorem is an immediate consequence of Proposition 3.3 and Remark 3.1.

**Theorem 3.1.** *Let $U\in C([0,T]\times\mathbb{R}^d)$ be given.*

(i) *Suppose $U(T,x)\leq\phi(x), \forall x\in\mathbb{R}^d$.*
*Then the following statements are equivalent:*

(a) *$U(\cdot)$ is a viscosity subsolution of* (3.1)–(3.2).
(b) *Let $u\in C^{12}([0,T)\times\mathbb{R}^d)\cap C([0,T]\times\mathbb{R}^d)$ be given.*
*Suppose that $U(\cdot)-u(\cdot)$ attains its global maximum at $(\bar{t},\bar{x})\in[0,T)\times\mathbb{R}^d$. Then*

$$-\partial_t u(\bar{t},\bar{x})+F(\bar{t},\bar{x},\partial_{xx}u(\bar{t},\bar{x}),\partial_x u(\bar{t},\bar{x}),U(\bar{t},\bar{x}))\leq 0. \tag{3.19}$$

(ii) *Suppose $U(T,x)\geq\phi(x), \forall x\in\mathbb{R}^d$.*
*Then the following statements are equivalent:*

(a) *$U(\cdot)$ is a viscosity supersolution of* (3.1)–(3.2).
(b) *Let $u\in C^{12}([0,T)\times\mathbb{R}^d)\cap C([0,T]\times\mathbb{R}^d)$ be given.*
*Suppose that $U(\cdot)-u(\cdot)$ attains its global minimum at $(\underline{t},\underline{x})\in[0,T)\times\mathbb{R}^d$. Then*

$$-\partial_t u(\underline{t},\underline{x})+F(\underline{t},\underline{x},\partial_{xx}u(\underline{t},\underline{x}),\partial_x u(\underline{t},\underline{x}),U(\underline{t},\underline{x}))\geq 0. \tag{3.20}$$

(iii) *If $U(\cdot)$ satisfies the polynomial growth condition* (3.11)*, then we may assume that $u(\cdot)$ in* (b) *satisfies*

$$u(s,y) = k(1+|y|^{2p}) \quad \textit{for } |y|\geq R, \quad s\in[0,T] \tag{3.21}$$

*for some large $R$, where $k=K$ in* (i) *and $k=-K$ in* (ii).

Set

$$\begin{aligned}\Sigma_{2p}(k) =&\{u\in C^{12}([0,T)\times\mathbb{R}^d)\cap C([0,T]\times\mathbb{R}^d);\\&\exists R>0, \text{such that } u(s,y)=k(1+|y|^{2p}) \text{ on } [0,T]\times S_R^c\}\end{aligned}\tag{3.22}$$

and

$$\Sigma_{2p} = \bigcup_{k>0} \Sigma_{2p}(k). \tag{3.23}$$

Using Theorem 3.1, we study the HJB equation.

**Theorem 3.2.** *Under conditions* $(b_1)$–$(b_5)$, *the value function* $v(T,t,x;\phi)$ $(= V_{tT}\phi(x))$ *is a viscosity solution of the HJB equation*

$$\begin{cases} 0 = -\partial_t U(t,x) + \mathrm{H}(t,x,\partial_{xx}U(t,x),\partial_x U(t,x),U(t,x)), & (t,x) \in [0,T) \times \mathbb{R}^d, \\ U(T,x) = \phi(x), \quad x \in \mathbb{R}^d. \end{cases} \tag{3.24}$$

*Proof.* Put $U(t,x) = V_{tT}\phi(x)$. Then Theorem 2.1 (ii) and (2.15) yield

$$U \in C([0,T] \times \mathbb{R}^d) \quad \text{with } |U(t,x)| \le \bar{K}(1+|x|^2). \tag{3.25}$$

Firstly we will show that $U(\cdot)$ is a viscosity subsolution of (3.24). Let $u \in \Sigma_2(\bar{K})$. Suppose that $(\bar{t},\bar{x}) \in [0,T) \times \mathbb{R}^d$ is a global maximizer of $U(\cdot) - u(\cdot)$ and $U(\bar{t},\bar{x}) = u(\bar{t},\bar{x})$. Fix a arbitrary reference probability system $(\Omega,\mathcal{F},(\mathcal{F}_t),P,W)$. Then Proposition 2.5 and DPP yield

$$\begin{aligned} u(\bar{t},\bar{x}) &= U(\bar{t},\bar{x}) \\ &= \inf_{\gamma(\cdot)\in\Gamma^W} E_{\bar{t}\bar{x}}\Big[\int_{\bar{t}}^{t} e^{-\int_{\bar{t}}^{s}\kappa\,d\lambda} f(s,X(s),\gamma(s))\,ds + e^{-\int_{\bar{t}}^{t}\kappa\,d\lambda} U(t,X(t))\Big] \\ &\le \inf_{\gamma(\cdot)\in\Gamma^W} E_{\bar{t}\bar{x}}\Big[\int_{\bar{t}}^{t} e^{-\int_{\bar{t}}^{s}\kappa\,d\lambda} f(s,X(s),\gamma(s))\,ds + e^{-\int_{\bar{t}}^{t}\kappa\,d\lambda} u(t,X(t))\Big], \end{aligned} \tag{3.26}$$

where for simplicity we put $X = X^{\gamma(\cdot)}$.

Since $u(\cdot)$ is smooth, (3.26) and Itô's formula yield

$$\begin{aligned} 0 \le \inf_{\gamma(\cdot)\in\Gamma^W} E_{\bar{t}\bar{x}}\Big[\int_{\bar{t}}^{t} e^{-\int_{\bar{t}}^{s}\kappa\,d\lambda}\big(f(s,X(s),\gamma(s)) + \partial_t u(s,X(s)) \\ + \mathrm{G}_s^{\gamma(s)} u(s,X(s)) - \kappa(s,X(s),\gamma(s))u(s,X(s))\big)\,ds\Big]. \end{aligned} \tag{3.27}$$

The same arguments as in Proposition 2.8 show that

$$0 \le \partial_t u(\bar{t},\bar{x}) + \mathrm{G}_{\bar{t}} u(\bar{t},\bar{x}), \tag{3.28}$$

i.e., $U(\cdot)$ is a viscosity subsolution of (3.24).

Since we can prove the supersolution part in the same fashion, this completes the proof. □

### 3.1.3 Viscosity Solutions via Semigroups

Let us revisit Theorem 3.2, according to [LN83]. Set

$$\tilde{\Sigma}_2 = \{\psi \in C^2(\mathbb{R}^d); \exists \text{ positive constants } K \text{ and } R, \\ \text{such that } |\psi(x)| \le K(1+|x|^2) \quad \text{for } |x| \ge R\}. \tag{3.29}$$

**Theorem 3.3.** *Let* $(\mathrm{T}_{\theta t}, 0 \le \theta \le t \le T)$ *be a 2-parameter semigroup on* $\tilde{C}$ *(see* (2.31)*) with* (i)–(v) *of Proposition* 2.7*. Suppose that, for* $\psi \in \tilde{\Sigma}_2$ *and* $\varphi \in \tilde{C}$*,*

$$\lim_{s\to 0} \frac{1}{s}\{\mathrm{T}_{\theta\theta+s}(\psi + s\varphi)(x) - \psi(x)\} = G_\theta\psi(x) + \varphi(x), \\ \forall(\theta, x) \in [0,T) \times \mathbb{R}^d. \tag{3.30}$$

*Then* $U(t,x) := \mathrm{T}_{tT}\phi(x)$ *is a viscosity solution of* (3.24).

*Proof.* $U(\cdot)$ is continuous on $[0,T]\times\mathbb{R}^d$ by (v).

Since (iv) yields

$$\|U(t,\cdot)\|_{\tilde{C}} \le \|\mathrm{T}_{tT}\phi - \mathrm{T}_{tT}0\|_{\tilde{C}} + \|T_{tT}0\|_{\tilde{C}} \le \tilde{K}(\|\phi\|_{\tilde{C}} + 1), \tag{3.31}$$

we have

$$|U(t,x)| \le \tilde{K}(\|\phi\|_{\tilde{C}} + 1)(1+|x|^2), \quad \forall t, x. \tag{3.32}$$

Now let us show that $U(\cdot)$ is a viscosity subsolution of (3.24). Let $u(\cdot) \in \Sigma_2$. Suppose that $U(\cdot) - u(\cdot)$ attains its global maximum at $(\bar{t},\bar{x}) \in [0,T)\times\mathbb{R}^d$ and $U(\bar{T},\bar{x}) = u(\bar{t},\bar{x})$. Then

$$\begin{aligned} 0 &= U(\bar{t},\bar{x}) - u(\bar{t},\bar{x}) = \mathrm{T}_{\bar{t}T}\phi(\bar{x}) - u(\bar{t},\bar{x}) \\ &= \mathrm{T}_{\bar{t}\bar{t}+s}U(\bar{t}+s,\cdot)(\bar{x}) - u(\bar{t},\bar{x}) \\ &\le \mathrm{T}_{\bar{t}\bar{t}+s}u(\bar{t}+s,\cdot)(\bar{x}) - u(\bar{t},\bar{x}) \quad (\text{by } U - u \le 0). \end{aligned} \tag{3.33}$$

Since $u(\cdot) \in \Sigma_2$ implies that there is $R > 0$, such that

$$e(s,x) := u(\bar{t}+s,x) - (u(\bar{t},x) + s\partial_t u(\bar{t},x)) = 0, \quad \forall |x| \ge R,$$

we have

$$\lim_{s\to 0}\frac{1}{s}e(s,x) = 0 \tag{3.34}$$

uniformly on $\mathbb{R}^d$.

On the other hand, (iv) yields

$$\|\mathrm{T}_{\bar{t}\,\bar{t}+s}u(\bar{t}+s,\cdot)-\mathrm{T}_{\bar{t}\,\bar{t}+s}(u(\bar{t},\cdot)+s\partial_t u(\bar{t},\cdot))\|_{\tilde{C}}\le\tilde{K}\|e(s,\cdot)\|_{\tilde{C}}. \tag{3.35}$$

From (3.34) and (3.35), it follows that

$$\lim_{s\to 0}\frac{1}{s}|\mathrm{T}_{\bar{t}\,\bar{t}+s}u(\bar{t}+s,\cdot)(x)-\mathrm{T}_{\bar{t}\,\bar{t}+s}(u(\bar{t},\cdot)+s\partial_t u(\bar{t},\cdot))(x)|=0 \tag{3.36}$$

uniformly on any bounded set.

Since $u(\bar{t},\cdot)\in\tilde{\Sigma}_2$ and $\partial_t u(\bar{t},\cdot)\in\tilde{C}$, (3.33) and (3.36) together with (3.30) yield that $U(\cdot)$ is a viscosity subsolution.

Since we can prove that $U(\cdot)$ is a viscosity supersolution in the same fashion, $U(\cdot)$ is a viscosity solution. □

Next we will prove that the semigroup $(V_{\theta t}, 0\le\theta\le t\le T)$ satisfies (3.30), which gives Theorem 3.2 in a different way.

**Proposition 3.4.** *For $\psi\in\tilde{\Sigma}_2$ and $\varphi\in\tilde{C}$,*

$$\begin{aligned}&\lim_{s\to 0}\frac{1}{s}\{V_{\theta\,\theta+s}(\psi+s\varphi)(x)-\psi(x)\}\\&=G_\theta\psi(x)+\varphi(x),\quad\forall(\theta,x)\in[0,T)\times\mathbb{R}^d.\end{aligned} \tag{3.37}$$

*Proof.* Fix a reference probability system $(\Omega,\mathcal{F},(\mathcal{F}_t),P,W)$. Then

$$\begin{aligned}&V_{\theta\,\theta+s}(\psi+s\varphi)(x)-\psi(x)-s\varphi(x)\\&=\inf_{\gamma(\cdot)\in\boldsymbol{\Gamma}^W}\{J(\theta+s,\theta,x,\gamma(\cdot);\psi)-\psi(x)\\&\quad+sE_{\theta x}[e^{-\int_\theta^{\theta+s}\kappa\,d\lambda}\varphi(X^{\gamma(\cdot)}(\theta+s))-\varphi(x)]\}.\end{aligned} \tag{3.38}$$

From (2.17) it follows that

$$\lim_{s\to 0}E_{\theta x}[e^{-\int_\theta^{\theta+s}\kappa\,d\lambda}\varphi(X^{\gamma(\cdot)}(\theta+s))-\varphi(x)]=0, \tag{3.39}$$

uniformly in $\gamma(\cdot)\in\boldsymbol{\Gamma}^W$.

Further, Proposition 2.8 yields

$$\lim_{s\to 0}\frac{1}{s}(V_{\theta\,\theta+s}\psi(x)-\psi(x))=G_\theta\psi(x),\quad\forall x\in\mathbb{R}^d. \tag{3.40}$$

By combining (3.39) and (3.40) with (3.38), we complete the proof. □

## 3.2 Uniqueness of Viscosity Solutions

This section deals mainly with the comparison principle between a viscosity subsolution $U(\cdot)$ and a viscosity supersolution $V(\cdot)$ of Eqs. (3.1)–(3.2).

The comparison principle asserts that

$$U(T, \cdot) \le V(T, \cdot) \Longrightarrow U(\cdot) \le V(\cdot).$$

Hence the uniqueness of the viscosity solution is immediate.

First we state the Crandall–Ishii Lemma, which provides a key tool for the proof of the comparison principle. Then, under the structural condition on $F$, we will establish the comparison principle, when both $U(\cdot)$ and $V(\cdot)$ are bounded and continuous (Theorem 3.5).

### 3.2.1 *Crandall–Ishii Lemma*

Let us recall Ishii's Lemma first, which states the maximum principle for elliptic differential equations. This result provides a fundamental tool for the theory of viscosity solutions.

**Definition 3.3.** 1. Let $u \in USC(\mathbb{R}^N)$. $\mathcal{J}^{2+}u(x)$ is defined as the superdifferential of $u(\cdot)$ at $x$,

$$\mathcal{J}^{2+}u(x) = \Big\{(p, A) \in \mathbb{R}^N \times S^N; \quad \limsup_{y \to x} \frac{u(y) - u(x) - p \cdot (y - x) - \frac{1}{2}(y - x)^\top A(y - x)}{|y - x|^2} \le 0\Big\}. \tag{3.41}$$

Its closure $\bar{\mathcal{J}}^{2+}u(x)$ is defined as

$$\bar{\mathcal{J}}^{2+}u(x) = \{(p, A) \in \mathbb{R}^N \times S^N; \quad \exists x_n \text{ and } (p_n, A_n) \in \mathcal{J}^{2+}u(x_n) \text{ such that } \lim_{n \to \infty}(x_n, u(x_n), p_n, A_n) = (x, u(x), p, A)\}. \tag{3.42}$$

2. Let $v \in LSC(\mathbb{R}^N)$. The subdifferential of $v(\cdot)$ at $x$, $\mathcal{J}^{2-}v(x)$, and its closure $\bar{\mathcal{J}}^{2-}v(x)$ are given by

$$\mathcal{J}^{2-}v(x) = -\mathcal{J}^{2+}(-v)(x)$$

and

$$\bar{\mathcal{J}}^{2-}v(x) = -\bar{\mathcal{J}}^{2+}(-v)(x),$$

respectively.

### Ishii's Lemma

Let $u_i \in USC(\mathbb{R}^N), i = 1, 2$, and $\phi \in C^2(\mathbb{R}^N \times \mathbb{R}^N)$. Suppose that $u_1(x_1) + u_2(x_2) - \phi(x_1, x_2)$ attains its local maximum at $(\bar{x}_1, \bar{x}_2)$. Then, for any $\varepsilon \in (0, 1)$, there exist $X_i \in S^N, i = 1, 2$, such that

$$(\partial_{x_i}\phi(\bar{x}_1, \bar{x}_2), X_i) \in \bar{\mathcal{J}}^{2+}u_i(\bar{x}_i), \quad i = 1, 2, \tag{3.43}$$

and

$$-\Big(\frac{1}{\varepsilon} + \|A\|\Big)I_{2N} \le \begin{pmatrix} X_1 & 0 \\ 0 & X_2 \end{pmatrix} \le A + \varepsilon A^2, \tag{3.44}$$

where

$$A = \begin{pmatrix} \partial_{x_1x_1}\phi(\bar{x}_1, \bar{x}_2) & \partial_{x_1x_2}\phi(\bar{x}_1, \bar{x}_2) \\ \partial_{x_2x_1}\phi(\bar{x}_1, \bar{x}_2) & \partial_{x_2x_2}\phi(\bar{x}_1, \bar{x}_2) \end{pmatrix} \quad \text{and} \quad \|A\| = \sup_{|y|=1} y^\top Ay. \tag{3.45}$$

See [CIL92], Appendix, [Ko04], p. 71, [Mo10], Theorem 4.4.6 for a proof.

For parabolic differential equations, Crandall and Ishii showed a similar fact, by using Ishii's lemma.

### Crandall-Ishii Lemma [CI90]

Let $u_i \in USC([0, T] \times \mathbb{R}^d), i = 1, 2$, and $\phi \in C^{12}([0, T] \times \mathbb{R}^d \times \mathbb{R}^d)$ .

Suppose that for any $M > 0$ there is a constant $c(M)$ such that, for $(t, x) \in [0, T) \times \mathbb{R}^d$,

$$\begin{aligned}&(q, p, A) \in \mathcal{P}^{2+}u_i(t, x)\\ &\text{and} \quad |x| + |u_i(t, x)| + |p| + |A| \le M \Longrightarrow q \le c(M).\end{aligned} \tag{3.46}$$

Then, for any $\varepsilon \in (0, 1)$ and any local maximizer $(\bar{t}, \bar{x}_1, \bar{x}_2)(\in (0, T) \times \mathbb{R}^d \times \mathbb{R}^d)$ of $u_1(t, x_1) + u_2(t, x_2) - \phi(t, x_1, x_2)$, there exist $q_i \in \mathbb{R}^1$ and $X_i \in S^d, i = 1, 2$, for which

$$(q_i, \partial_{xi}\phi(\bar{t}, \bar{x}_1, \bar{x}_2), X_i) \in \bar{\mathcal{P}}^{2+}u_i(\bar{t}, \bar{x}_i), \quad i = 1, 2, \tag{3.47}$$

$$q_1 + q_2 = \partial_t \phi(\bar{t}, \bar{x}_1, \bar{x}_2), \tag{3.48}$$

$$-\Big(\frac{1}{\varepsilon} + \|A\|\Big) I_{2d} \le \begin{pmatrix} X_1 & 0 \\ 0 & X_2 \end{pmatrix} \le A + \varepsilon A^2, \tag{3.49}$$

$$\text{where } A = \begin{pmatrix} \partial_{x_1x_1}\phi(\bar{t}, \bar{x}_1, \bar{x}_2) & \partial_{x_1x_2}\phi(\bar{t}, \bar{x}_1, \bar{x}_2) \\ \partial_{x_2x_1}\phi(\bar{t}, \bar{x}_1, \bar{x}_2) & \partial_{x_2x_2}\phi(\bar{t}, \bar{x}_1, \bar{x}_2) \end{pmatrix}.$$

We can see that condition (3.46) is satisfied when $u_i$ $(i = 1, 2)$ is a viscosity subsolution of (3.1)–(3.2).

Since we need the Crandall–Ishii lemma for $u_i \in C([0, T] \times \mathbb{R}^d)$ with the linear growth condition

$$|u_i(t, x)| \le K_0(1 + |x|), \quad \forall (t, x) \in [0, T] \times \mathbb{R}^d, \tag{3.50}$$

we will show the following theorem, following [CI90].

**Theorem 3.4.** *Let $u_i \in C([0, T] \times \mathbb{R}^d)$ satisfying (3.50) be given. Suppose that $u_i, i = 1, 2$, satisfy the condition (3.46). Put*

$$\phi(t, x_1, x_2) = \frac{\lambda}{2}|x_1 - x_2|^2 + \beta(T - t) + \frac{c}{2}(|x_1|^2 + |x_2|^2), \tag{3.51}$$

*with positive constants $\lambda, \beta$ and $c$.*

*Then for any $\varepsilon > 0$ and any global maximizer $(\bar{t}, \bar{x}_1, \bar{x}_2) \in (0, T) \times \mathbb{R}^d \times \mathbb{R}^d$ of $u_1(t, x_1) + u_2(t, x_2) - \phi(t, x_1, x_2)$, there exists $q_i \in \mathbb{R}^1$ and $X_i \in S^d, i = 1, 2$, satisfying*

$$\begin{cases} (q_1, \lambda(\bar{x}_1 - \bar{x}_2) + c\bar{x}_1, X_1) \in \bar{\mathcal{P}}^{2+} u_1(\bar{t}, \bar{x}_1), \\ (q_2, -\lambda(\bar{x}_1 - \bar{x}_2) + c\bar{x}_2, X_2) \in \bar{\mathcal{P}}^{2+} u_2(\bar{t}, \bar{x}_2), \end{cases} \tag{3.47$'$}$$

$$q_1 + q_2 = -\beta, \tag{3.48$'$}$$

$$-\Big(\frac{1}{\varepsilon} + \|A\|\Big) I_{2d} \le \begin{pmatrix} X_1 & 0 \\ 0 & X_2 \end{pmatrix} \le A + \varepsilon A^2, \tag{3.49$'$}$$

$$\textit{where } A = \begin{pmatrix} (\lambda + c)I_d & -\lambda I_d \\ -\lambda I_d & (\lambda + c)I_d \end{pmatrix}.$$

*Outline of Proof.* We divide the proof into three steps.

**Step 1.** ***Reductions***

Put

$$u_i(t, x_i) = -\infty \quad \text{for } t \notin [0, T], \quad i = 1, 2. \tag{3.52}$$

Set

$$\tilde{u}_i(t,x_i) = u_i(t,x_i) - \frac{c}{2}|x_i|^2 - \frac{\beta}{2}(T-t) \tag{3.53}$$

and

$$\psi(x_i,x_2) = \frac{\lambda}{2}|x_1-x_2|^2. \tag{3.54}$$

Define $\omega_i \in USC(\mathbb{R}^1 \times \mathbb{R}^d)$ by

$$\begin{aligned}\omega_i(t,x_i) =& \tilde{u}_i(t+\bar{t}, x_i+\bar{x}_i) - \tilde{u}_i(\bar{t},\bar{x}_i)\\ & - \partial_{x_i}\psi(\bar{x}_1,\bar{x}_2)\cdot x_i - \frac{r}{2}(|x_i|^4+|t|^4),\end{aligned} \tag{3.55}$$

with $r > 0$.

Then (3.50) and (3.55) imply

$$\omega_i(t,x_i) < -1 \quad \text{for } \forall t \text{ and } |x_i| \geq N_0,\ i=1,2, \tag{3.56}$$

with some $N_0 = N_0(c)$.

Further, $(0,0,0)$ is the unique maximizer of $\omega_1(t,x_1)+\omega_2(t,x_2)-\psi(x_1,x_2)$ and

$$\begin{cases}\omega_1(0,0) = \omega_2(0,0),\\ \omega_1(t,x_1)+\omega_2(t,x_2) < \dfrac{\lambda}{2}|x_1-x_2|^2 \quad \text{for } (t,x_1,x_2) \notin (0,0,0).\end{cases} \tag{3.57}$$

**Step 2.** ***Associated Ishii's Lemma***

For $\delta > 0$, we define $\varphi$ and $\Phi$ by

$$\varphi(t_1,x_1,t_2,x_2) = \frac{\lambda}{2}|x_1-x_2|^2 + \frac{1}{2\delta}|t_1-t_2|^2 \tag{3.58}$$

and

$$\Phi(t_1,x_1,t_2,x_2) = \omega_1(t_1,x_1)+\omega_2(t_2,x_2)-\varphi(t_1,x_1,t_2,x_2), \tag{3.59}$$

respectively. Then $\Phi$ attains its global maximum at some point $z(\delta) = (t_1(\delta), x_1(\delta), t_2(\delta), x_2(\delta))$.

Since $\Phi(0,0,0,0) = 0$, (3.52) and (3.56) yield

$$t_i(\delta) \in [-\bar{t}, T-\bar{t}], \quad x_i(\delta) \in S_{N_0}, \quad \forall\delta > 0. \tag{3.60}$$

By Ishii's Lemma, we have $X_i := X_i(\delta, \varepsilon) \in S^{1+d}$ such that

$$((\partial_{t_i}\varphi(z(\delta)), \partial_{x_i}\varphi(z(\delta))), X_i) \in \bar{\mathcal{J}}^{2+}\omega_i(t_i(\delta), x_i(\delta)) \tag{3.61}$$

and (3.44) hold.

Hence (3.3) indicates that $Z_i := Z_i(\delta, \varepsilon)$ (= minor matrix obtained by deleting the first row and the first column of $X_i$) gives

$$(\partial_{t_i}\varphi(z(\delta)), \partial_{x_i}\varphi(z(\delta)), Z_i) \in \bar{\mathcal{P}}^{2+}\omega_i(t_i(\delta), x_i(\delta)) \tag{3.62}$$

and

$$-\Big(\frac{1}{\varepsilon} + 2\lambda\Big) I_{2d} \le \begin{pmatrix} Z_1 & 0 \\ 0 & Z_2 \end{pmatrix} \le (1 + 2\lambda\varepsilon)\hat{A}, \tag{3.63}$$

with $\hat{A} = \lambda \begin{pmatrix} I_d & -I_d \\ -I_d & I_d \end{pmatrix}$.

**Step 3.** ***Limit of*** $z(\delta)$ ***as*** $\delta \to 0$

We can easily see that $\omega_i(\cdot)$ also satisfies (3.46), with a different $c(M)$. Hence, from (3.60) and (3.46) it follows that there is a positive constant $c_0$ such that, for any $\delta > 0$,

$$\begin{cases} \partial_{t_1}\varphi(z(\delta)) = \dfrac{1}{\delta}(t_1(\delta) - t_2(\delta)) < c_0, \\ \partial_{t_2}\varphi(z(\delta)) = -\dfrac{1}{\delta}(t_1(\delta) - t_2(\delta)) < c_0, \end{cases} \tag{3.64}$$

that is

$$\frac{1}{\delta}|t_1(\delta) - t_2(\delta)| < c_0, \quad \forall \delta > 0. \tag{3.65}$$

Next, taking into account that $\Phi(z(\delta)) \ge \Phi(0, 0, 0, 0) = 0$, we have

$$\begin{aligned} &\omega_1(t_1(\delta), x_1(\delta)) + \omega_2(t_2(\delta), x_2(\delta)) \\ &\ge \frac{\lambda}{2}|x_1(\delta) - x_2(\delta)|^2 + \frac{1}{2\delta}|t_1(\delta) - t_2(\delta)|^2. \end{aligned} \tag{3.66}$$

On the other hand, (3.60) implies that

$$\bar{K} := \sup_{\delta > 0} \omega_i(t_i(\delta), x_i(\delta)) < \infty. \tag{3.67}$$

Combining (3.60), (3.65), and (3.63) together, we can choose $\delta_j, j = 1.2, \ldots$ tending to 0, so that

$$\lim_{j\to\infty} t_1(\delta_j) = \lim_{j\to\infty} t_2(\delta_j) =: \theta(\in [-\bar{t}, T-\bar{t}]), \tag{3.68}$$

$$\lim_{j\to\infty} x_i(\delta_j) =: \xi_j(\in S_{N_0}), \quad i = 1,2, \tag{3.69}$$

$$\lim_{j\to\infty} \frac{1}{\delta_j}(t_1(\delta_j) - t_2(\delta_j)) =: q \quad (|q| \le c_0), \tag{3.70}$$

and

$$\lim_{j\to\infty} Z_i(\delta_j, \varepsilon) =: Z_i(\varepsilon), \quad i = 1,2, \tag{3.71}$$

where $Z_i(\varepsilon), i = 1.2$, satisfy (3.63).

Finally we show that

$$\theta = 0, \quad \xi_i = 0, \quad i = 1,2. \tag{3.72}$$

Indeed, (3.68) and (3.69) together with (3.66) yield

$$\omega_1(\theta, \xi_1) + \omega_2(\theta, \xi_2) \ge \frac{\lambda}{2}|\xi_1 - \xi_2|^2. \tag{3.73}$$

However, (3.73) contradicts (3.57) if $(\theta, \xi_i, \xi_2) \ne (0,0,0)$. Thus (3.72) is valid.

Collecting the above results, we see that

$$\begin{cases} (q, 0, Z_1(\varepsilon)) \in \bar{\mathcal{P}}^{2+}\omega_1(0,0), \\ (-q, 0, Z_2(\varepsilon)) \in \bar{\mathcal{P}}^{2+}\omega_2(0,0). \end{cases} \tag{3.74}$$

Coming back to $u_i(\cdot)$, we conclude the proof of the theorem. □

### 3.2.2 *Structural Condition*

Let us introduce the structural condition on $F$ related to (3.63).

#### Structural Condition

If $A, B \in S^d$ and $\mu > 1$ satisfy

$$-3\mu \begin{pmatrix} I_d & 0 \\ 0 & I_d \end{pmatrix} \le \begin{pmatrix} A & 0 \\ 0 & -B \end{pmatrix} \le 3\mu \begin{pmatrix} I_d & -I_d \\ -I_d & I_d \end{pmatrix}, \tag{3.75}$$

then there is a modulus function $m_F$, such that

$$\begin{aligned}&F(t,y,B,\mu(x-y),z)-F(t,x,A,\mu(x-y),z)\\ \leq& m_F(\mu|x-y|^2+|x-y|+|z||x-y|), \quad \forall t,x,y,z. \end{aligned} \tag{3.76}$$

We give two examples.

*Example 3.1 (HJB equation).* Let $\Gamma$ be a parameter set. Suppose

$$\begin{aligned}\alpha &: [0,T]\times\mathbb{R}^d\times\Gamma\longmapsto\mathbb{R}^d\otimes\mathbb{R}^m,\\ b &: [0,T]\times\mathbb{R}^d\times\Gamma\longmapsto\mathbb{R}^d,\\ \kappa &: [0,T]\times\mathbb{R}^d\times\Gamma\longmapsto[0,c_0],\\ f &: [0,T]\times\mathbb{R}^d\times\Gamma\longmapsto\mathbb{R}^1\end{aligned} \tag{3.77}$$

are Lipschitz continuous w.r.t. $x\in\mathbb{R}^d$, uniformly on $[0,T]\times\Gamma$. Put $a=\alpha\alpha^\top$. Then $F$ given by

$$\begin{aligned}&F(t,x,A,p,z)\\ =&\sup_{\gamma\in\Gamma}\Big(-\frac{1}{2}\mathrm{tr}(a(t,x,\gamma)A)-b(t,x,\gamma)\cdot p+\kappa(t,x,\gamma)z-f(t,x,\gamma)\Big)\end{aligned} \tag{3.78}$$

satisfies the structural condition.

*Example 3.2 (Isaacs equation).* Replacing $\Gamma$ of Example 3.1 by $\Gamma^I\times\Gamma^{II}$, and giving $\alpha, b, \kappa, f$ and $a$ in the same way, we define $F$ by

$$\begin{aligned}&F(t,x,A,p,z)\\ =&\inf_{\gamma^I\in\Gamma^I}\sup_{\gamma^{II}\in\Gamma^{II}}\Big\{-\frac{1}{2}\mathrm{tr}(a(t,x,\gamma^I,\gamma^{II})A)\\ &-b(t,x,\gamma^I,\gamma^{II})\cdot p+\kappa(t,x,\gamma^I,\gamma^{II})z-f(t,x,\gamma^I,\gamma^{II})\Big\}.\end{aligned} \tag{3.79}$$

Then $F$ satisfies the structural condition.

Indeed, we denote the $k$-th column vectors of $\alpha(t,x,\gamma)$ and $\alpha(t,y,\gamma)$ by $\xi_k^\gamma$ and $\eta_k^\gamma$, respectively ($\gamma$ stands for $(\gamma^I,\gamma^{II})$ in Example 3.42). Then the Lipschitz condition ensures that there is a constant $c_1$, such that

$$\sum_{k=1}^m|\xi_k^\gamma-\eta_k^\gamma|^2\leq c_1|x-y|^2, \quad \forall\gamma. \tag{3.80}$$

Observing that

$$
\begin{aligned}
&\operatorname{tr}(a(t,x,\gamma)A) - \operatorname{tr}(a(t,y,\gamma)B) \\
&= \sum_{k=1}^{m} (\xi_k^{\gamma\top} A \xi_k^{\gamma} - \eta_k^{\gamma\top} B \eta_k^{\gamma}) \\
&\leq 3\mu \sum_{k=1}^{m} \begin{pmatrix} \xi_k^{\gamma} \\ \eta_k^{\gamma} \end{pmatrix}^{\top} \begin{pmatrix} I_d & -I_d \\ -I_d & I_d \end{pmatrix} \begin{pmatrix} \xi_k^{\gamma} \\ \eta_k^{\gamma} \end{pmatrix} \quad \text{(by (3.75))} \\
&= 3\mu \sum_{k=1}^{m} |\xi_k^{\gamma} - \eta_k^{\gamma}|^2 \leq 3\mu c_1 |x-y|^2 \quad \text{(by (3.80))},
\end{aligned}
$$

we conclude that the structural condition is satisfied.

### 3.2.3 Comparison Principle

In this subsection we compare a viscosity subsolution and a viscosity supersolution of (3.1)–(3.2). Since we are mainly concerned with the HJB equation and the Isaacs equation, we assume, besides ($d_1$) and ($d_2$), that the following condition is satisfied.

($d_3$) $F(t,x,A,p,z)$ is Lipschitz continuous in $p, A$ and $z$, uniformly in $(t,x)$, say

$$
\begin{aligned}
&|F(t,x,A,p,z) - F(t,x,\hat{A},\hat{p},\hat{z})| \\
&\leq l_0\{|A-\hat{A}| + |p-\hat{p}| + |z-\hat{z}|\}, \quad \forall t,x, 
\end{aligned} \tag{3.81}
$$

with a constant $l_0 > 0$.

**Theorem 3.5 (Comparison Principle).** *Assume* ($d_1$)–($d_3$) *and the structural condition. Let $U$ and $V \in C_b([0,T]\times\mathbb{R}^d)$ be a viscosity subsolution and supersolution of* (3.1)–(3.2), *respectively. Then*

$$
U(t,x) \leq V(t,x), \quad \forall (t,x) \in [0,T]\times\mathbb{R}^d. \tag{3.82}
$$

*Proof.* We divide the proof into three steps. Suppose

$$
|U(t,x)| + |V(t,x)| \leq K, \quad \forall (t,x,y) \in [0,T]\times\mathbb{R}^d\times\mathbb{R}^d. \tag{3.83}
$$

**Step 1.** ***Preparation***

Let us suppose the contrary, namely there is $(\theta,z) \in (0,T)\times\mathbb{R}^d$, such that

$$
4\Delta := U(\theta,z) - V(\theta,z) > 0. \tag{3.84}
$$

For $\rho > 0$, we put $\frac{\rho}{0} = \infty$ and

$$U^{\rho}(t,x) := U(t,x) - \frac{\rho}{t}, \quad (t,x) \in [0,T] \times \mathbb{R}^d. \tag{3.85}$$

Then $U^{\rho} \in USC([0,T] \times \mathbb{R}^d)$ and, for $(t,x) \in (0,T) \times \mathbb{R}^d$,

$$(q,p,A) \in \mathcal{P}^{2+}U(t,x) \Longleftrightarrow \left(q + \frac{\rho}{t^2}, p, A\right) \in \mathcal{P}^{2+}U^{\rho}(t,x).$$

Thus, putting $q^{\rho} = q + \frac{\rho}{t^2}$, condition $(\mathrm{d}_1)$ yields

$$\begin{aligned} & -q^{\rho} + F(t,x,A,p,U^{\rho}(t,x)) \\ & \le -q^{\rho} + F(t,x,A,p,U(t,x)) \le -\frac{\rho}{t^2} < 0. \end{aligned} \tag{3.86}$$

Hence $U^{\rho}$ is a viscosity subsolution of (3.1)–(3.2).

Next we define $\tilde{U}^{\rho}$ and $\tilde{V}$ by

$$\tilde{U}^{\rho}(t,x) = U^{\rho}(t,x) - \varepsilon(1 + |x|^2) \tag{3.87}$$

and

$$\tilde{V}(t,x) = V(t,x) + \varepsilon(1 + |x|^2), \tag{3.88}$$

respectively, where $\rho$ and $\varepsilon$ are positive constants satisfying

$$\frac{\rho}{\theta \wedge 1} \in (0,\Delta) \text{ and } 2\varepsilon(1 + |z|^2) \in (0,\Delta). \tag{3.89}$$

Then from (3.84) and (3.89) it follows that

$$\tilde{U}^{\rho}(\theta,z) - \tilde{V}(\theta,z) > 2\Delta > 0. \tag{3.90}$$

For $\beta \in (0, \frac{\Delta}{T})$ and $\mu > 1$, we put

$$\varphi(t,x,y) = \frac{\mu}{2}|x-y|^2 + \beta(T-t) \ (\ge 0) \tag{3.91}$$

and

$$\Phi(t,x,y) = \tilde{U}^{\rho}(t,x) - \tilde{V}(t,y) - \varphi(t,x,y). \tag{3.92}$$

Then there exists a global maximizer of $\Phi$, say $(\bar{t}_{\mu\beta\varepsilon\rho}, \bar{x}_{\mu\beta\varepsilon\rho}, \bar{y}_{\mu\beta\varepsilon\rho})$, because (3.83) leads to

$$\lim_{|x|+|y|\to\infty} \tilde{U}^{\rho}(t,x) - \tilde{V}(t,y) = -\infty$$

uniformly in $t$.

**Step 2.** ***Limit of the maximizer*** $(\bar{t}_{\mu\beta\varepsilon\rho}, \bar{x}_{\mu\beta\varepsilon\rho}, \bar{y}_{\mu\beta\varepsilon\rho})$ ***as*** $\mu \to \infty$

We omit the subscripts $\mu, \beta, \varepsilon$ and $\rho$, if no confusion occurs.
First we consider $\bar{x}$ and $\bar{y}$. Noting that

$$\tilde{U}^\rho(\bar{t}, \bar{x}) - \tilde{V}(\bar{t}, \bar{y}) \geq \Phi(\bar{t}, \bar{x}, \bar{y}) \geq \Phi(\theta, z, z) \geq \Delta \tag{3.93}$$

and

$$\tilde{U}^\rho(t, x) - \tilde{V}(t, y) \leq K - \varepsilon(|x|^2 + |y|^2), \tag{3.94}$$

we have

$$|\bar{x}|^2 + |\bar{y}|^2 \leq \frac{K}{\varepsilon} =: C^2. \tag{3.95}$$

On the other hand, since

$$0 \leq \Phi(\bar{t}, \bar{x}, \bar{y}) - \Phi(\bar{t}, \bar{x}, \bar{x}) = -\tilde{V}(\bar{t}, \bar{y}) + \tilde{V}(\bar{t}, \bar{x}) - \frac{\mu}{2}|\bar{x} - \bar{y}|^2, \tag{3.96}$$

(3.83) and (3.95) yield

$$\frac{\mu}{2}|\bar{x} - \bar{y}|^2 \leq -V(\bar{t}, \bar{y}) + V(\bar{t}, \bar{x}) + \varepsilon(|\bar{x}|^2 + |\bar{y}|^2) \leq 3K. \tag{3.97}$$

Next we consider $\bar{t}$. Since

$$\lim_{t \to 0} \tilde{U}^\rho(t, x) - \tilde{V}(t, y) = -\infty \quad \text{uniformly on } S_C \times S_C,$$

there is $t_0 = t_0(\varepsilon, \rho) > 0$ such that

$$\bar{t}_{\mu\beta\varepsilon\rho} \geq t_0, \quad \forall \mu. \tag{3.98}$$

Let us take $t_1 = t_1(\varepsilon, \rho) \in (\frac{T}{2}, T)$, such that, for $t > t_1$

$$|U(t, x) - \phi(x)| + |V(t, y) - \phi(y)| < \frac{\rho}{2T}, \quad \forall x, y \in S_C. \tag{3.99}$$

Then for $t > t_1$,

$$\Phi(t, \bar{x}, \bar{y}) < \phi(\bar{x}) - \phi(\bar{y}) + \frac{\rho}{2T} - \frac{\rho}{t} < \phi(\bar{x}) - \phi(\bar{y}). \tag{3.100}$$

Since $\phi(\cdot)$ is uniformly continuous on $S_C$, (3.97) provides a large $\mu_0(\varepsilon)$, such that

$$|\phi(\bar{x}) - \phi(\bar{y})| < \Delta \quad \text{for } \mu > \mu_0(\varepsilon). \tag{3.101}$$

Hence (3.100), (3.101), and (3.93) yield

$$\bar{t}_{\mu\beta\varepsilon\rho} \le t_1 \quad \text{for } \mu > \mu_0(\varepsilon). \tag{3.102}$$

Finally, we consider the limit of the maximizer. Fix $\beta, \varepsilon$, and $\rho$. By (3.95), (3.97), (3.98), and (3.102), we can take $\mu_n, n = 1, 2 \ldots$, tending to $\infty$, so that

$$\lim_{n\to\infty} \bar{x}_{\mu_n\beta\varepsilon\rho} = \lim_{n\to\infty} \bar{y}_{\mu_n\beta\varepsilon\rho} \in S_C \tag{3.103}$$

and

$$\lim_{n\to\infty} \bar{t}_{\mu_n\beta\varepsilon\rho} =: \bar{t}_{\beta\varepsilon\rho} \in [t_0, t_1]. \tag{3.104}$$

By using (3.96), (3.103) and (3.104), we have

$$\frac{\mu_n}{2}|\bar{x}_{\mu_n\beta\varepsilon\rho} - \bar{y}_{\mu_n\beta\varepsilon\rho}|^2 \le \tilde{V}(\bar{t}_{\mu_n\beta\varepsilon\rho}, \bar{x}_{\mu_n\beta\varepsilon\rho}) - \tilde{V}(\bar{t}_{\mu_n\beta\varepsilon\rho}, \bar{y}_{\mu_n\beta\varepsilon\rho}) \longrightarrow 0$$
$$\text{as } \mu_n \to \infty. \tag{3.105}$$

**Step 3.** ***Contradiction to the hypotheses (3.84)***

Let $(t, x) \in (0, T) \times \mathbb{R}^d$. We note the following facts:

$$\begin{aligned} &(q, p, A) \in \bar{\mathcal{P}}^{2+}\tilde{U}^\rho(t, x) \\ \Longleftrightarrow\ &(q, p + 2\varepsilon x, A + 2\varepsilon I_d) \in \bar{\mathcal{P}}^{2+}U^\rho(t, x), \end{aligned} \tag{3.106}$$

$$\begin{aligned} &|F(t, x, A, p, U^\rho(t, x)) - F(t, x, A + 2\varepsilon I_d, p + 2\varepsilon x, U^\rho(t, x))| \\ &\le 2l_0\varepsilon(\sqrt{d} + |x|) \quad (\text{by } (3.81)), \end{aligned} \tag{3.107}$$

and

$$F(t, x, A, p, \tilde{U}^\rho(t, x)) \le F(t, x, A, p, U^\rho(t, x)) \quad (\text{by } (\text{d}_1)). \tag{3.108}$$

Since $U^\rho(\cdot)$ is a viscosity subsolution, (3.106)–(3.108) yield

$$-q + F(t, x, A, p, \tilde{U}^\rho(t, x)) \le 2l_0\varepsilon(\sqrt{d} + |x|). \tag{3.109}$$

Similarly, for $(\hat{q}, \hat{p}, \hat{A}) \in \bar{\mathcal{P}}^{2-}\tilde{V}(t, y)$,

$$-\hat{q} + F(t, y, \hat{A}, \hat{p}, \tilde{V}(t, y)) \ge -2l_0\varepsilon(\sqrt{d} + |y|). \tag{3.110}$$

Now we apply Theorem 3.4 to $\Phi(t,x,y)$. For $\bar{t} = \bar{t}_{\mu\beta\varepsilon\rho}, \bar{x} = \bar{x}_{\mu\beta\varepsilon\rho}$, and $\bar{y} = \bar{y}_{\mu\beta\varepsilon\rho}$, we take $q, \hat{q} \in \mathbb{R}^1$ and $A, \hat{A} \in S^d$ such that

$$
\begin{aligned}
&(q, \mu(\bar{x}-\bar{y}), A) \in \bar{\mathcal{P}}^{2+}\tilde{U}^{\rho}(\bar{t},\bar{x}),\\
&(\hat{q}, \mu(\bar{x}-\bar{y}), \hat{A}) \in \bar{\mathcal{P}}^{2-}\tilde{V}(\bar{t},\bar{y}),\\
&q-\hat{q} = -\beta,\\
&-3\mu\begin{pmatrix} I_d & 0\\ 0 & I_d\end{pmatrix} \le \begin{pmatrix} A & 0\\ 0 & -\hat{A}\end{pmatrix} \le 3\mu\begin{pmatrix} I_d & -I_d\\ -I_d & I_d\end{pmatrix}.
\end{aligned}
$$

Referring to (3.109) and (3.110), we have

$$
-q + F(\bar{t},\bar{x},A,\mu(\bar{x}-\bar{y}),\tilde{U}^{\rho}(\bar{t},\bar{x})) \le 2l_0\varepsilon(\sqrt{d}+|\bar{x}|) \tag{3.111}
$$

and

$$
-\hat{q} + F(\bar{t},\bar{y},\hat{A},\mu(\bar{x}-\bar{y}),\tilde{V}(\bar{t},\bar{y}) \ge -2l_0\varepsilon(\sqrt{d}+|\bar{y}|)). \tag{3.112}
$$

Subtracting (3.111) from (3.112) we get

$$
\begin{aligned}
&F(\bar{t},\bar{y},\hat{A},\mu(\bar{x}-\bar{y}),\tilde{V}(\bar{t},\bar{y})) - F(\bar{t},\bar{x},A,\mu(\bar{x}-\bar{y}),\tilde{U}^{\rho}(\bar{t},\bar{x}))\\
&\ge -2l_0\varepsilon(2\sqrt{d}+|\bar{x}|+|\bar{y}|)+\beta.
\end{aligned} \tag{3.113}
$$

On the other hand, the structural condition implies that

$$
\begin{aligned}
&\text{LHS of (3.113)}\\
&\le F(\bar{t},\bar{y},\hat{A},\mu(\bar{x}-\bar{y}),\tilde{V}(\bar{t},\bar{y})) - F(\bar{t},\bar{x},A,\mu(\bar{x}-\bar{y}),\tilde{V}(\bar{t},\bar{y}))\\
&\quad \text{(by (3.93) and (d}_1\text{))}\\
&\le m_F(\mu|\bar{x}-\bar{y}|^2 + |\bar{x}-\bar{y}| + |\tilde{V}(\bar{t},\bar{y})||\bar{x}-\bar{y}|).
\end{aligned} \tag{3.114}
$$

Thus, we obtain, letting $\mu \to \infty$ in (3.113),

$$
0 \ge -4l_0\varepsilon(\sqrt{d}+C)+\beta, \tag{3.115}
$$

thanks to (3.113), (3.114), and (3.103). Since $C = \sqrt{\frac{K}{\varepsilon}}$, letting $\varepsilon \to 0$ yields "$0 \ge \beta$". Since $\beta > 0$, this completes the proof. □

The following uniqueness result is immediate from Theorem 3.5.

**Theorem 3.6.** *For the* HJB *equation* (3.24), *we assume that* $\alpha, b, \kappa$ *and* $f$ *are bounded, continuous, and Lipschitz continuous in* $x$, *uniformly in* $(t,\gamma)$. *Then, for* $\phi \in C_b(\mathbb{R}^d)$, *the value function gives the unique bounded viscosity solution of* (3.24).

## 3.3 HJB Equations for Control-Stopping Problems

In Sect. 3.3, we control not only the dynamics of stochastic system, but also the terminal time of its evolution. Accordingly, here an admissible system consists a pair of admissible control $A = (\Omega, \mathcal{F}, (\mathcal{F}_t), P, W, \gamma(\cdot))$ and an $(\mathcal{F}_t)$-stopping time $\tau$. By $\mathfrak{A}$, we denote the set of all admissible systems. We will deal with this problem in the same framework as the control problem.

Firstly we formulate control-stopping problems in Sect. 3.3.1. In Sects. 3.3.2 and 3.3.3, we study the DPP and the viscosity solutions for the HJB equations via semigroup arguments respectively. Section 3.3.4 deals with the American option price problem as an example.

### 3.3.1 *Formulations*

We assume that all coefficients $\alpha, b, \kappa$ and $f$ are time independent and that conditions $(b_1)$–$(b_5)$ in Sect. 2.1 are satisfied. $(A, \tau)$ is called an admissible system if $A = (\Omega, \mathcal{F}, (\mathcal{F}_t), P, W, \gamma(\cdot))$ is an admissible control and $\tau$ is a $[0, T]$-valued $(\mathcal{F}_t)$-stopping time. $\mathfrak{A}$ denotes the set of all admissible systems.

For $(A, \tau) \in \mathfrak{A}$, we have the following SDE (3.116) for the response $X^A$ and the payoff $j(\cdot)$ in the control-stopping problem:

$$dX(t) = b(X(t), \gamma(t))\, dt + \alpha(X(t), \gamma(t))\, dW(t), \quad t \in (0, T], \tag{3.116}$$

and

$$j(t, x, (A, \tau); \phi) := J(t \wedge \tau, 0, x, A; \phi) = E_{0x} C(t \wedge \tau, A; \phi), \tag{3.117}$$

where the cost functional $C(\cdot)$ is given by

$$\begin{aligned} &C(s, A; \phi) \\ &= \int_0^s \exp\{-\int_0^\theta \kappa(X^A(h), \gamma(h))\, dh\} f(X^A(\theta), \gamma(\theta))\, d\theta \\ &\quad + \exp\{-\int_0^s \kappa(X^A(h), \gamma(h))\, dh\} \phi(X^A(s)). \end{aligned}$$

The value function is defined by

$$v(t, x; \phi) = \inf_{(A,\tau)\in\mathfrak{A}} j(t, x, (A, \tau); \phi). \tag{3.118}$$

Thus, we aim to analyze the value function and find an optimal admissible system. For simplicity, we assume that T is an integer.

Before we study the control-stopping problems, we shall summarize the properties of $v(\cdot)$. Let $\phi \in \tilde{C}$ and $\varepsilon, R > 0$ be given. Then there are positive constants $\tilde{K}, \delta_{\varepsilon R}$ and $\Delta_{\varepsilon R}$, such that

$$\sup_{0 \le t \le T} |v(t, x; \phi)| \le \tilde{K}(1 + |x|^2), \tag{3.119}$$

$$\sup_{0 \le t \le T} |v(t, x; \phi) - v(t, y; \phi)| < \varepsilon \tag{3.120}$$

for $|x - y| < \delta_{\varepsilon R}$ and $x, y \in S_R$, and

$$|v(t, x; \phi) - v(\theta, x; \phi)| < \varepsilon \tag{3.121}$$

for $|t - \theta| < \Delta_{\varepsilon R}$ and $x \in S_R$.

From (3.119)–(3.121), it is follows that

$$v(t, \cdot; \phi) \in \tilde{C}, \quad \forall t \in [0, T]. \tag{3.122}$$

**Proposition 3.5 (Dependence on the terminal cost function).** *Suppose that $\phi, \phi_n \in \tilde{C}$ satisfy $\|\phi\|_{\tilde{C}}, \|\phi_n\|_{\tilde{C}} \le K$ $(n = 1, 2, \dots)$ and $\phi_n$ converges to $\phi$ uniformly on any bounded set. Then, for any $R > 0$,*

$$\lim_{n \to \infty} \sup_{0 \le t \le T, |x| \le R} |v(t, x; \phi_n) - v(t, x; \phi)| = 0. \tag{3.123}$$

From now on, we put $E_x = E_{0x}$, for simplicity.

*Proof.* We have

$$\begin{aligned} |v(t, x; \phi_n) - v(t, x; \phi)| &\le \sup_{(A\tau) \in \mathfrak{A}} E_x |\phi_n(X^A(t \wedge \tau)) - \phi(X^A(t \wedge \tau))| \\ &\le \sup_{A \in \mathbb{A}} E_x \Big[ \sup_{0 \le s \le T} |\phi_n(X^A(s)) - \phi(X^A(s))| \Big]. \end{aligned} \tag{3.124}$$

By the assumptions, we can choose $N(\varepsilon, R)$ so that

$$\sup_{y \in S_R} |\phi_n(y) - \phi(y)| < \varepsilon \quad \text{for } n > N(\varepsilon, R).$$

Hence

$$\begin{aligned} &E_x \Big[ \sup_{0 \le s \le T} |\phi_n(X^A(s)) - \phi(X^A(s))| \Big] \\ &\le \varepsilon + 2K E_x \Big[ 1 + \sup_{0 \le s \le T} |X^A(s)|^2; \sup_{0 \le s \le T} |X^A(s)| \ge R \Big] \end{aligned}$$

$$\leq \varepsilon + c_1 \sqrt{E_x\Big(1 + \sup_{0\leq s\leq T} |X^A(s)|^2\Big)^2}\, \frac{\sqrt{E_x[\sup_{0\leq s\leq T} |X^A(s)|^2]}}{R}$$

$$\leq \varepsilon c_2 \frac{1 + |x|^3}{R} \quad \text{(by (2.9))} \tag{3.125}$$

with constants $c_1$ and $c_2$ independent of $A$.

Taking $R$ so that "$c_2 \frac{1+\sqrt{R}}{R} < \varepsilon$", we conclude that, for $n > N(\varepsilon, R)$,

$$\sup_{0\leq t\leq T} |v(t, x; \phi_n) - v(t, x; \phi)| < 2\varepsilon \quad \text{for } |x| < R^{\frac{1}{6}}.$$

This completes the proof. □

## 3.3.2 DPP

Let us consider discrete-time DPP firstly. By $\mathfrak{A}_N$, we denote the set of $(A, \tau) \in \mathfrak{A}$ whose control process is switching at $\{k2^{-N}; k = 1, 2, \ldots, 2^N T - 1\}$ and with stopping time $\tau$ taking values in $\{k2^{-N}; k = 0, 1, \ldots, 2^N T\}$. Put

$$v_N(t, x; \phi) = \inf_{(A\tau)\in\mathfrak{A}_N} j(t, x, (A, \tau); \phi) \quad \text{for } \phi \in \tilde{C}. \tag{3.126}$$

We sometimes write $(\gamma(\cdot), \tau)$ instead of $(A, \tau)$, when there is no danger of confusion.

**Proposition 3.6.** *Let $\Gamma$ be convex and compact. Then*

$$\begin{aligned} v_N(t + \theta, x; \phi) &= \inf_{(A,\tau)\in\mathfrak{A}_N} j(t, x, (A, \tau); v_N(\theta, \cdot; \phi)) \\ &= v_N(t, x; v_N(s, \cdot; \phi)) \quad \text{for } t = 2^{-N} i, \quad \theta = 2^{-N} j, \quad t + \theta \leq T. \end{aligned} \tag{3.127}$$

Before we start the proof, we prepare two lemmas. First of all, we introduce the mapping $j_N : \tilde{C} \to \tilde{C}$, according to [S08], (2.6), [N81], 2.4.1 [N78]. Put $\Delta = 2^{-N}$ and fix a reference probability system $(\Omega, \mathcal{F}, (\mathcal{F}_t), P, W)$.

Let us define $\nu_N$ and $j_N : \tilde{C} \to \tilde{C}$ by

$$\nu_N \phi(x) = \inf_{\gamma\in\Gamma} E_x C(\Delta, \gamma; \phi) \tag{3.128}$$

and

$$j_N \phi(x) = \phi(x) \wedge \nu_N \phi(x). \tag{3.129}$$

That is to say, $j_N \phi$ is the expected optimal value at one step.

Put $j_N^0 =$ identity and $j_N^{k+1}\phi = j_N(j_N^k\phi), k = 0, 1, \ldots, 2^N T - 1$.

**Lemma 3.1.**

$$j_N^{k+1}\phi = \phi \wedge \nu_N(j_N^k\phi), \quad k = 0, 1, 2, \cdots. \tag{3.130}$$

*Proof.* We use induction on $k$. For $k = 0$, (3.130) is the definition of $j_N$. Suppose that (3.130) holds for $k$. Then

$$\begin{aligned} j_N^{k+2}\phi &= j_N(j_N^{k+1}\phi) = j_N^{k+1}\phi \wedge \nu_N(j_N^{k+1}\phi) \\ &= \phi \wedge \nu_N(j_N^k\phi) \wedge \nu_N(j_N^{k+1}\phi) \end{aligned} \tag{3.131}$$

by the hypothesis for $k$.

Observing that $j_N^{k+1}\phi \le j_N^k\phi$ and $\nu_N$ is monotone, we obtain (3.131) for $k + 1$. This completes the proof. □

In a way similar to that described in Sect. 2.2.4, we have

**Lemma 3.2.** *Let $\phi \in \tilde{C}$, an integer $k$, and $x \in \mathbb{R}^d$ be given. Then there exist $(\mathcal{F}_t^W)$-progressively measurable switching control process $\gamma^*(\cdot)$ and an $(\mathcal{F}_t^W)$-stopping time $\tau^*$ such that $(\gamma^*(\cdot), \tau^*) \in \mathfrak{A}_N$ and*

$$j_N^k\phi(x) = j(k2^{-N}, x, (\gamma^*(\cdot), \tau^*); \phi) = J(k2^{-N} \wedge \tau^*, x, \gamma^*(\cdot); \phi). \tag{3.132}$$

*Proof.* We divide the proof into three steps. Fix a reference probability system $(\Omega, \mathcal{F}, (\mathcal{F}_t), P, W)$. By $\mathbb{S}^W$ we denote the set of all $(\mathcal{F}_t^W)$-stopping times with values in $[0, T]$.

**Step 1.** ***Construction of $\gamma^*(\cdot)$ on the reference probability system***

For $\Psi \in \tilde{C}$, $J(\triangle, y, \gamma; \Psi)$ is continuous w.r.t. $(y, \gamma) \in \mathbb{R}^d \times \Gamma$. Since $\Gamma$ is compact, there exists a minimum selector $\hat{\gamma} : \mathbb{R}^d \to \Gamma$, such that

$$J(\triangle, y, \hat{\gamma}(y; \Psi); \Psi) = \inf_{\gamma \in \Gamma} J(\triangle, y, \gamma; \Psi) = \nu_N\Psi(y). \tag{3.133}$$

Define $\gamma^*(\cdot)$ by

$$\gamma^*(t) = \hat{\gamma}(x; j_N^{k-1}\phi), \quad \forall t \in [0, \triangle). \tag{3.134}$$

Denoting by $X^*$ its response, we have

$$\begin{cases} dX^*(t) = b(X^*(t), \gamma^*(0))\, dt + \alpha(X^*(t), \gamma^*(0))\, dW(t), & t \in (0, \triangle], \\ X^*(0) = x. \end{cases} \tag{3.135}$$

Putting $\gamma^*(t) = \hat{\gamma}(X^*(\triangle); j_N^{k-2}\phi)$, for $t \in [\triangle, 2\triangle)$, we have its response $X^*$ on $[0, 2\triangle]$.

Repeating this procedure successively, we obtain a switching control $\gamma^*(\cdot) \in \boldsymbol{\Gamma}^W$ and its response $X^*$ on the time interval $[0, k\triangle]$.

From (3.133), it follows that

$$J(\triangle, x, \gamma^*(\cdot), j_N^{k-1}\phi) = v_N(j_N^{k-1}\phi)(x). \tag{3.136}$$

**Step 2.** ***Construction of*** $\tau^* \in \mathbb{S}^W$

Put

$$D(l) = \{z \in \mathbb{R}^d; \phi(z) = j_N^l\phi(z)\}, \quad l = 0, 1, \ldots, k. \tag{3.137}$$

Then $D(0) = \mathbb{R}^d$ and the closed set $D(l)$ is decreasing. Thinking of $D(k-l)$ as the stop region at time $l\triangle$, we define $\tau^*$ by

$$\begin{aligned}\tau^*(\omega) &= \min\{i\triangle; X^*(i\triangle, \omega) \in D(k-i)\}\\ &= \min\{i\triangle; \phi(X^*(i\triangle, \omega)) = j_N^{k-i}\phi(X^*(i\triangle, \omega))\}.\end{aligned} \tag{3.138}$$

Since $j_N^0\phi = \phi$, $\tau^*$ is an $(\mathcal{F}_t^W)$-stopping time taking values in $\{0, \triangle, \ldots, k\triangle\}$.

**Step 3.** $(\gamma^*(\cdot), \tau^*) \in \boldsymbol{\Gamma}^W \times \mathbb{S}^W$ satisfies (3.132). Note that the triviality of $\mathcal{F}_0^W$ yields

$$P_x(\tau^* = 0) = 1 \text{ or } 0. \tag{3.139}$$

For $x \in D(k)$, we have $P_x(\tau^* = 0) = 1$ and

$$j_N^k\phi(x) = \phi(x) = j(k\triangle, x, (\gamma^*(\cdot), \tau^*); \phi). \tag{3.140}$$

For $x \notin D(k)$, it holds that

$$P_x(\tau^* \geq \triangle) = 1. \tag{3.141}$$

Let us compute the RHS of (3.132):

$$\begin{aligned}J &:= E_x[C(\tau^*, \gamma^*(\cdot); \phi)]\\ &= \sum_{l=1}^{k} E_x[C(l\triangle, \gamma^*(\cdot); \phi); \tau^* = l\triangle]\\ &=: \sum_{l=1}^{k} J_l.\end{aligned} \tag{3.142}$$

Observing that $(\tau^* = k\Delta) = (\tau^* \leq (k-1)\Delta)^c \in \mathcal{F}^W_{(k-1)\Delta}$ and

$$\begin{aligned}
&E_x\Big(\int_{(k-1)\Delta}^{k\Delta} \exp\{-\int_{(k-1)\Delta}^{s} \kappa(\lambda)\,d\lambda\} f(X^*(s), \hat{\gamma}(X^*((k-1)\Delta);\phi))\,ds \\
&\quad + \exp\{-\int_{(k-1)\Delta}^{k\Delta} \kappa(\lambda)\,d\lambda\}\phi(X^*(k\Delta))|\mathcal{F}^W_{(k-1)\Delta}\Big) \\
=&\nu_N\phi(X^*((k-1)\Delta)),
\end{aligned} \tag{3.143}$$

where $\kappa(\lambda)$ denotes $\kappa(X^*(\lambda), \hat{\gamma}(X^*((k-1)\Delta);\phi))$ for simplicity, we have

$$J_k = E_x[C((k-1)\Delta, \gamma^*(\cdot); \nu_N\phi); \tau^* = k\Delta]. \tag{3.144}$$

Therefore,

$$J_k + J_{k-1} = E_x[C((k-1)\Delta, \gamma^*(\cdot); j_N\phi); \tau^* \geq (k-1)\Delta]. \tag{3.145}$$

By using the same arguments successively, we obtain

$$\begin{aligned}
J &= J_k + J_{k-1} + \cdots + J_1 \\
&= E_x[C(\Delta, \gamma^*(\cdot), j_N^{k-1}\phi); \tau^* \geq \Delta] \\
&= \nu_N(j_N^{k-1}\phi)(x) = j_N^k\phi(x),
\end{aligned} \tag{3.146}$$

because $x \notin D(k)$ and $P_x(\tau^* \geq \Delta) = 1$.

Now (3.146) and (3.140) complete the proof of the lemma. □

Now we are in the position to prove Proposition 3.6.

*Proof.* First we will show that, for $(A, \tau) \in \mathfrak{A}_N$,

$$j_N^k\phi(x) \leq j(k\Delta, x, (A, \tau); \phi), \quad k = 1, 2, \ldots. \tag{3.147}$$

We use induction on $k$. It holds that

$$j(\Delta, x, (A, \tau); \phi) = E_x(C(\Delta, \gamma(0); \phi); \tau \geq \Delta] + \phi(x)P_x(\tau = 0). \tag{3.148}$$

Since $(\tau \geq \Delta) = (\tau = 0)^c \in \mathcal{F}_0$ and $\gamma(0)$ is $\mathcal{F}_0$-measurable,

$$\begin{aligned}
\text{1st term of RHS of (3.148)} &= E_x[E(C(\Delta, \gamma(0); \phi)|\mathcal{F}_0); \tau \geq \Delta] \\
&\geq E_x[\nu_N\phi(x); \tau \geq \Delta] = \nu_N\phi(x)P_x(\tau \geq \Delta).
\end{aligned} \tag{3.149}$$

Substituting (3.149) into (3.148), we obtain (3.147) for $k = 1$.

Suppose that, for any $\phi \in \tilde{C}$, (3.147) is valid for $k$. Noting that

$$\begin{aligned} & j((k+1)\triangle, x, (A,\tau); \phi) \\ &= E_x[C((k+1)\triangle, A; \phi); \tau \geq (k+1)\triangle] + E_x[C(\tau, A; \phi); \tau \leq k\triangle] \end{aligned} \tag{3.150}$$

and recalling that $(\tau \geq (k+1)\triangle) = (\tau \leq k\triangle)^c \in \mathcal{F}_{k\triangle}$ and $\gamma(k\triangle)$ is $\mathcal{F}_{k\triangle}$-measurable, we have

$$\begin{aligned} \text{1st term of RHS of (3.150)} &\geq E_x[C(k\triangle, A; v_N\phi); \tau \geq (k+1)\triangle] \\ &\geq E_x[C(k\triangle, A; j_N\phi); \tau \geq (k+1)\triangle]. \end{aligned} \tag{3.151}$$

From (3.150) and (3.151) it follows that

$$\begin{aligned} j((k+1)\triangle, x, (A,\tau); \phi) &\geq j(k\triangle, x, (A,\tau); j_N\phi) \\ &\geq j_N^k(j_N\phi)(x) \quad \text{(by the induction hypothesis)} \\ &= j_N^{k+1}\phi(x), \end{aligned} \tag{3.152}$$

which yields (3.147) for $k+1$.

Taking the infimum of the RHS of (3.147) over $\mathfrak{A}_N$, we have

$$j_N^k\phi(x) \leq v_N(k\triangle, x; \phi). \tag{3.153}$$

Since (3.132) yields the opposite inequality, we obtain

$$j_N^k\phi(x) = v_N(k\triangle, x; \phi) \tag{3.154}$$

which in turn yields (3.127). □

*Remark.* By referring to Lemma 3.2, we can assert that there is a Brownian adapted pair $(\gamma^*(\cdot), \tau^*)$, that is optimal in $\mathfrak{A}_N$.

Next we need the following approximation result.

**Lemma 3.3.** *Let $\Gamma$ be convex and compact. Then for binary rational $t$*

$$v(t, x : \phi) = \lim_{N\to\infty} v_N(t, x : \phi), \quad \forall x \in \mathbb{R}^d. \tag{3.155}$$

*Proof.* By the definitions of $v_N(\cdot)$ and $v(\cdot)$, $v_N(\cdot)$ is decreasing, as $N \to \infty$, and, for a binary rational $t$,

$$v(t, x; \phi) \leq \lim_{N\to\infty} v_N(t, x; \phi). \tag{3.156}$$

For the converse inequality, it is enough to prove that

$$j(t, x, (A,\tau); \phi) \geq \lim_{N\to\infty} v_N(t, x; \phi), \quad \forall (A,\tau) \in \mathfrak{A}. \tag{3.157}$$

Let a reference probability system $(\Omega, \mathcal{F}, (\mathcal{F}_t), P, W)$ be given. For an $(\mathcal{F}_t)$-progressively measurable control process $\gamma(\cdot)$, we can take an approximate switching control $\gamma_N(\cdot)$ $(N = 1, 2, \ldots)$ by Theorem 2.3, such that

$$\gamma_N(t) = \gamma_N(([2^N t]2^{-N}) \wedge T) \tag{3.158}$$

and

$$\gamma_N(\cdot) \longrightarrow \gamma(\cdot) \tag{3.159}$$

in $L^2([0, T] \times \Omega)$ and a.e. in $[0, T]$ $P$-a.s.

Moreover, its response $X_N$ converge to $X(=$ response for $\gamma(\cdot))$,

$$\sup_{0 \leq t \leq T} |X_N(t) - X(t)| \longrightarrow 0 \tag{3.160}$$

in $L^2(\Omega)$ and $P$-a.s.

Let $\tau$ be an $(\mathcal{F}_t)$-stopping time with values in $[0, T]$. Put $\tau_N(\omega) = T \wedge [1 + 2^N \tau(\omega)]2^{-N}$. Then $\tau_N \searrow \tau$ $P$-a.s. and $(\gamma_N(\cdot), \tau_N) \in \mathfrak{A}_N$.

By noticing that $C(t \wedge \tau_N, \gamma_N(\cdot); \phi)$ converges to $C(t \wedge \tau, \gamma(\cdot); \phi)$ $P$-a.s., the convergence theorem together with $(b_4)$ and (2.9) shows that

$$\begin{aligned} j(t, x, (\gamma(\cdot), \tau); \phi) &= \lim_{N \to \infty} j(t, x, (\gamma_N(\cdot), \tau_N); \phi) \\ &\geq \lim_{N \to \infty} v_N(t, x; \phi). \end{aligned} \tag{3.161}$$

This completes the proof of the lemma. □

From the Remark after the proof of Lemma 3.2 it follows that

$$\begin{aligned} v_N(t, x; \phi) &= \inf_{\mathfrak{A}_N \cap (\Gamma^W \times \mathbb{S}^W)} j(t, x, (\gamma(\cdot), \tau); \phi) \\ &\geq \inf_{\Gamma^W \times \mathbb{S}^W} j(t, x, (\gamma(\cdot), \tau); \phi). \end{aligned} \tag{3.162}$$

Consequently, Lemma 3.3 together with (3.162) yields

$$v(t, x; \phi) \geq \inf_{\Gamma^W \times \mathbb{S}^W} j(t, x, (\gamma(\cdot), \tau); \phi).$$

Since the opposite inequality is clear, we have

$$v(t, x; \phi) = \inf_{\Gamma^W \times \mathbb{S}^W} j(t, x, (\gamma(\cdot), \tau); \phi) \tag{3.163}$$

for a binary rational $t$.

Finally we prove DPP for the value function $v(\cdot)$.

**Theorem 3.7.** *Let $\Gamma$ be convex and $\sigma$-compact. Suppose* (b$_1$)–(b$_5$) *holds. Then for* $\phi \in \tilde{C}$,

$$v(t+s, x; \phi) = v(t, x; v(s, \cdot; \phi)) \tag{3.164}$$

*for* $0 \le s, t \le s+t \le T$, $x \in \mathbb{R}^d$.

*Proof.* We divide the proof into three steps.

**Step 1.** Let $\Gamma$ be convex and compact. Let $t$ and $s$ be binary, say, $t = i2^{-p}$ and $s = j2^{-p}$.

Since $v_N(\theta, y; \phi)$ is decreasing to $v(\theta, y; \phi)$ as $N \to \infty$, Proposition 3.6 and the monotonicity property of $v_N(\cdot)$ yield

$$\begin{aligned} v_N(t+s, x; \phi) &= v_N(t, x; v_N(s, \cdot; \phi)) \\ &\ge v_N(t, x; v(s, \cdot; \phi)) \quad \text{for } N \ge p, \end{aligned} \tag{3.165}$$

and, as $N \to \infty$,

$$v(t+s, x; \phi) \ge v(t, s; v(s, \cdot; \phi)). \tag{3.166}$$

Hence we need to establish the opposite inequality of (3.166). Observing

$$v_{N+m}(t+s, x; \phi) \le v_{N+m}(t, x; v_m(s, \cdot; \phi)) \tag{3.167}$$

for $N \ge p, m = 1, 2, \ldots$, and, letting $N \to \infty$, we have

$$\begin{aligned} v(t+s, x; \phi) &\le v(t, x; v_m(s, \cdot; \phi)) \\ &\le j(t, x, (\gamma(\cdot), \tau); v_m(s, \cdot; \phi)), \quad \forall (\gamma(\cdot), \tau) \in \boldsymbol{\Gamma}^W \times \mathbb{S}^W, \end{aligned} \tag{3.168}$$

by (3.163). Again, using that

$$\lim_{m \to \infty} C(t \wedge \tau, \gamma(\cdot); v_m(s, \cdot; \phi)) = C(t \wedge \tau, \gamma(\cdot); v(s, \cdot \phi)) \quad P\text{-a.s.} \tag{3.169}$$

and

$$|C(t \wedge \tau, \gamma(\cdot); v_m(s, \cdot; \phi)| \le c_1 \Big(1 + \sup_{0 \le t \le T} |X(t)|^2\Big)$$

with a constant $c_1$, independent of $m$, we obtain

$$\lim_{m \to \infty} j(t, x, (\gamma(\cdot), \tau); v_m(s, \cdot; \phi)) = j(t, x, (\gamma(\cdot), \tau), v(s, \cdot; \phi)), \tag{3.170}$$

by the dominated convergence theorem. Taking the infimum over $(\gamma(\cdot), \tau) \in \boldsymbol{\Gamma}^W \times \mathbb{S}^W$ yields the opposite of inequality (3.166), by (3.163).

**Step 2.** Let $t$ and $s$ be given. Suppose that binary $t_m$ and $s_n$ approach $t$ and $s$, respectively. Setting $\phi_n = v(s_n, \cdot; \phi)$ in Proposition 3.5 and using (3.119) and (3.121), we get

$$\lim_{n\to\infty} \sup_{|x|\le R} |v(t_m, x; v(s_n, \cdot; \phi)) - v(t_m, x; v(s, \cdot; \phi))| = 0. \tag{3.171}$$

Now

$$v(t_m, x; v(s, \cdot; \phi)) = v(t_m + s, x; \phi) \tag{3.172}$$

follows from Step 1. Letting $m \to \infty$, we obtain (3.164).

**Step 3.** $\Gamma$ is convex and $\sigma$- compact. Take a sequence of convex and compact sets $\Gamma_n, n = 1, 2, \ldots$, so that

$$\Gamma_n \subset \Gamma_{n+1}, n = 1, 2, \ldots \quad \text{and} \quad \bigcup_n \Gamma_n = \Gamma. \tag{3.173}$$

Let $\mathfrak{A}_N$ denote the set of all $(A, \tau)$ with control region $\Gamma_N$. Then $\mathfrak{A} = \bigcup_N \mathfrak{A}_N$ by Definition 2.1. Set

$$v_N(t, x; \phi) := \inf_{(A,\tau)\in\mathfrak{A}_N} j(t, x, (A, \tau); \phi).$$

Since $v_N(t, x; \phi)$ is decreasing as $N \to \infty$, and

$$j(t, x, (A, \tau); \phi) \ge \lim_{N\to\infty} v_N(t, x; \phi) \ge v(t, x; \phi), \quad \forall (A, \tau) \in \mathfrak{A}, \tag{3.174}$$

by taking the infimum of LHS of (3.174) over $\mathfrak{A}$ we have

$$v(t, x; \phi) = \lim_{N\to\infty} v_N(t, x; \phi), \quad \forall t, x. \tag{3.175}$$

From DPP for $v_N$ and (3.175) we obtain (3.164) by using the same arguments as in Step 2 of the proof of Theorem 2.4.

This completes the proof. □

### 3.3.3 Semigroups Associated with DPP

Let us define a mapping $\mathrm{V}_t : \tilde{C} \mapsto \tilde{C}$, by

$$\mathrm{V}_t \phi(x) = v(t, x; \phi). \tag{3.176}$$

Then by DPP $(\mathrm{V}_t; t \in [0, T])$ is a one-parameter semigroup on $\tilde{C}$.

The following properties are easily verified.

**Proposition 3.7.**

(i) ***Semigroup*** $\mathrm{V}_0 =$ *identity,* $\mathrm{V}_{t+s} = \mathrm{V}_t\mathrm{V}_s = \mathrm{V}_s\mathrm{V}_t$*;*
(ii) ***Monotonicity***

$$\phi \le \psi \Longrightarrow \mathrm{V}_t\phi \le \mathrm{V}_t\psi, \quad \forall t \in [0,T];$$

(iii) ***Lipschitz condition***
$\exists K^* > 0$ *such that*

$$\|\mathrm{V}_t\phi - \mathrm{V}_t\psi\|_{\tilde{C}} \le K^*\|\phi - \psi\|_{\tilde{C}}, \quad \forall t \in [0,T], \quad \forall \phi,\psi \in \tilde{C}; \tag{3.177}$$

(iv) ***Continuity in the time parameter***

$$\lim_{t\to 0}\sup_{|x|\le R}|\mathrm{V}_t\phi(x) - \phi(x)| = 0, \quad \forall R > 0; \tag{3.178}$$

(v)

$$\mathrm{V}_t\phi \le \phi \quad \text{and} \quad \mathrm{V}_t\phi \le \mathrm{H}_t^\gamma\phi, \quad \forall t \in [0,T], \quad \forall \gamma \in \Gamma,$$

*where* $\mathrm{H}_t^\gamma$ *is the semigroup on* $\tilde{C}$ *given by*

$$\mathrm{H}_t^\gamma\phi(x) = E_x C(t,\gamma;\phi),$$

(vi) ***Maximality***
*Suppose that* $(\mathrm{U}_t; t \in [0,T])$ *is a semigroup on* $\tilde{C}$*, satisfying* (i)–(v). *Then*

$$\mathrm{U}_t\phi \le \mathrm{V}_t\phi, \quad \forall t \in [0,T], \quad \forall \phi \in \tilde{C}.$$

For the generator of $\mathrm{V}_t$, we have

**Proposition 3.8.** *Put*

$$\hat{\mathrm{G}}\phi(x) = 0 \wedge \inf_{\gamma\in\Gamma}\mathrm{h}^\gamma\phi(x), \tag{3.179}$$

*where* $\mathrm{h}^\gamma$ *is the generator of* $\mathrm{H}_t^\gamma$*. Then for any* $R > 0$*,*

$$\lim_{t\to 0}\sup_{|x|\le R}\left|\frac{1}{t}(\mathrm{V}_t\phi(x) - \phi(x)) - \hat{\mathrm{G}}\phi(x)\right| = 0 \tag{3.180}$$

*for any* $\phi \in \tilde{C} \cap C_p^2(\mathbb{R}^d)$*.*

*Proof.* We use the same arguments as in Proposition 2.8. Fix a reference probability system $(\Omega, \mathcal{F}, (\mathcal{F}_t), P, W)$. For simplicity, we put

$$I(\gamma(\cdot), \tau) = j(t, x, (\gamma(\cdot), \tau); \phi) - \phi(x) \quad \text{for } (\gamma(\cdot), \tau) \in \boldsymbol{\Gamma}^W \times \mathbb{S}^W.$$

For $\phi \in C_p^2(\mathbb{R}^d)$, Itô's formula implies

$$\begin{aligned} I(\gamma(\cdot), \tau) &\geq E_x \int_0^{t\wedge\tau} \exp\{-\int_0^s \kappa(X(\lambda), \gamma(\lambda))\, d\lambda\}\big(0 \wedge \mathrm{h}^{\gamma(s)}\phi(X(s))\big)\, ds \\ &\geq E_x \int_0^t \big(0 \wedge \mathrm{h}^{\gamma(s)}\phi(X(s))\big)\, ds. \end{aligned} \tag{3.181}$$

On the other hand, Lemma 2.1 asserts that, for $\varepsilon$ and $R > 0$, there is $t_0 = t_0(\varepsilon, R)$ such that

$$\sup_{|x|\leq R} E_x|\mathrm{h}^{\gamma(s)}\phi(X(s)) - \mathrm{h}^{\gamma(s)}\phi(x)| < \varepsilon, \quad \forall \gamma(\cdot) \in \boldsymbol{\Gamma}^W \tag{3.182}$$

whenever $s < t_0$.

Plugging (3.182) into the RHS of (3.181), we have

$$\begin{aligned} I(\gamma(\cdot), \tau) &\geq (0 \wedge \inf_{\gamma\in\Gamma} \mathrm{h}^\gamma\phi(x))t - \varepsilon t \\ &= (\hat{\mathrm{G}}\phi(x) - \varepsilon)t \quad \text{for } x \in S_R,\ t < t_0. \end{aligned} \tag{3.183}$$

Taking the infimum of the LHS of (3.183) over $(\gamma(\cdot), \tau) \in \boldsymbol{\Gamma}^W \times \mathbb{S}^W$ and dividing both sides by $t$, we obtain

$$\frac{1}{t}(\mathrm{V}_t\phi(x) - \phi(x)) - \hat{\mathrm{G}}\phi(x) > -\varepsilon, \quad \forall x \in S_R, \quad t < t_0. \tag{3.184}$$

For the converse inequality, we notice that

$$\mathrm{V}_t\phi(x) \leq \phi(x) \wedge \inf_{\gamma\in\Gamma} j(t, x, (\gamma, t); \phi). \tag{3.185}$$

Hence, for $x \in S_R$ and $t < t_0$,

$$\begin{aligned} &\mathrm{V}_t\phi(x) - \phi(x) \\ &\leq 0 \wedge \inf_{\gamma\in\Gamma} E_x \int_0^t \exp\{-\int_0^s \kappa(X(\lambda), \gamma)\, d\lambda\}\mathrm{h}^\gamma\phi(X(s))\, ds \\ &\leq 0 \wedge \inf_{\gamma\in\Gamma} E_x\Big[\int_0^t \exp\{-\int_0^s \kappa(X(\lambda), \gamma)\, d\lambda\}\mathrm{h}^\gamma\phi(x)\, ds\Big] + \varepsilon t \quad \text{(by (3.182))} \\ &\leq \Big(0 \wedge \inf_{\gamma\in\Gamma} \mathrm{h}^\gamma\phi(x)t\Big) + \varepsilon t + \sup_{\gamma\in\Gamma} |\mathrm{h}^\gamma\phi(x)|t^2 c_0 \quad \text{(by } (b_5)\text{)}. \end{aligned} \tag{3.186}$$

Now (3.184) and (3.186) yield (3.180). □

The HJB equation associated with the control-stopping problem reads and is considered

$$\partial_t v(t,x) + 0 \vee \mathrm{H}(x, \partial_{xx} v(t,x), \partial_x v(t,x), v(t,x)) = 0,$$
$$(t,x) \in (0,T] \times \mathbb{R}^d, \tag{3.187}$$

with the initial condition

$$v(0,x) = \phi(x), \quad x \in \mathbb{R}^d, \tag{3.188}$$

where

$$\mathrm{H}(x,Q,p,z) = \sup_{\gamma \in \Gamma}\Big(-\frac{1}{2}\mathrm{tr}(a(x,\gamma)Q) - b(x,\gamma)\cdot p + \kappa(x,\gamma)z - f(x,\gamma)\Big).$$

**Proposition 3.9.** *Suppose that the coefficients $\alpha, b, \kappa$, and $f$ are bounded, continuous, and Lipschitz continuous w.r.t. $x$ uniformly on $\Gamma$. Then the value function is the unique bounded continuous viscosity solution of* (3.187)–(3.188), *for $\phi \in C_b(\mathbb{R}^d)$.*

Proof mimics Theorem 3.3, because $\mathrm{V}_{\theta t} := \mathrm{V}_{t-\theta}$ is a 2 parameter semigroup satisfying (3.30).

### 3.3.4 American Option Price

Let us apply previous results to the HJB equation arising from the problem of American option price. Following [Ma00], we consider a time-homogeneous one-dimensional model.

The asset price $\xi(\cdot)$ evolves according to the linear SDE

$$d\xi(t) = \rho\xi(t)\,dt + \sigma\xi(t)\,dW(t), \tag{3.189}$$

where $\rho$ and $\sigma$ are positive constants and $W$ is a real Wiener process. Let $T > 0$ denote the maturity time. For a Lipschitz continuous function $g$, the American option gives an asset holder the right to get the amount of money $g(\xi(\tau))$ with the discount rate $\rho$ at his/her chosen stopping time $\tau$ $(\leq T)$. Thus the asset holder wants to maximize its expectation by choosing a suitable stopping time.

First we formulate the problem and recall its HJB equation. Put

$$\mathrm{S}_\theta = \{\tau; \tau - \theta \text{ is an}(\mathcal{F}_t^{W_\theta^+})\text{-stopping time valued in } [0, T-\theta]\} \tag{3.190}$$

and

$$u(\theta, x) = \sup_{\tau \in \mathrm{S}_\theta} E_{\theta x}[e^{-\rho(\tau-\theta)} g(\xi(\tau))], \quad \theta \in [0, T], \quad x > 0. \tag{3.191}$$

$u(\theta, x)$ is called the price of American option for $g$ at $\theta$. We assume that

$g : (0, \infty) \mapsto [0, \infty)$ is Lipschitz continuous.

Put

$$\hat{u}(t, x) = \sup_{\tau \in \mathrm{S}_0^t} E_{0x}[e^{-\rho\tau} g(\xi(\tau))] \tag{3.192}$$

where

$$\mathrm{S}_0^t = \{\tau \in \mathrm{S}_0; \tau \in [0, t] \ P\text{-a.s.}\}$$

is the set of $(\mathcal{F}_t^W)$-stopping times taking values in $[0, t]$.

Since the coefficients are independent of the time variable, we have

$$u(\theta, x) = \hat{u}(T - \theta, x). \tag{3.193}$$

Hence, $u(\cdot)$ is continuous on $[0, T] \times (0, \infty)$ and (3.187) yields that, when $u(\cdot)$ is smooth,

$$0 = -\partial_t u + 0 \wedge \Big(-\frac{\sigma^2}{2} x^2 \frac{\partial^2 u}{\partial x^2} - \rho\Big(x \frac{\partial u}{\partial x} - u\Big)\Big) \quad \text{on } [0, T) \times (0, \infty) \tag{3.194}$$

with the lateral boundary condition

$$u(T, x) = g(x), \quad x \in (0, \infty). \tag{3.195}$$

Next we revisit the problem, by changing the space variable $x$ to $y = \log x$. Put

$$f(y) = g(e^y), \quad \eta(t) = \log \xi(t). \tag{3.196}$$

Then $\eta(\cdot)$ satisfies

$$d\eta(t) = \Big(\rho - \frac{\sigma^2}{2}\Big) dt + \sigma \, dW(t). \tag{3.197}$$

Namely, $\eta(\cdot)$ is a Gaussian diffusion with generator

$$\mathrm{G} = \frac{\sigma^2}{2} \frac{\partial^2}{\partial x^2} + \Big(\rho - \frac{\sigma^2}{2}\Big) \frac{\partial}{\partial x}.$$

Since

$$E_{\theta x}[e^{-\rho(\tau-\theta)}g(\xi(\tau))] = E_{\theta \log x}[e^{-\rho(\tau-\theta)}f(\eta(\tau))] \tag{3.198}$$

holds, putting

$$v(\theta, y) = \sup_{\tau \in \mathrm{S}_\theta} E_{\theta y}[e^{-\rho(\tau-\theta)}f(\eta(\tau))], \tag{3.199}$$

we have

$$v(\theta, y) = u(\theta, e^y), \quad y \in \mathbb{R}^1. \tag{3.200}$$

By (3.197) and (3.199), the HJB equation for $v(\cdot)$ is

$$0 = -\partial_t v + 0 \wedge (-\mathrm{G}v(t,\cdot) + \rho v) \quad \text{on } [0,T) \times \mathbb{R}^1 \tag{3.201}$$

with

$$v(T, y) = g(e^y), \quad y \in \mathbb{R}^1.$$

Now Proposition 3.9 yields

**Proposition 3.10.** *When $g$ is bounded and Lipschitz continuous, $v(\cdot)$ is bounded and continuous and is the unique viscosity solution of* (3.201).

Finally we note the variational inequality for $v(\cdot)$, according to [JLL90]. When $g$ is convex, $v(t,\cdot)$ is also convex. Referring to [JLL90], Theorem 3.6 and Corollary 3.7, we have

**Proposition 3.11.** *Suppose that $g$ is convex and Lipschitz continuous. Then $v(\cdot)$ has a bounded continuous derivative $\frac{\partial v}{\partial x}$ and bounded generalized derivatives $\frac{\partial v}{\partial t}$ and $\frac{\partial^2 v}{\partial x^2}$. Moreover,*

$$\partial_t v + \mathrm{G}v(t,\cdot) - \rho v \le 0 \quad \textit{a.e. on } [0,T) \times \mathbb{R}^1 \tag{3.202}$$

*and*

$$(\partial_t v + \mathrm{G}v(t,\cdot) - \rho v)(f - v) = 0 \quad \textit{a.e. on } [0,T) \times \mathbb{R}^1. \tag{3.203}$$

Let $(\theta, y)$ be given. $\eta(\cdot)$ denotes the solution of (3.197) with $\eta(\theta) = y$. Noticing that $v \ge f$ and $v(T,\cdot) = f(\cdot)$, and using (3.202) and (3.203), we can take an optimal stopping time in $\mathrm{S}_\theta$ such that

$$\tau^* = \inf\{s \in [\theta, T]; v(s, \eta(s)) = f(\eta(s))\}. \tag{3.204}$$

Indeed, the Itô–Krylov formula together with (3.202) and (3.203) yields

$$E_{\theta y}[e^{-\rho(\tau^* \wedge \tau_R - \theta)} v(\tau^* \wedge \tau_R, \eta(\tau^* \wedge \tau_R)] = v(\theta, y), \tag{3.205}$$

where $\tau_R$ denotes the exit time of $\eta(\cdot)$ from $[-R, R]$. Since $v(\cdot)$ is bounded, the bounded convergence theorem yields

$$E_{\theta y}[e^{-\rho(\tau^* - \theta)} v(\tau^*, \eta(\tau^*))] = v(\theta, y),$$

which establishes the assertion.

Regarding the martingale approach to American option price problem, the reader is referred to [KS98], Appendix D. [CPY09] and [St11] treat other topics on optimal stopping problems.

# Chapter 4
# Stochastic Differential Games

**Abstract** In this chapter, we will deal with zero-sum two-player time-homogeneous stochastic differential games and viscosity solutions of the Isaacs equations arising from such games, via the dynamic programming principle.

In Sect. 4.1, we are concerned with basic concepts and definitions and we introduce stochastic differential games, referring to (Controlled MarkovProcesses and viscosity solutions, 2nd edn. Springer, New York 2006), XI. Then, using a semi-discretization argument, we study the DPP for lower- and upper-value functions in Sect. 4.2. In Sect. 4.3, we will consider the Isaacs equations, via semigroups related to DPP. In Sect. 4.4, we consider a link between stochastic controls and differential games via risk sensitive controls.

## 4.1 Formulations

In this section we introduce some basic concepts for stochastic differential games and formulate problems. When players I and II choose Wiener-adapted control processes $\mathbf{y} := \mathbf{y}(\cdot)$ and $\mathbf{z} := \mathbf{z}(\cdot)$ respectively, the response $X$ evolves according to SDE and the cost functional is given by the usual form:

$$C(t, \theta, x, \mathbf{y}, \mathbf{z}; \phi) = \int_\theta^t f(X(s), \mathbf{y}(s), \mathbf{z}(s))\, ds + \phi(X(t))$$

for $\mathbf{y}, \mathbf{z}$ with $X(\theta) = x$. Its expectation is called the payoff. Now player I wants to maximize the payoff and player II wants to minimize it.

In Sect. 4.1.1, we define control processes and strategies, based on the concepts in [FS06], XI. 4, and study properties of responses. In Sect. 4.1.2 we formulate two kinds of stochastic differential games and investigate lower- and upper value functions.

### *4.1.1 Admissible Controls and Strategies*

Let $W$ be an $m$-dimensional Wiener process, defined on $(\Omega, \mathcal{F}, P)$. Let Y and Z be convex and compact subsets of $\mathbb{R}^{q_1}$ and $\mathbb{R}^{q_2}$, respectively. Put

© Springer Japan 2015 
M. Nisio, *Stochastic Control Theory*, Probability Theory and Stochastic Modelling 72, DOI 10.1007/978-4-431-55123-2_4

$$\begin{cases} \mathbb{Y}_0^W[\theta,t] = L^\infty([\theta,t],(\mathcal{F}_t^W);\mathrm{Y}), & \mathbb{Y}_0 = \mathbb{Y}_0^W[0,T], \\ \mathbb{Z}_0^W[\theta,t] = L^\infty([\theta,t],(\mathcal{F}_t^W);\mathrm{Z}), & \mathbb{Z}_0 = \mathbb{Z}_0^W[0,T]. \end{cases} \tag{4.1}$$

Let $\theta \in (0,T)$ be given. Since $\mathbf{y} \in \mathbb{Y}_0$ is given by (4.2), with a progressively measurable map $\Phi : [0,T] \times C([0,T-\theta] : \mathbb{R}^m) \times C([0,\theta];\mathbb{R}^m) \mapsto \mathrm{Y}$,

$$\mathbf{y}(t,W) = \Phi(t, W_\theta^+, W_\theta^-) \quad P\text{-a.s.} \tag{4.2}$$

under $P(\cdot|\mathcal{F}_\theta^W)$, we freeze $W_\theta^-$ in (4.2) and $(\mathbf{y}(t), t \in [\theta,T])$ can be regarded as an element of $L^\infty([0,T-\theta],(\mathcal{F}_t^{W_\theta^+});\mathrm{Y})$.

Denote by $\mathbb{Y}_{\mathrm{SC}}$ the set of all switching processes of $\mathbb{Y}_0$, namely $\mathbf{y}(t) = \mathbf{y}(t_i), t \in [t_i, t_{i+1}), i = 0,\ldots,p$ with some $0 = t_0 < t_1 < \cdots < t_p < t_{p+1} = T$. For $\mathbb{Y} \subset \mathbb{Y}_0$ and $t > 0$, $\mathbb{Y}_t$ denotes the set of restriction of $\mathbf{y} \in \mathbb{Y}$ to $[0,t]$.

Next we introduce admissible controls, by referring to [FS06], X1.4. Let $\mathbb{Y}$ be a subset of $\mathbb{Y}_0$ satisfying

(a) $\mathbb{Y} \supset \mathbb{Y}_{\mathrm{SC}}$;
(b) For $\mathbf{y}_1, \mathbf{y}_2 \in \mathbb{Y}$ and $\theta \in (0,T)$, the concatenation $\mathbf{y}$ given by

$$\mathbf{y}(t) = \mathbf{y}_1(t)\chi_{[0,\theta)}(t) + \mathbf{y}_2(t)\chi_{[\theta,T]}(t)$$

is in $\mathbb{Y}$;
(c) Under $P(\cdot|\mathcal{F}_\theta^W), (\mathbf{y}(t), t \in [\theta,T])$ can be regarded as an element of $\mathbb{Y}_{T-\theta}^{W_\theta^+}$ (= the set given by $\mathbb{Y}_{T-\theta}$ with $W$ replaced by $W_\theta^+$).

Clearly $\mathbb{Y}_{\mathrm{SC}}$ and $\mathbb{Y}_0$ satisfy (a)–(c) and

$$\mathbb{Y}_{\mathrm{SC}} \subset \mathbb{Y} \subset \mathbb{Y}_0. \tag{4.3}$$

Replacing Y by Z, we define $\mathbb{Z}, \mathbb{Z}_t$, and $\mathbb{Z}_{\mathrm{SC}}$ in the same way.

**Definition 4.1.** $\mathbf{y} \in \mathbb{Y}$ (resp. $\mathbf{z} \in \mathbb{Z}$) is called an admissible control for player I (resp. II).

**Definition 4.2.** An admissible strategy for player I is a map $\eta : \mathbb{Z} \mapsto \mathbb{Y}$ with the property that if $P(\mathbf{z}(s) = \tilde{\mathbf{z}}(s)) = 1, \forall s \in [0,t]$, then $P(\eta(\mathbf{z})(s) = \eta(\tilde{\mathbf{z}})(s)) = 1, \forall s \in [0,t]$, for any $t \in [0,T]$.

The set of these $\eta$ is denoted by $\mathscr{Y}$. For player II, an admissible strategy is defined in the similar way and $\mathscr{Z}$ denotes the corresponding set of strategies. When $\mathbb{Y} = \mathbb{Y}_0$ and $\mathbb{Z} = \mathbb{Z}_0$, the strategy is called Elliott–Kalton strategy and the sets of such strategies is denoted by $\mathscr{Y}_{\mathrm{EK}}$ and $\mathscr{Z}_{\mathrm{EK}}$.

Now let us apply the time discretization to controls and strategies. Let $\mathcal{D} = (t_1,\ldots,t_N)$ be a division of $[0,T], 0 = t_0 < t_1 < \cdots < t_N < t_{N+1} = T$, and put $|\mathcal{D}| = \max_i(t_{i+1} - t_i)$.

**Definition 4.3.** $\mathbf{y} \in \mathbb{Y}$ (resp. $\mathbf{z} \in \mathbb{Z}$) is called $\mathcal{D}$-admissible for player I (resp. II), if $\mathbf{y}(t) = \mathbf{y}(t_j)$ (resp $\mathbf{z}(t) = \mathbf{z}(t_j)$) on $[t_j, t_{j+1})$, $j = 0, 1, \ldots, N$, $P$-a.s. $\mathbb{Y}^{\mathcal{D}}$ (resp. $\mathbb{Z}^{\mathcal{D}}$) denotes the set of all $\mathcal{D}$-admissible controls for player I (resp. II).

**Definition 4.4.** $\eta \in \mathscr{Y}$ is called $\mathcal{D}$-admissible for player I, if $\eta : \mathrm{Z} \mapsto \mathbb{Y}^{\mathcal{D}}$ satisfies

$$\eta(\mathbf{z})(s) = \eta(\mathbf{z})(0), \quad s < t_1 \text{ is a constant } (\in \mathrm{Y}) \text{ independent of } \mathbf{z} (\in \mathbb{Z}) \tag{4.4}$$

and

$$\begin{aligned} &P(\mathbf{z}(s) = \tilde{\mathbf{z}}(s)) = 1 \quad \text{for } s < t_k \\ \Longrightarrow\ &P(\eta(\mathbf{z})(t_k) = \eta(\tilde{\mathbf{z}})(t_k)) = 1, \quad k = 1, \ldots, N. \end{aligned} \tag{4.5}$$

We note that (4.5) and $\mathbb{Y}^{\mathcal{D}}$ yield that

$$\begin{aligned} &P(\mathbf{z}(s) = \tilde{\mathbf{z}}(s)) = 1, \quad \forall s < t_k \\ \Longrightarrow\ &P(\eta(\mathbf{z})(s) = \eta(\tilde{\mathbf{z}})(s)) = 1 \quad \text{for } s < t_{k+1}. \end{aligned} \tag{4.5$'$}$$

$\mathscr{Y}^{\mathcal{D}}$ denotes the set of all $\mathcal{D}$-admissible strategies. $\mathscr{Z}^{\mathcal{D}}$ is given in the similar way for player II.

When players I and II choose $\mathbf{y} \in \mathbb{Y}$ and $\mathbf{z} \in \mathbb{Z}$, respectively, the response $X = X^{\mathbf{yz}}$ evolves according to the $d$-dimensional SDE

$$dX(t) = b(X(t), \mathbf{y}(t), \mathbf{z}(t))\, dt + \alpha(X(t), \mathbf{y}(t), \mathbf{z}(t))\, dW(t), \quad t \in (0, T], \tag{4.6}$$

with the initial condition

$$X(0) = x \ (\in \mathbb{R}^d). \tag{4.7}$$

We always assume that the coefficients $\alpha, b$, and $f$(= running cost) are bounded and Lipschitz continuous, say

(d)

$$|\alpha(x, y, z)| + |b(x, y, z)| + |f(x, y, z)| \le K_0$$

and

$$\begin{aligned} &|\alpha(x, y, z) - \alpha(\tilde{x}, \tilde{y}, \tilde{z})| + |b(x, y, z) - b(\tilde{x}, \tilde{y}, \tilde{z})| + |f(x, y, z) - f(\tilde{x}, \tilde{y}, \tilde{z})| \\ &\le l_0(|x - \tilde{x}| + |y - \tilde{y}| + |z - \tilde{z}|). \end{aligned}$$

Then the following result is immediate.

**Proposition 4.1.** *The* SDE (4.6)–(4.7) *admits a unique* $(\mathcal{F}_t^W)$*-adapted solution. Moreover, the following estimates hold:*

$$E_x\Big[\sup_{0\le s\le T}|X(s)|^{2p}\Big]\le K_p(1+|x|^{2p}),\quad p\ge 1, \tag{4.8}$$

$$E_x\Big[\sup_{\theta\le s\le t}|X(s)-X(\theta)|^{2p}\Big]\le K_p|t-\theta|^p,\quad p\ge 1, \tag{4.9}$$

$$E\Big[\sup_{0\le s\le T}|X_{x_1}(s)-X_{x_2}(s)|^2\Big]\le K_1|x_1-x_2|^2, \tag{4.10}$$

*where* $x_i$ *denotes the initial state of the solution, and*

$$\begin{aligned}&E_x\Big[\sup_{0\le s\le t}|X^{\mathbf{yz}}(s)-X^{\tilde{\mathbf{y}}\tilde{\mathbf{z}}}(s)|^2\Big]\\ \le &K_1\int_0^t E(|\mathbf{y}(s)-\tilde{\mathbf{y}}(s)|^2+|\mathbf{z}(s)-\tilde{\mathbf{z}}(s)|^2)\,ds,\end{aligned} \tag{4.11}$$

*where the constants* $K_1$ *and* $K_p$ *depend only on* $l_0$ *and* $K_0$.

### 4.1.2 Formulation of Stochastic Differential Games

We are concerned with two kinds of stochastic differential games, lower games and upper games. In the lower game, the player I chooses an admissible control $\mathbf{y}$ first and $\mathbf{y}$ is known when the player II chooses an element of $\mathbb{Z}$. Hence player II chooses an admissible strategy from $\mathscr{Z}$. When $\mathbf{y}\in\mathbb{Y}$ and $\zeta\in\mathscr{Z}$ are chosen, the response $X^{\mathbf{y}\zeta}$ evolves according to (4.6) with $\mathbf{z}=\zeta(\mathbf{y})$. Thus, when game stops at $t$, the payoff is given by

$$J(t,x,\mathbf{y},\zeta;\phi)=E_x\Big[\int_0^t f(X(s),\mathbf{y}(s),\zeta(\mathbf{y})(s))\,ds+\phi(X(t))\Big] \tag{4.12}$$

with $\phi\in C_{\mathrm{bu}}(\mathbb{R}^d)$.

The $(\mathbb{Y},\mathscr{Z})$-lower value $V^-$ is defined by

$$V^-(t,x;\phi)=\inf_{\zeta\in\mathscr{Z}}\sup_{\mathbf{y}\in\mathbb{Y}}J(t,x,\mathbf{y},\zeta;\phi). \tag{4.13}$$

Hence the minimizing player II has information advantage. When $\mathbb{Y}=\mathbb{Y}_0$ and $\mathbb{Z}=\mathbb{Z}_0$, $V^-$ is called the Elliott–Kalton lower value and is denoted by $V^-_{\mathrm{EK}}$. We also write $\mathscr{Z}_{\mathrm{EK}}$ and $\mathscr{Y}_{\mathrm{EK}}$, instead of $\mathscr{Z}$ and $\mathscr{Y}$.

The upper game is defined in the similar way and the $(\mathbb{Z}, \mathscr{Y})$-upper value $V^+$ is given by

$$V^+(t,x;\phi) = \sup_{\eta\in\mathscr{Y}} \inf_{\mathbf{z}\in\mathbb{Z}} J(t,x,\eta,\mathbf{z};\phi). \tag{4.14}$$

The Elliott–Kalton upper value. is denoted by $V^+_{\mathrm{EK}}$. However, it seems to be unfair that only the stronger player chooses a strategy. Now following [FH11] we reduce the class of strategies to a smaller one which eliminates the information advantage.

We will treat mainly the lower game for a while.

**Definition 4.5.** $\zeta \in \mathscr{Z}_{\mathrm{EK}}$ is called a strictly progressively measurable strategy, if for each $\eta \in \mathscr{Y}_{\mathrm{EK}}$ the equations

$$\mathbf{y} = \eta(\mathbf{z}), \quad \mathbf{z} = \zeta(\mathbf{y}) \tag{4.15}$$

have a solution, namely $\eta\zeta$ and $\zeta\eta$ have fixed points. $\mathscr{Z}_{\mathrm{s}}$ denotes the set of strictly progressively measurable strategies.

Let $\zeta \in \mathscr{Z}_{\mathrm{s}}$ and $\eta \in \mathscr{Y}_{\mathrm{EK}}$ be given. Then, it is immediate that

$$J(t,x,\hat{\mathbf{y}},\zeta;\phi) = J(t,x,\eta,\hat{\mathbf{z}};\phi), \quad \forall (t,x) \tag{4.16}$$

where $(\hat{\mathbf{y}}, \hat{\mathbf{z}})$ is a solution of (4.15). Hence we have

$$\inf_{\zeta\in\mathscr{Z}_s} \sup_{\mathbf{y}\in\mathbb{Y}_0} J(t,x,\mathbf{y},\zeta;\phi) \ge V^+_{\mathrm{EK}}(t,x,\phi), \tag{4.17}$$

which says that $\mathscr{Z}_{\mathrm{s}}$ eliminates the advantage of player II.

Indeed, for $\varepsilon > 0$, there is $\zeta_\varepsilon \in \mathscr{Z}_{\mathrm{s}}$, such that

$$\text{LHS of (4.17)} + \varepsilon \ge \sup_{\mathbf{y}\in\mathbb{Y}_0} J(t,x,\mathbf{y},\zeta_\varepsilon;\phi). \tag{4.18}$$

For any $\eta \in \mathscr{Y}_{\mathrm{EK}}$,

$$\begin{aligned}\text{RHS of (4.18)} &\ge J(t,x,\hat{\mathbf{y}},\zeta_\varepsilon;\phi) = J(t,x,\eta,\hat{\mathbf{z}};\phi)\\ &\ge \inf_{\mathbf{z}\in\mathbb{Z}_0} J(t,x,\eta,\mathbf{z};\phi),\end{aligned} \tag{4.19}$$

where $(\hat{\mathbf{y}}, \hat{\mathbf{z}})$ is a solution of (4.15) for $(\eta, \zeta_\varepsilon)$.

Taking the supremum $\eta$ over $\mathscr{Y}_{\mathrm{EK}}$ yields (4.17).

*Example 4.1.* $\mathscr{Z}^{\mathcal{D}} \subset \mathscr{Z}_{\mathrm{s}}$.

Indeed, for $\zeta \in \mathscr{Z}^{\mathcal{D}}$ and $\eta \in \mathscr{Y}_{\mathrm{EK}}$, we construct a solution of (4.15) in the following way. Let $\mathcal{D} = (t_1, \ldots, t_n)$. Fix $\mathrm{z} \in \mathrm{Z}$ arbitrarily. Regarding $\mathrm{z}$ as the constant control process, we define $\mathbf{y}_k \in \mathbb{Y}_0$ and $\mathbf{z}_k \in \mathbb{Z}^{\mathcal{D}}$ by the following recursive procedure: put $\mathbf{y}_0 = \eta(\mathrm{z}),\ \mathbf{z}_0 = \zeta(\mathbf{y}_0)$, and

$$\mathbf{y}_k = \eta(\mathbf{z}_{k-1}), \quad \mathbf{z}_k = \zeta(\mathbf{y}_k), \quad k = 1, 2, \ldots. \tag{4.20}$$

Then, we can show that

$$\mathbf{y}_{j+1} = \mathbf{y}_j \text{ on } [0, t_j), \quad \mathbf{z}_{j+1} = \mathbf{z}_j \text{ on } [0, t_{j+1}), \quad j = 1, 2, \ldots, n. \tag{4.21}$$

Since (4.4) yields that, for $s \in [0, t_1)$, $\zeta(\mathbf{y})(s) = \zeta(\mathbf{y})(0) =$ constant independent of $\mathbf{y}(=: \tilde{\mathrm{z}})$, we see that

$$\mathbf{z}_k(s) = \mathbf{z}_k(0) = \tilde{\mathrm{z}} \quad \text{for } s \in [0, t_1), \quad k = 1, 2, \ldots$$

and

$$\mathbf{y}_k(s) = \eta(\mathbf{z}_{k-1})(s) = \eta(\tilde{\mathrm{z}})(s), \quad s \in [0, t_1), \quad k = 1, 2, \ldots$$

holds by Definition 4.2. Hence, $\mathbf{z}_k(t_1) = \zeta(\mathbf{y}_k)(t_1)(k = 1, 2, \ldots)$ does not depend on $k$, and (4.21) holds for $j = 1$.

If (4.21) holds for $j$, then the same arguments lead to

$$\mathbf{y}_{j+2}(s) = \eta(\mathbf{z}_{j+1})(s) = \eta(\mathbf{z}_j)(s) = \mathbf{y}_{j+1}(s) \quad \text{for } s < t_{j+1} \tag{4.22}$$

and

$$\mathbf{z}_{j+2}(s) = \mathbf{z}_{j+1}(s) \quad \text{on } [0, t_{j+1}). \tag{4.23}$$

Since $\mathbf{z}_{j+2}$ and $\mathbf{z}_{j+1}$ are in $\mathbb{Z}^{\mathcal{D}}$, we have

$$\mathbf{z}_{j+2} = \mathbf{z}_{j+1} \text{ on } [0, t_{j+2}) \tag{4.24}$$

and

$$\mathbf{y}_{j+2} = \mathbf{y}_{j+1} \text{ on } [0, t_{j+1}), \tag{4.25}$$

which yield (4.21) for $j + 1$.

Finally, we have $\mathbf{z}_{n+1}(t_n) = \mathbf{z}_n(t_n)$. Hence,

$$\mathbf{z}_{n+1} = \mathbf{z}_n \text{ on } [0, T] \quad \text{and} \quad \mathbf{y}_{n+1} = \eta(\mathbf{z}_n) = \eta(\mathbf{z}_{n+1})$$

show that $(\mathbf{y}_{n+1}, \mathbf{z}_{n+1})$ is a solution.

*Example 4.2 (Constant strategy).* Let $\mathbf{z}_0 \in \mathbb{Z}_0$ be given. Define $\zeta_{\mathbf{z}_0} \in \mathscr{Z}_{\mathrm{EK}}$ by $\zeta_{\mathbf{z}_0}(\mathbf{y}) = \mathbf{z}_0$ for any $\mathbf{y} \in \mathbb{Y}_0$. Then $\zeta_{\mathbf{z}_0} \in \mathscr{Z}_{\mathrm{s}}$ and the solution of (4.15) for $(\eta, \zeta_{\mathbf{z}_0})$ is unique.

Indeed, for $\eta \in \mathscr{Y}_{\rm EK}$, we put $\mathbf{y}_0 = \eta(\mathbf{z}_0)$. Then $\zeta_{\mathbf{z}_0}(\mathbf{y}_0) = \mathbf{z}_0$ yields that $(\mathbf{y}_0, \mathbf{z}_0)$ is a solution. If $(\hat{\mathbf{y}}, \hat{\mathbf{z}})$ is a solution, then $\zeta_{\mathbf{z}_0}(\hat{\mathbf{y}}) = \hat{\mathbf{z}}$ shows that $\hat{\mathbf{z}} = \mathbf{z}_0$. Hence $\hat{\mathbf{y}} = \eta(\mathbf{z}_0) = \mathbf{y}_0$.

Regarding the solution of (4.15) and related topics in deterministic differential games, [FKSh10], Sect. 4 presents interesting results.

The case when the equality holds in (4.17), i.e.,

$$\inf_{\zeta \in \mathscr{Z}_{\rm s}} \sup_{\mathbf{y} \in \mathbb{Y}_0} J(t,x,\mathbf{y},\zeta;\phi) = \sup_{\eta \in \mathscr{Y}_{\rm EK}} \inf_{\mathbf{z} \in \mathbb{Z}_0} J(t,x,\eta,\mathbf{z};\phi) \tag{4.26}$$

suggests a saddle point property for the upper game.

By (4.17) and (4.26), we recall the Definition and Theorem from [FH11], Sect. 3.

**Definition.** The saddle point property for the upper game is said to hold, if there exists a real valued function $V(t,x)$ such that for each $\varepsilon > 0$ and $(t,x) \in [0,T] \times \mathbb{R}^d$ there exist $\zeta_\varepsilon \in \mathscr{Z}_{\rm s}$ and $\eta_\varepsilon \in \mathscr{Y}_{\rm EK}$ such that

(a) $V(t,x) - \varepsilon \le \inf_{\mathbf{z} \in \mathbb{Z}_0} J(t,x,\eta_\varepsilon,\mathbf{z};\phi)$,
(b) $\sup_{\mathbf{y} \in \mathbb{Y}_0} J(t,x,\mathbf{y},\zeta_\varepsilon;\phi) \le V(t,x) + \varepsilon$.

**Theorem.** *The saddle point property holds with $V = V^+_{\rm EK} = v$, where $v$ is the unique bounded, uniformly continuous viscosity solution to the upper Isaacs equation with the initial state $\phi$.*

Finally, we revisit the saddle point property (4.26) in terms of strategies of the two players. For $\zeta \in \mathscr{Z}_{\rm s}$ and $\eta \in \mathscr{Y}_{\rm EK}$, denote

$$\mathbb{Y}(\eta,\zeta) = \{\hat{\mathbf{y}};\, (\hat{\mathbf{y}}, \zeta(\hat{\mathbf{y}})) \text{ is a solution}\}, \tag{4.27}$$

$$\mathbb{Z}(\eta,\zeta) = \{\hat{\mathbf{z}};\, (\eta(\hat{\mathbf{z}}), \hat{\mathbf{z}}) \text{ is a solution}\}, \tag{4.27$'$}$$

$$J^\zeta(t,x,\eta,\zeta;\phi) = \sup_{\mathbf{y} \in \mathbb{Y}(\eta,\zeta)} J(t,x,\mathbf{y},\zeta(\mathbf{y});\phi), \tag{4.28}$$

$$J_\eta(t,x,\eta,\zeta;\phi) = \inf_{\mathbf{z} \in \mathbb{Z}(\eta,\zeta)} J(t,x,\eta(\mathbf{z}),\mathbf{z};\phi), \tag{4.28$'$}$$

$$J(t,x,\eta,\zeta;\phi) = \{J(t,x,\hat{\mathbf{y}},\hat{\mathbf{z}};\phi);\, (\hat{\mathbf{y}},\hat{\mathbf{z}}) \text{ is a solution}\}. \tag{4.29}$$

If the solution is unique, we simply write

$$J(t,x,\eta,\zeta;\phi) = J(t,x,\hat{\mathbf{y}},\hat{\mathbf{z}};\phi). \tag{4.30}$$

From (4.27) and Example 4.2, it follows that

$$\begin{aligned}\sup_{\mathbf{y} \in \mathbb{Y}_0} J(t,x,\mathbf{y},\zeta;\phi) &= \sup_{\mathbf{y} \in \mathbb{Y}_0} J(t,x,\eta_{\mathbf{y}},\zeta;\phi)\\ &\le \sup_{\eta \in \mathscr{Y}_{\rm EK}} J^\zeta(t,x,\eta,\zeta;\phi) \le \sup_{\mathbf{y} \in \mathbb{Y}_0} J(t,x,\mathbf{y},\zeta;\phi).\end{aligned}$$

Consequently,

$$\sup_{\mathbf{y}\in\mathbb{Y}_0} J(t,x,\mathbf{y},\zeta;\phi) = \sup_{\eta\in\mathscr{Y}_{\mathrm{EK}}} J^{\zeta}(t,x,\eta,\zeta;\phi).$$

Taking the infimum $\zeta$ over $\mathscr{Z}_{\mathrm{s}}$ leads to

$$\text{LHS of (4.26)} = \inf_{\zeta\in\mathscr{Z}_{\mathrm{s}}} \sup_{\eta\in\mathscr{Y}_{\mathrm{EK}}} J^{\zeta}(t,x,\eta,\zeta;\phi). \tag{4.31}$$

Let $\eta \in \mathscr{Y}_{\mathrm{EK}}$ be given. By using similar arguments, we have

$$\inf_{\mathbf{z}\in\mathbb{Z}_0} J(t,x,\eta,\mathbf{z};\phi) = \inf_{\zeta\in\mathscr{Z}_{\mathrm{s}}} J_{\eta}(t,x,\eta,\zeta;\phi),$$

and so

$$\text{RHS of (4.26)} = \sup_{\eta\in\mathscr{Y}_{\mathrm{EK}}} \inf_{\zeta\in\mathscr{Z}_{\mathrm{s}}} J_{\eta}(t,x,\eta,\zeta;\phi). \tag{4.32}$$

Hence (4.26) implies that

$$\inf_{\zeta\in\mathscr{Z}_{\mathrm{s}}} \sup_{\eta\in\mathscr{Y}_{\mathrm{EK}}} J^{\zeta}(t,x,\eta,\zeta;\phi) = \sup_{\eta\in\mathscr{Y}_{\mathrm{EK}}} \inf_{\zeta\in\mathscr{Z}_{\mathrm{s}}} J_{\eta}(t,x,\eta,\zeta;\phi).$$

Before we end Sect. 4.1, let us list some basic properties of payoff, by using Proposition 4.1.1.

**Proposition 4.2.** *Let* $\phi \in C_{\mathrm{bu}}(\mathbb{R}^d)$.

(i) ***Bound***

$$|J(t,x,\mathbf{y},\mathbf{z};\phi)| \le K_0 t + \|\phi\|_{\mathrm{C}}, \tag{4.33}$$

(ii) ***Continuous dependence on*** $t$
*For* $\varepsilon > 0$, *there is* $\Delta_{\varepsilon\phi} > 0$, *such that*

$$|J(t_1,x,\mathbf{y},\mathbf{z};\phi) - J(t_2,x,\mathbf{y},\mathbf{z};\phi)| < \varepsilon, \quad \forall x,\mathbf{y},\mathbf{z}, \tag{4.34}$$

*whenever* $|t_1 - t_2| < \Delta_{\varepsilon\phi}$.

(iii) ***Continuous dependence on the initial state***
*For* $\varepsilon > 0$, *there is* $\delta_{\varepsilon,\phi} > 0$, *such that*

$$|J(t,x_1,\mathbf{y},\mathbf{z};\phi) - J(t,x_2,\mathbf{y},\mathbf{z};\phi)| < \varepsilon, \quad \forall t,\mathbf{y},\mathbf{z}, \tag{4.35}$$

*whenever* $|x_1 - x_2| < \delta_{\varepsilon\phi}$.

(iv) ***Continuous dependence on admissible controls***
*For $\varepsilon > 0$, there is $\Lambda_{\varepsilon\phi} > 0$, such that*

$$|J(t,x,\mathbf{y}_1,\mathbf{z}_1;\phi) - J(t,x,\mathbf{y}_2,\mathbf{z}_2;\phi)| < \varepsilon, \quad \forall t,x, \tag{4.36}$$

*whenever* $E[\int_0^T (|\mathbf{y}_1(t) - \mathbf{y}_2(t)|^2 + |\mathbf{z}_1(t) - \mathbf{z}_2(t)|^2)\,dt] < \Lambda_{\varepsilon\phi}$,

(v) ***Monotonicity***

$$\phi \le \psi \Longrightarrow J(t,x,\mathbf{y},\mathbf{z};\phi) \le J(t,x,\mathbf{y},\mathbf{z};\psi), \quad \forall t,x,\mathbf{y},\mathbf{z}, \tag{4.37}$$

(iv) ***Contractiveness***

$$|J(t,x,\mathbf{y},\mathbf{z};\phi) - J(t,x,\mathbf{y},\mathbf{z},\psi)| \le \|\phi - \psi\|_C, \quad \forall t,x,\mathbf{y},\mathbf{z}.$$

## 4.2 DPP

We have already emphasized the importance of DPP for stochastic control problems in Chap. 2. Here we will consider DPP (sometimes we say minimax principle) for stochastic differential games. In this section, we deal with stochastic games where the player with information advantageous uses Elliott–Kalton strategies and the other player uses switching controls. Hence, it is dealing with an unfair game. By using semi-discretization arguments, introduced in [N88] and [FSo89], we firstly consider $\mathcal{D}$-lower and $\mathcal{D}$-upper value functions in Sect. 4.2.1. These cases admit the discrete time DPP. Taking finer and finer divisions of $[0,T]$, we obtain the DPP for the limit games in Sect. 4.2.2 (see Theorem 4.2).

### *4.2.1 $\mathcal{D}$-Lower and $\mathcal{D}$-Upper Value Functions*

For $\mathcal{D} = (t_1,\ldots,t_n)$, we define the $\mathcal{D}$-lower value function $v^{\mathcal{D}}$ by replacing $(\mathbb{Y},\mathscr{Z})$ by $(\mathbb{Y}^{\mathcal{D}},\mathscr{Z}_{\mathrm{EK}})$ in (4.13), namely

$$v^{\mathcal{D}}(t,x;\phi) = \inf_{\zeta\in\mathscr{Z}_{\mathrm{EK}}} \sup_{\mathbf{y}\in\mathbb{Y}^{\mathcal{D}}} J(t,x,\mathbf{y},\zeta;\phi). \tag{4.38}$$

Similarly, $\mathcal{D}$-upper value function $u^{\mathcal{D}}$ is given by

$$u^{\mathcal{D}}(t,x;\phi) = \sup_{\eta\in\mathscr{Y}_{\mathrm{EK}}} \inf_{\mathbf{z}\in\mathbb{Z}^{\mathcal{D}}} J(t,x,\eta,\mathbf{z};\phi). \tag{4.38$'$}$$

Here we are concerned with DPP for $v^{\mathcal{D}}$ and $u^{\mathcal{D}}$. From now on, we mainly consider $v^{\mathcal{D}}$, because $u^{\mathcal{D}}$ can be treated by the same arguments.

Recalling Proposition 4.2, we easily obtain;

**Proposition 4.3.** $v^{\mathcal{D}}(t,\cdot;\phi) \in C_{\text{bu}}(\mathbb{R}^d)$.

(i) $|v^{\mathcal{D}}(t,x;\phi)| \leq K_0 t + \|\phi\|_C$.
(ii) *With the same constant* $\Delta_{\varepsilon\phi} > 0$, *as in* (4.34),

$$|v^{\mathcal{D}}(t_1,x;\phi) - v^{\mathcal{D}}(t_2,x;\phi)| < \varepsilon, \quad \forall x, \tag{4.39}$$

*whenever* $(t_1 - t_2) < \Delta_{\varepsilon\phi}$.
(iii) *With the same constant* $\delta_{\varepsilon\phi} > 0$, *as in* (4.35),

$$|v^{\mathcal{D}}(t,x_1;\phi) - v^{\mathcal{D}}(t,x_2;\phi)| < \varepsilon, \quad \forall t, \tag{4.40}$$

*whenever* $|x_1 - x_2| < \delta_{\varepsilon\phi}$.
(iv) ***Monotonicity***

$$\phi \leq \psi \Longrightarrow v^{\mathcal{D}}(t,\cdot;\phi) \leq v^{\mathcal{D}}(t,\cdot;\psi), \quad \forall t.$$

(v) ***Contractiveness***

$$\|v^{\mathcal{D}}(t,\cdot;\phi) - v^{\mathcal{D}}(t,\cdot;\psi)\|_C \leq \|\phi - \psi\|_C, \quad \forall t.$$

**Corollary 4.1.** $v^{\mathcal{D}}(\cdot;\phi)$ *has* $\mathcal{D}$*-independent bound and uniform continuity on* $[0,T] \times \mathbb{R}^d$. *Moreover, the sequence* $\{v^{\mathcal{D}_n}(\cdot;\phi), n = 1,2,\ldots\}$ *has a subsequence which converges uniformly on any bounded subset of* $[0,T] \times \mathbb{R}^d$.

Considering $\mathrm{y} \in \mathrm{Y}$ as a constant control in $\mathbb{Y}_0$, we introduce two notations;

$$\begin{cases} I(t,x,\mathrm{y};\phi) = \inf\limits_{\mathbf{z}\in\mathbb{Z}_0} J(t,x,\mathrm{y},\mathbf{z};\phi), \\ \mathrm{v}(t,x;\phi) = \sup\limits_{\mathrm{y}\in\mathrm{Y}} I(t,x,\mathrm{y};\phi). \end{cases} \tag{4.41}$$

Let us define an operator $\nu(t) : C_{\text{bu}}(\mathbb{R}^d) \mapsto C_{\text{bu}}(\mathbb{R}^d)$ by

$$\nu(t)\phi(x) = \mathrm{v}(t,x;\phi). \tag{4.42}$$

We aim to prove the following form which leads to DPP.

**Theorem 4.1.** *Let* $\tau \in (t_p, t_{p+1}]$ *be given. Then*

$$v^{\mathcal{D}}(t,x;\phi) = \nu(t_1)\nu(t_2 - t_1)\cdots\nu(\tau - t_p)\phi(x). \tag{4.43}$$

*Note.* Let $k < p$ and $\theta \in [t_k, t_{k+1})$ be given. By $\mathcal{D}_\theta$ we denote the division $0 < t_{k+1} - \theta < t_{k+2} - \theta < \cdots < t_n - \theta < t_{n+1} - \theta = T - \theta$ on $[0, T - \theta]$.

We define $v^{\mathcal{D}_\theta}$ in the same way as $v^{\mathcal{D}}$. Then Theorem 4.1 leads to the DPP

$$v^{\mathcal{D}}(\tau, x; \phi) = v^{\mathcal{D}}(t_k, x; v^{\mathcal{D}_{t_k}}(\tau - t_k, \cdot; \phi)). \tag{4.44}$$

*Proof.* We divide the proof into five steps.

**Step 1.** Construct $\mathrm{y}^*_{\varepsilon t\phi}(x) \in \mathrm{Y}$, which yields a nearly optimal control process:

$$\mathrm{v}(t, x; \phi) \le J(t, x, \mathrm{y}^*_{\varepsilon t\phi}(x), \mathbf{z}; \phi) + 2\varepsilon, \quad \forall \mathbf{z} \in \mathbb{Z}_0. \tag{4.45}$$

Since $I(t, x, \mathrm{y}; \phi)$ is uniformly continuous in $(x, \mathrm{y}) \in \mathbb{R}^d \times \mathrm{Y}$ and $\mathrm{Y}$ is compact, there is a maximum selector $\hat{\mathrm{y}}_{t\phi} : \mathbb{R}^d \mapsto \mathrm{Y}$, such that

$$\mathrm{v}(t, x; \phi) = I(t, x, \hat{\mathrm{y}}_{t\phi}(x); \phi) \le J(t, x, \hat{\mathrm{y}}_{t\phi}(x), \mathbf{z}; \phi), \quad \forall \mathbf{z} \in \mathbb{Z}_0, \tag{4.46}$$

where $\hat{\mathrm{y}}_{t\phi}(x) \in \mathrm{Y}$ is regarded as a constant control process in $\mathbb{Y}_0$. On the other hand, Proposition 4.2 says that for a division of $\mathbb{R}^d$ $(\Xi_j, j = 1, 2, \dots)$ with diameter of $\Xi_j < \delta_{\varepsilon\phi}, j = 1, 2, \dots$ one has

$$\begin{aligned} &|J(t, x_1, \mathbf{y}, \mathbf{z}; \phi) - J(t, x_2, \mathbf{y}, \mathbf{z}; \phi)| \\ &\vee |I(t, x_1, \mathrm{y}; \phi) - I(t, x_2, \mathrm{y}; \phi)| \vee |\mathrm{v}(t, x_1; \phi) - \mathrm{v}(t, x_2; \phi)| < \varepsilon, \quad \forall t, \mathbf{y}, \mathbf{z}, \mathrm{y} \end{aligned} \tag{4.47}$$

whenever $x_1, x_2 \in \Xi_j$ $(j = 1, 2, \dots)$.
We fix $x_j \in \Xi_j$ arbitrarily and put

$$\mathrm{y}^*_{\varepsilon t\phi}(x) = \sum_{j=1}^{\infty} \hat{\mathrm{y}}_{t\phi}(x_j) \chi_{\Xi_j}(x). \tag{4.48}$$

Then (4.46) and (4.47) yield (4.45).
**Step 2.** Using $\mathrm{y}^*_{\varepsilon t\phi}$, we define $\eta_\varepsilon \in \mathscr{Y}^{\mathcal{D}}$ as

$$\eta_\varepsilon(\mathbf{z})(t) = \sum_{i=0}^{p-1} \chi_{[t_i, t_{i+1})}(t) \mathrm{y}^*_{\varepsilon, t_{i+1} - t_i, \phi_{i+1}}(X(t_i)) + \chi_{[t_p, T]}(t) \mathrm{y}^*_{\varepsilon, \tau - t_p, \phi}(X(t_p)) \tag{4.49}$$

where

$$\begin{aligned} &\phi_{p+1} = \phi, \quad \phi_p = \nu(\tau - t_p)\phi_{p+1}, \dots, \\ &\phi_l = \nu(t_{l+1} - t_l)\phi_{l+1}, \dots, \phi_0 = \nu(t_1)\phi_1, \end{aligned} \tag{4.50}$$

and $X$ is the response for $(\eta_\varepsilon, \mathbf{z})$ constructed by the following stepwise procedure. Let $x$ be the initial state. Fix $\mathbf{z} \in \mathbb{Z}_0$ arbitrarily and define $\eta_\varepsilon$ by

$$\eta_\varepsilon(\mathbf{z})(s) = \mathrm{y}^*_{\varepsilon t_1 \phi_1}(x), \quad s \in [0, t_1). \tag{4.51}$$

$(X(s), s \leq t_1)$ denotes the response for $(\eta_\varepsilon, \mathbf{z})$ with $X(0) = x$. Next we define $(\eta_\varepsilon(\mathbf{z})(s), s \in [t_1, t_2))$ by

$$\eta_\varepsilon(\mathbf{z})(s) = \eta_\varepsilon(\mathbf{z})(t_1) = \mathrm{y}^*_{\varepsilon, t_2 - t_1, \phi_2}(X(t_1)). \tag{4.52}$$

Thus, we have $\eta_\varepsilon(\mathbf{z})$ on $[0, t_2)$ and its response $(X(s), s \in [0, t_2])$. Repeating this procedure, we obtain $\eta_\varepsilon(\mathbf{z})$ on $[0, t_p)$ and its response $(X(s), s \in [0, t_p])$. Finally, we put

$$\eta_\varepsilon(\mathbf{z})(s) = \eta_\varepsilon(\mathbf{z})(t_p) = \mathrm{y}^*_{\varepsilon, \tau - t_p, \phi}(X(t_p)), \quad s \in [t_p, T]. \tag{4.53}$$

Then $\eta_\varepsilon \in \mathscr{Y}^{\mathcal{D}}$ satisfies (4.49).

**Step 3.** One has

$$J(\tau, x, \eta_\varepsilon, \mathbf{z}; \phi) \geq \nu(t_1) \cdots \nu(t_p - t_{p-1}) \nu(\tau - t_p) \phi(x) - 2\varepsilon p. \tag{4.54}$$

The reason is following. Under $P(\cdot | \mathcal{F}^W_{t_p})$, $(\mathbf{z}(s), s \in [t_p, T])$ can be regarded as an element of $\mathbb{Z}_0^{W^+_{t_p}}$ and $\eta_\varepsilon(\mathbf{z})(t_p)$ as constant $(\in \mathrm{Y})$ $P$-a.s.
Hence (4.45) yields

$$\begin{aligned} &E\Big(\int_{t_p}^{\tau} f(X(s), \eta_\varepsilon(\mathbf{z})(t_p), \mathbf{z}(s))\, ds + \phi(X(\tau)) | \mathcal{F}^W_{t_p}\Big) \\ &= E(C(\tau, t_p, X(t_p), \mathrm{y}^*_{\varepsilon, \tau - t_p, \phi}(X(t_p)), \mathbf{z}; \phi) | \mathcal{F}^W_{t_p}) \\ &\geq \mathrm{v}(\tau - t_p, X(t_p); \phi) - 2\varepsilon, \quad P\text{-a.s.} \\ &= \nu(\tau - t_p)\phi(X(t_p)) - 2\varepsilon, \quad P\text{-a.s.} \end{aligned} \tag{4.55}$$

Further, similar estimates yield

$$\begin{aligned} &E(C(\tau, t_{p-1}, X(t_{p-1}), \eta_\varepsilon, \mathbf{z}; \phi) | \mathcal{F}^W_{t_{p-1}}) \\ &\geq \nu(t_p - t_{p-1}) \nu(\tau - t_p) \phi(X(t_{p-1})) - 4\varepsilon \quad P\text{-a.s.} \end{aligned} \tag{4.56}$$

Repeating the arguments, we obtain (4.54).
Now taking the infimum over $\mathbb{Z}_0$, in (4.56), we conclude

$$\inf_{\mathbf{z} \in \mathbb{Z}_0} J(\tau, x, \eta_\varepsilon, \mathbf{z}; \phi) \geq \nu(t_1) \cdots \nu(\tau - t_p) \phi(x) - 2\varepsilon p. \tag{4.57}$$

Observing $\eta_\varepsilon \in \mathscr{Y}^{\mathcal{D}}$ and letting $\varepsilon \to 0$, we have

$$\sup_{\eta \in \mathscr{Y}^{\mathcal{D}}} \inf_{\mathbf{z} \in \mathbb{Z}_0} J(\tau, x, \eta, \mathbf{z}; \phi) \geq \nu(t_1) \cdots \nu(\tau - t_p) \phi(x). \tag{4.58}$$

Next we will show that

$$v^{\mathcal{D}}(\tau, x; \phi) \leq v(t_1) \cdots v(\tau - t_p)\phi(x). \tag{4.59}$$

**Step 4.** Construct a nearly optimal $\zeta_\varepsilon \in \mathscr{Z}_{\mathrm{EK}}$ by using arguments similar to those in Steps 1 and 2.

Let $\varepsilon > 0$ be given. By (4.36), we can take $d_{\varepsilon\phi} > 0$, such that

$$|J(t,x,\mathrm{y}_1,\mathbf{z};\phi) - J(t,x,\mathrm{y}_2,\mathbf{z};\phi)| + |I(t,x,\mathrm{y}_1;\phi) - I(t,x,\mathrm{y}_2;\phi)| < \varepsilon, \quad \forall t,x,\mathbf{z}, \tag{4.60}$$

whenever $|\mathrm{y}_1 - \mathrm{y}_2| < d_{\varepsilon\phi}$.

Let $\mathrm{Y}^\varepsilon = (\mathrm{Y}_1, \ldots, \mathrm{Y}_l)$ be a division of Y with the diameter of $\mathrm{Y}_j < d_{\varepsilon\phi}, j = 1, 2, \ldots, l$. We fix $\mathrm{y}_j \in \mathrm{Y}_j$ and $x_i \in \Xi_i$ $(i = 1, 2, \ldots, j = 1, \ldots, l)$ arbitrarily and take $\mathbf{z}_{\epsilon t \phi_{ij}} \in \mathbb{Z}_0$ so that

$$J(t, x_i, \mathrm{y}_j, \mathbf{z}_{\epsilon t \phi_{ij}}; \phi) \leq I(t, x_i, \mathrm{y}_j; \phi) + \varepsilon. \tag{4.61}$$

Emphasizing the dependence on the Wiener path $w$, we put

$$\mathbf{z}_{\epsilon t \phi}(\mathrm{y}, x)(s, w) = \sum_{i=1}^{\infty} \sum_{j=1}^{l} \chi_{\Xi_i}(x) \chi_{\mathrm{Y}_j}(\mathrm{y}) \mathbf{z}_{\varepsilon t \phi_{ij}}(s, w). \tag{4.62}$$

Then the inequality

$$J(t, x, \mathrm{y}, \mathbf{z}_{\varepsilon t \phi}(\mathrm{y}, x); \phi) \leq I(t, x, \mathrm{y}; \phi) + 5\varepsilon, \quad \forall x \in \mathbb{R}^d, \quad \mathrm{y} \in \mathrm{Y}, \tag{4.63}$$

follows from (4.60), (4.61) and (4.47).

Now we construct $\zeta_\varepsilon$ by using $\mathbf{z}_{\varepsilon t \phi}$. Let $\mathbf{y} \in \mathbb{Y}_0$ be given. Since $\mathbf{y}(0)$ is constant, say $\mathrm{y}_0$, we define $\zeta_\varepsilon$ by

$$\zeta_\varepsilon(\mathbf{y})(s, w) = \mathbf{z}_{\varepsilon t_1 \phi_1}(\mathrm{y}_0, x)(s, w) \quad \text{for } s \in [0, t_1]. \tag{4.64}$$

$(X(s), s \in [0, t_1])$ denotes the response for $(\mathbf{y}, \zeta_\varepsilon)$ with $X(0) = x$. Then we put

$$\zeta_\varepsilon(\mathbf{y})(s, w) = \mathbf{z}_{\varepsilon, t_2 - t_1, \phi_2}(\mathbf{y}(t_1, w), X(t_1, w))(s - t_1, w_{t_1}^+) \quad \text{for } s \in (t_1, t_2]. \tag{4.65}$$

Hence we obtain $\zeta_\varepsilon(\mathbf{y})$ on $[0, t_2]$ and its response $X$. Repeating this procedure, we construct $\zeta_\varepsilon(\mathbf{y})$ on $[0, t_p]$. Now put

$$\zeta_\varepsilon(\mathbf{y})(s, w) = \mathbf{z}_{\varepsilon, \tau - t_p, \phi}(\mathbf{y}(t_p, w), X(t_p, w))(s - t_p, w_{t_p}^+) \quad \text{for } s \in (t_p, T]. \tag{4.66}$$

Then $\zeta_\varepsilon$ is in $\mathscr{Z}_{\mathrm{EK}}$.

**Step 5.** We will show that

$$J(\tau, x, \mathbf{y}, \zeta_\varepsilon; \phi) \leq \nu(t_1) \cdots \nu(\tau - t_p)\phi(x) + 5\varepsilon p \quad \text{for } \mathbf{y} \in \mathbb{Y}^{\mathcal{D}}. \tag{4.67}$$

Using the same arguments as in Step 3, we deduce from (4.63) that

$$\begin{aligned} &E\Big(\int_{t_p}^{\tau} f(X(s), \mathbf{y}(t_p), \zeta_\varepsilon(\mathbf{y})(s))\, ds + \phi(X(\tau)) \,|\mathcal{F}_{t_p}^W\Big) \\ &\leq I(\tau - t_p, X(t_p), \mathbf{y}(t_p); \phi) + 5\varepsilon \\ &\leq \nu(\tau - t_p)\phi(X(t_p)) + 5\varepsilon. \end{aligned} \tag{4.68}$$

Again repeating the same arguments, we obtain (4.67).
Since $\mathbf{y} \in \mathbb{Y}^{\mathcal{D}}$ is arbitrary, (4.67) yields

$$\sup_{\mathbf{y} \in \mathbb{Y}^{\mathcal{D}}} J(\tau, x, \mathbf{y}, \zeta_\varepsilon; \phi) \leq \nu(t_1) \cdots \nu(\tau - t_p)\phi(x) + 5\varepsilon p. \tag{4.69}$$

Observing that $\zeta_\varepsilon \in \mathscr{Z}_{\mathrm{EK}}$ and letting $\varepsilon \to 0$ in (4.69) we conclude (4.59).

Finally we show the opposite of inequality (4.59) by using (4.57). Since $\eta_\varepsilon$ of (4.57) is in $\mathscr{Y}^{\mathcal{D}}$, Example 4.1 says that, for any $\zeta \in \mathscr{Z}_{\mathrm{EK}}$, there exist $\hat{\mathbf{y}} \in \mathbb{Y}^{\mathcal{D}}$ and $\hat{\mathbf{z}} \in \mathbb{Z}_0$ such that

$$\eta_\varepsilon(\hat{\mathbf{z}}) = \hat{\mathbf{y}}, \quad \zeta(\hat{\mathbf{y}}) = \hat{\mathbf{z}}.$$

Therefore,

$$J(t, x, \eta_\varepsilon, \hat{\mathbf{z}}; \phi) = J(t, x, \hat{\mathbf{y}}, \zeta; \phi). \tag{4.70}$$

From (4.57) and (4.70), it follows that

$$\begin{aligned} \text{LHS of (4.57)} &\leq J(\tau, x, \eta_\varepsilon, \hat{\mathbf{z}}; \phi) \\ &= J(\tau, x, \hat{\mathbf{y}}, \zeta; \phi) \\ &\leq \sup_{\mathbf{y} \in \mathbb{Y}^{\mathcal{D}}} J(\tau, x, \mathbf{y}, \zeta; \phi) \quad (\text{by } \hat{\mathbf{y}} \in \mathbb{Y}^{\mathcal{D}}). \end{aligned} \tag{4.71}$$

Since $\zeta \in \mathscr{Z}_{\mathrm{EK}}$ is arbitrary, (4.71) yields

$$\text{LHS of (4.57)} \leq \nu^{\mathcal{D}}(\tau, x; \phi), \tag{4.72}$$

which concludes the opposite inequality in (4.59). This completes the proof of theorem. □

### 4.2.2 DPP *for Lower- and Upper Value Functions*

Let $\mathcal{D}_n, n = 1, 2, \ldots$ be a sequence of division of $[0, T]$. Suppose $\mathcal{D}_n \subset \mathcal{D}_{n+1}, n = 1, 2, \ldots$. Then $v^{\mathcal{D}_n}(t, x; \phi)$ is increasing uniformly on any bounded set of $[0, T] \times \mathbb{R}^d$, by Corollary 4.1. Put

$$v^{(\mathcal{D}_n)}(t, x; \phi) = \lim_{n\to\infty} v^{\mathcal{D}_n}(t, x; \phi). \tag{4.73}$$

The following result holds true.

**Proposition 4.4.** *Suppose that $\mathcal{D}_n \subset \mathcal{D}_{n+1}, n = 1, 2, \ldots$ and $\lim_{n\to} |\mathcal{D}_n| = 0$. Then $v^{(\mathcal{D}_n)}$ does not depend on the sequence $\mathcal{D}_n, n = 1, 2, \ldots$.*

**Definition 4.6.** Put $v(t, x; \phi) = v^{(\mathcal{D}_n)}(t, x; \phi)$. The function $v(\cdot)$ is called the lower value function.

First let us prove the following Lemma.

**Lemma 4.1.** *Let $\mathcal{D} = (t_1, \ldots, t_n)$ and $\tilde{\mathcal{D}} = (t_1, \ldots, t_p, \tau, t_{p+1}, \ldots, t_n)$. For $t \in (t_j, t_{j+1}],\ j \geq p + 1,$ we put*

$$\Phi(x) := \Phi_t(x) = \nu(t_{p+2} - t_{p+1}) \cdots \nu(t - t_j)\phi(x).$$

*Then*

$$|v^{\mathcal{D}}(t, x; \phi) - v^{\tilde{\mathcal{D}}}(t, x; \phi)| < 2\varepsilon, \quad \forall x, \quad \text{if } t_{p+1} - t_p < \Delta_{\varepsilon\Phi}, \tag{4.74}$$

*where $\Delta_{\varepsilon\Phi}$ is the constant given by* (4.34) *for $\Phi$.*

*Proof of Lemma.* Theorem 4.1 implies that

$$v^{\mathcal{D}}(t, x; \phi) = \nu(t_1) \cdots \nu(t_{p+1} - t_p)\Phi(x) \tag{4.75}$$

and

$$v^{\tilde{\mathcal{D}}}(t, x; \phi) = \nu(t_1) \cdots \nu(t_p - t_{p-1})\nu(\tau - t_p)\nu(t_{p+1} - \tau)\Phi(x). \tag{4.76}$$

Put

$$v^*(t, x) = \nu(t_1) \cdots \nu(t_p - t_{p-1})\Phi(x). \tag{4.77}$$

From (4.39) and (4.43) it follows that for $\theta \in [0, s]$,

$$\begin{aligned} &|\nu(\theta)\nu(s - \theta)\Phi(x) - \Phi(x)| \\ &\leq \sup_{\mathbf{z}\in\mathbb{Z}_0} \sup_{\mathbf{y}\in\mathbb{Y}_0} |J(s, x, \mathbf{y}, \mathbf{z}; \Phi) - \Phi(x)| < \varepsilon, \quad \forall x \end{aligned} \tag{4.78}$$

whenever $s < \Delta_{\varepsilon\Phi}$.

We notice that, if $t_{p+1} - t_p < \Delta_{\varepsilon\Phi}$, then

$$|v^{\mathcal{D}}(t,x;\phi) - v^*(t,x)| \leq \|\nu(t_{p+1}-t_p)\Phi - \Phi\|_C < \varepsilon$$

and

$$|v^{\tilde{\mathcal{D}}}(t,x;\phi) - v^*(t,x)| \leq \|\nu(\tau - t_p)\nu(t_{p+1}-\tau)\Phi - \Phi\|_C < \varepsilon,$$

which yields (4.74). □

*Proof of Proposition 4.4.* Let $\mathcal{D} \cup \tilde{\mathcal{D}}$ be the division with $\mathscr{P}_{\mathcal{D}\cup\tilde{\mathcal{D}}} = \mathscr{P}_{\mathcal{D}} \cup \mathscr{P}_{\tilde{\mathcal{D}}}$. Suppose that $(\mathcal{D}_n, n = 1,2,\ldots)$ and $(\tilde{\mathcal{D}}_n, n = 1,2,\ldots)$ satisfy the condition of the proposition. Put

$$v(t,x) = v^{(\mathcal{D}_n)}(t,x;\phi),\ \tilde{v}(t,x) = v^{(\tilde{\mathcal{D}}_n)}(t,x;\phi),$$

and $\mathcal{D}_{nk} = \mathcal{D}_n \cup \tilde{\mathcal{D}}_k$.

Then for all $n,k = 1,2,\ldots,$

$$v^{\mathcal{D}_{nk}}(t,x;\phi) \geq v^{\mathcal{D}_n}(t,x;\phi), \tag{4.79}$$

$$v^{\mathcal{D}_{nk}}(t,x;\phi) \geq v^{\tilde{\mathcal{D}}_k}(t,x;\phi). \tag{4.80}$$

On the other hand, by Lemma 4.1,

$$\lim_{n\to\infty} v^{\mathcal{D}_{nk}}(t,x;\phi) = \lim_{n\to\infty} v^{\mathcal{D}_n}(t,x;\phi) = v(t,x), \quad k = 1,2,\ldots \tag{4.81}$$

and

$$\lim_{k\to\infty} v^{\mathcal{D}_{nk}}(t,x;\phi) = \tilde{v}(t,x), \quad n = 1,2,\ldots. \tag{4.82}$$

Now letting $k \to \infty$ in (4.79) and then letting $n \to \infty$, we obtain $\tilde{v}(\cdot) \geq v(\cdot)$ by (4.82). Since in the same way we have $v(\cdot) \geq \tilde{v}(\cdot)$, we conclude that $v(\cdot) = \tilde{v}(\cdot)$. □

We are ready to prove DPP.

**Theorem 4.2.** *For* $0 \leq t, s, t+s \leq T$,

$$v(t+s,x;\phi) = v(t,x;v(s,\cdot;\phi)), \quad \forall x. \tag{4.83}$$

*Proof.* We divide the proof into three steps.

**Step 1.** We prove

$$v(t+s,x;\phi) \leq v(t,x;v(s,\cdot;\phi)). \tag{4.84}$$

Suppose that $\mathcal{D}_n \subset \mathcal{D}_{n+1}, n = 1,2,\dots, \lim_{n\to\infty}|\mathcal{D}_n| = 0$ and $t \in \mathscr{P}_{\mathcal{D}_1}$. Then (4.44) yields

$$v^{\mathcal{D}_n}(t+s,x;\phi) = v^{\mathcal{D}_n}(t,x;v^{\mathcal{D}_{nt}}(s,\cdot;\phi)). \tag{4.85}$$

Since $(\mathcal{D}_{nt}, n = 1,2,\dots)$ satisfies the condition of Proposition 4.4, by replacing $T$ with $T-t$, $v^{\mathcal{D}_{nt}}$ is increasing to $v$, as $n \to \infty$. Hence the monotonicity provides

$$v^{\mathcal{D}_n}(t+s,x;\phi) \le v^{\mathcal{D}_n}(t,x;v(s,\cdot;\phi)) \le v(t,x;v(s,\cdot;\phi)), \quad n = 1,2,\dots. \tag{4.86}$$

Letting $n \to \infty$, we obtain (4.84).

**Step 2.** For the opposite inequality, we first need

**Lemma 4.2.** *Suppose that* $\psi_l \in C_{\mathrm{bu}}(\mathbb{R}^d), l = 1,2,\dots$ *satisfy* $K_0 := \sup_l \|\psi_l\|_C < \infty$ *and* $\psi_l$ *converges to* $\psi_\infty (\in C_{\mathrm{bu}}(\mathbb{R}^d))$ *uniformly on any bounded set. Then*

$$\lim_{l\to\infty} |v^{\mathcal{D}_n}(t,x;\psi_l) - v^{\mathcal{D}_n}(t,x;\psi_\infty)| = 0 \tag{4.87}$$

*uniformly in n.*

*Proof of Lemma.* Let $X$ be the response for $\mathbf{y} \in \mathbb{Y}_0$ and $\mathbf{z} \in \mathbb{Z}_0$. Then (4.8) implies

$$E_x[|\psi_l(X(t)\chi(|X(t)| > \lambda)|] \le K_0 P_x(|X(t)| > \lambda) < \varepsilon \tag{4.88}$$

whenever $\lambda > \lambda_\varepsilon := \left(\frac{K_0K_1(1+|x|^2)}{\varepsilon}\right)^{\frac{1}{2}}$.

Let $\rho_\varepsilon \in C(\mathbb{R}^d;[0,1])$ satisfy

$$\rho_\varepsilon(y) = \begin{cases} 1, \text{ for } |y| \le \lambda_\varepsilon, \\ 0, \text{ for } |y| \ge \lambda_\varepsilon + 1. \end{cases} \tag{4.89}$$

Then we have

$$\begin{aligned} |J(t,x,\mathbf{y},\mathbf{z};\psi_l) - J(t,x,\mathbf{y},\mathbf{z};\psi_l\rho_\varepsilon)| &\le E_x|\psi_l(X(t)) - (\psi_l\rho_\varepsilon)(X(t))| \\ &\le 2K_0P_x(|X(t)| > \lambda_\varepsilon) < 2K_0\varepsilon, \end{aligned} \tag{4.90}$$

for $l = 1,2,\dots,\infty$. Since $\psi_l\rho_\varepsilon$ converges to $\psi_\infty\rho_\varepsilon$ uniformly,

$$\sup_{\mathbf{y},\mathbf{z}} |J(t,x,\mathbf{y},\mathbf{z};\psi_l\rho_\varepsilon) - J(t,x,\mathbf{y},\mathbf{z};\psi_\infty\rho_\varepsilon)| \le \|\psi_l\rho_\varepsilon - \psi_\infty\rho_\varepsilon\|_C \longrightarrow 0 \tag{4.91}$$

as $l \to \infty$. Combining (4.90) and (4.91), we obtain

$$|v^{\mathcal{D}_n}(t,x;\psi_l) - v^{\mathcal{D}_n}(t,x;\psi_\infty)| \leq 4K_0\varepsilon + \|\psi_l\rho_\varepsilon - \psi_\infty\rho_\varepsilon\|_C, \quad n = 1,2,\ldots. \tag{4.92}$$

Now letting $l \to \infty$, and then letting $\varepsilon \to 0$, we get (4.87). □

**Step 3.** ***Opposite of inequality* (4.84)**
From (4.44), we deduce that, for $n$ and $l$,

$$v(t+s,x;\phi) \geq v^{\mathcal{D}_{n+l}}(t+s,x;\phi) \geq v^{\mathcal{D}_n}(t,x;v^{\mathcal{D}_{lt}}(s,\cdot;\phi)). \tag{4.93}$$

Putting $\psi_l = v^{\mathcal{D}_{lt}}(s,\cdot;\phi)$ and $\psi_\infty = v(s,\cdot;\phi)$, we have

$$v^{\mathcal{D}_n}(t,x;\psi_l) \longrightarrow v^{\mathcal{D}_n}(t,x;\psi_\infty) \quad \text{as } l \to \infty, \tag{4.94}$$

by Lemma 4.2. Thus, (4.93) and (4.94) give the opposite inequality.

This completes the proof of the theorem. □

To conclude Sect. 4.2, we introduce the upper value function. Under the same conditions as in Proposition 4.4, the $\mathcal{D}_n$-upper value function $u^{\mathcal{D}_n}(\cdot;\psi)$ is decreasing uniformly on any bounded set of $[0,T] \times \mathbb{R}^d$, as $n \to \infty$. Its limit $u(\cdot;\phi)$, called the upper value function, is independent of $\mathcal{D}_n, n = 1,2,\ldots$ and the DPP holds:

$$u(t+s,x;\phi) = u(t,x;u(s,\cdot;\phi)), \quad \forall x, \tag{4.95}$$

for $0 \leq t, s \leq t+s \leq T$.
Using (4.38′), (4.58), and Theorem 4.1, we have

$$u^{\mathcal{D}}(t,x;\phi) \geq \sup_{\eta \in \mathcal{Y}^{\mathcal{D}}} \inf_{\mathbf{z} \in \mathbb{Z}_0} J(t,x,\eta,\mathbf{z};\phi) \geq v^{\mathcal{D}}(t,x;\phi). \tag{4.96}$$

Hence,

$$u(t,x;\phi) \geq v(t,x;\phi). \tag{4.97}$$

When the two value functions coincide, we use the name value function.

## 4.3 Isaacs Equations

In this section, we study the dynamics of lower- and upper value functions, via the semigroup formulation of DPP. Referring to Sect. 2.2, we introduce semigroups on $C_{\mathrm{bu}}(\mathbb{R}^d)$ related to DPP for lower (resp. upper) value function in Sect. 4.3.1. Their generators lead to a (nonlinear) parabolic equation, called the lower (resp. upper) Isaacs equation. In Sect. 4.3.2, we prove that the lower value function is the unique viscosity solution of the lower Isaacs equation, by using the same arguments as in Sect. 3.1.3. Similar results are valid for the upper value function.

### 4.3.1 *Semigroups Related to the* DPP

Let us define $\mathrm{V}_t; C_{\mathrm{bu}}(\mathbb{R}^d) \mapsto C_{\mathrm{bu}}(\mathbb{R}^d)$, by

$$\mathrm{V}_t\phi(x) = v(t, x; \phi). \tag{4.98}$$

Then, the following properties are clear.

**Proposition 4.5.**

(i) ***Semigroup*** $\mathrm{V}_0 =$ *identity map,*

$$\mathrm{V}_{t+s}\phi = \mathrm{V}_t(\mathrm{V}_s\phi) = \mathrm{V}_s(\mathrm{V}_t\phi), \quad 0 \le t, \quad s \le t+s \le T.$$

(ii) ***Monotonicity***

$$\phi \le \psi \Longrightarrow \mathrm{V}_t\phi \le \mathrm{V}_t\psi, \quad \forall t.$$

(iii) ***Contractiveness***

$$\|\mathrm{V}_t\phi - \mathrm{V}_t\psi\|_C \le \|\phi - \psi\|_C, \quad \forall t.$$

(iv) ***Continuity***

$$\lim_{t\to 0} \|\mathrm{V}_{t+\theta}\phi - \mathrm{V}_\theta\phi\|_C \le \lim_{t\to 0} \|\mathrm{V}_t\phi - \phi\|_C = 0, \quad \forall \theta.$$

Fix $\mathrm{y} \in \mathrm{Y}$ arbitrarily. From the definition of $I(t, x, \mathrm{y}; \phi)$ (see (4.41)) the maps $\mathrm{I}_t^{\mathrm{y}}; C_{\mathrm{bu}}(\mathbb{R}^d) \mapsto C_{\mathrm{bu}}(\mathbb{R}^d)$ given by

$$\mathrm{I}_t^{\mathrm{y}}\phi(x) = I(t, x, \mathrm{y}; \phi), \quad t \in [0, T], \tag{4.99}$$

form a semigroup, which formulates the DPP for the value function of stochastic control (cf. Sect. 2.2). Hence, putting

$$\mathrm{G}(\mathrm{y},\mathrm{z})\psi(x) = \frac{1}{2}\mathrm{tr}(a(x,\mathrm{y},\mathrm{z})\partial_{xx}\psi(x)) + b(x,\mathrm{y},\mathrm{z})\cdot\partial_x\psi(x) + f(x,\mathrm{y},\mathrm{z}) \tag{4.100}$$

for $\psi \in C_{\mathrm{b}}^2(\mathbb{R}^d)$, we have

**Proposition 4.6.** *For any* $\mathrm{y} \in \mathrm{Y}$,

(i) ***Semigroup***
$\mathrm{I}_0^{\mathrm{y}} =$*identity map,*

$$\mathrm{I}_{t+s}^{\mathrm{y}}\phi = \mathrm{I}_t^{\mathrm{y}}(\mathrm{I}_s^{\mathrm{y}}\phi) = \mathrm{I}_s^{\mathrm{y}}(I_t^{\mathrm{y}}\phi), \quad 0 \le t, \quad s \le t+s \le T.$$

(ii) ***Monotonicity***

$$\phi \leq \psi \Longrightarrow \mathrm{I}_t^{\mathrm{y}}\phi \leq \mathrm{I}_t^{\mathrm{y}}\psi, \quad \forall t.$$

(iii) ***Contractiveness***

$$\|\mathrm{I}_t^{\mathrm{y}}\phi - \mathrm{I}_t^{\mathrm{y}}\psi\|_C \leq \|\phi - \psi\|_C, \quad \forall t.$$

(iv) ***Continuity***

$$\lim_{t\to 0}\|\mathrm{I}_{t+\theta}^{\mathrm{y}}\phi - \mathrm{I}_\theta^{\mathrm{y}}\phi\|_C \leq \lim_{t\to 0}\|\mathrm{I}_t^{\mathrm{y}}\phi - \phi\|_C = 0, \quad \forall\theta,$$

(v) ***Generator***
*For* $\psi \in C_{\mathrm{b}}^2(\mathbb{R}^d)$,

$$\lim_{t\to 0}\frac{1}{t}(\mathrm{I}_t^{\mathrm{y}}\psi(x) - \psi(x)) = \inf_{\mathrm{z}\in Z}\mathrm{G}(\mathrm{y},\mathrm{z})\psi(x).$$

By using the same arguments as in Sect. 2.2.5, the following proposition is also immediate.

**Proposition 4.7.** *Let* $\mathbb{Q}$ *denote the set of all continuous, monotone, and contractive semigroup* $(\mathrm{Q}_t) := (\mathrm{Q}_t; t \in [0,T])$ *satisfying*

$$\mathrm{Q}_t\phi \geq \mathrm{I}_t^{y}\phi, \quad \forall\phi \in C_{bu}(\mathbb{R}^d) \tag{4.101}$$

*for any* $t$ *and* y. *Then* $(\mathrm{V}_t)$ *is the minimal element of* $\mathbb{Q}$, *that is* $(\mathrm{V}_t) \in \mathbb{Q}$ *and for any* $(\mathrm{Q}_t) \in \mathbb{Q}$,

$$\mathrm{Q}_t\phi \geq \mathrm{V}_t\phi, \quad \forall\phi \in C_{bu}(\mathbb{R}^d), \quad \forall t. \tag{4.102}$$

Next we compute the generator $\mathscr{G}$ of $(\mathrm{V}_t)$.

**Theorem 4.3.** *For* $\psi \in C_b^2(\mathbb{R}^d)$

$$\mathscr{G}\psi(x) = \lim_{t\to 0}\frac{1}{t}(\mathrm{V}_t\psi(x) - \psi(x)) = \sup_{\mathrm{y}\in\mathrm{Y}}\inf_{\mathrm{z}\in Z}\mathrm{G}(\mathrm{y},\mathrm{z})\psi(x) \ (:= \mathrm{G}\psi(x)). \tag{4.103}$$

*Proof.* We divide the proof into three steps.

**Step 1.** Let $\mathbf{y} \in \mathbb{Y}_0, \mathbf{z} \in \mathbb{Z}_0$ be given. $X$ denotes the corresponding response. For $R, \varepsilon > 0$, the uniform continuity of all coefficients and (4.9) provide a constant $\Delta_{\varepsilon\psi_R} > 0$, independent of **y** and **z**, such that

$$E_x|\mathrm{G}(\mathbf{y}(s),\mathbf{z}(s))\psi(X(s)) - \mathrm{G}(\mathbf{y}(s),\mathbf{z}(s))\psi(x)| < \varepsilon \tag{4.104}$$

for any $x \in \mathrm{S}_R$, whenever $s < \Delta_{\varepsilon\psi_R}$.

Hence Itô's formula yields that for $t < \Delta_{\varepsilon\psi_R}$,

$$\left| J(t,x,\mathbf{y},\zeta;\psi) - \psi(x) - E_x \int_0^t \mathrm{G}(\mathbf{y}(s),\zeta(\mathbf{y})(s))\psi(x)\,ds \right| < \varepsilon t,$$
$$\forall x \in S_R, \quad \mathbf{y} \in \mathbb{Y}_0, \quad \zeta \in \mathscr{Z}_{\mathrm{EK}}. \tag{4.105}$$

**Step 2.** For $x \in \mathrm{S}_R$, we compute

$$c(t) = c(t,x) := v^{\mathcal{D}}(t,x;\psi) - \psi(x) \quad \text{for } t < \Delta_{\varepsilon\psi_R}.$$

Let $\mathcal{D} = (t_1,\ldots,t_n)$ and $t \in (t_p, t_{p+1}]$. For simplicity, we put $t = t_{p+1}$. From (4.105) it follows that

$$c(t) \le \inf_{\zeta\in\mathscr{Z}_{\mathrm{EK}}} \sup_{\mathbf{y}\in\mathbb{Y}^{\mathcal{D}}} \sum_{j=0}^{p} E \int_{t_j}^{t_{j+1}} \mathrm{G}(\mathbf{y}(t_j),\zeta(\mathbf{y})(s))\psi(x)\,ds + \varepsilon t. \tag{4.106}$$

Since $\mathrm{G}(\mathrm{y},\mathrm{z})\psi(x)$ is continuous in $(\mathrm{y},\mathrm{z})$ and Z is compact, there is a minimum selector $\hat{\zeta} : \mathrm{Y} \mapsto \mathrm{Z}$, such that

$$\mathrm{G}(\mathrm{y},\hat{\zeta}(\mathrm{y})) = \inf_{\mathrm{z}\in\mathrm{Z}} \mathrm{G}(\mathrm{y},\mathrm{z})\psi(x). \tag{4.107}$$

Define $\zeta^* \in \mathscr{Z}_{\mathrm{EK}}$ by

$$\zeta^*(\mathbf{y})(s) = \hat{\zeta}(\mathbf{y}(t_j)) \quad \text{for } s \in [t_j, t_{j+1}), \quad j = 0,1,\ldots,n. \tag{4.108}$$

Then (4.107) and (4.108) yield

$$\text{RHS of (4.106)}$$
$$\le \sup_{\mathbf{y}\in\mathbb{Y}^{\mathcal{D}}} \sum_{j=0}^{p} E \int_{t_j}^{t_j+1} \mathrm{G}(\mathbf{y}(t_j),\zeta^*(\mathbf{y})(s))\psi(x)\,ds + \varepsilon t \tag{4.109}$$

and

$$\int_{t_j}^{t_{j+1}} \mathrm{G}(\mathbf{y}(t_j),\zeta^*(\mathbf{y})(s))\psi(x)\,ds = \inf_{\mathrm{z}\in\mathrm{Z}} \mathrm{G}(\mathbf{y}(t_j),\mathrm{z})\psi(x)(t_{j+1}-t_j)$$
$$\le \mathrm{G}\psi(x)(t_{j+1}-t_j). \tag{4.110}$$

Thus (4.106), (4.109), and (4.110) imply that

$$v^{\mathcal{D}}(t,x;\psi) - \psi(x) \le t(\mathrm{G}\psi(x)+\varepsilon). \tag{4.111}$$

**Step 3.** ***Conclusion***
Suppose that $\mathcal{D}_n, n = 1, 2, \ldots$ satisfy the conditions of Proposition 4.4. Then $v^{\mathcal{D}_n}(\cdot; \psi)$ is increasing to $v(\cdot; \psi)$ uniformly on any bounded subset of $[0, T] \times \mathbb{R}^d$, and so (4.111) yields

$$\limsup_{t \to 0} \frac{1}{t}(v(t, x; \psi) - \psi(x)) \leq \mathrm{G}\psi(x). \tag{4.112}$$

For the converse inequality of (4.112), we recall

$$\mathrm{V}_t \psi(x) \geq \mathrm{I}_t^{\mathrm{y}} \psi(x), \quad \forall \mathrm{y}, t, x. \tag{4.113}$$

Now it follows from Proposition 4.6 (v) that

$$\liminf_{t \to 0} \frac{1}{t}(\mathrm{V}_t \psi(x) - \psi(x)) \geq \mathrm{G}\psi(x). \tag{4.114}$$

Inequalities (4.112) and (4.114) complete the proof. □

## 4.3.2 *Viscosity Solutions of the Isaacs Equations*

In the case when the lower value function $v(\cdot; \psi)$ is smooth, Theorem 4.3 says that $v(\cdot; \psi)$ satisfies

$$\partial_t v(t, x) - \sup_{\mathrm{y} \in \mathrm{Y}} \inf_{\mathrm{z} \in \mathrm{Z}} \mathrm{G}(\mathrm{y}, \mathrm{z}) v(t, x) = 0, \quad t \in (0, T], \quad x \in \mathbb{R}^d, \tag{4.115}$$

with the initial condition

$$v(0, x) = \psi(x), \quad x \in \mathbb{R}^d. \tag{4.116}$$

Now we intend to prove that the lower value function satisfies (4.115)–(4.116) in the viscosity sense. Define the Hamiltonian $\mathcal{H} : \mathbb{R}^d \times S^d \times \mathbb{R}^d \mapsto \mathbb{R}^1$ by

$$\mathcal{H}(x, A, p) = \inf_{\mathrm{y} \in \mathrm{Y}} \sup_{\mathrm{z} \in \mathrm{Z}} \Big( -\frac{1}{2} \mathrm{tr}(a(x, \mathrm{y}, \mathrm{z}) A) - b(x, \mathrm{y}, \mathrm{z}) \cdot p - f(x, \mathrm{y}, \mathrm{z}) \Big), \tag{4.117}$$

where $a = \alpha \alpha^\top$. Then (4.115) can be recast as

$$\partial_t v(t, x) + \mathcal{H}(x, \partial_{xx} v(t, x), \partial_x v(t, x)) = 0. \tag{4.118}$$

This parabolic equation (4.118) is called the lower Isaacs equation.

First we will show

**Theorem 4.4.** *Let* $(\mathrm{S}_t, t \in [0, T])$ *be a continuous semigroup on* $C_{\mathrm{bu}}(\mathbb{R}^d)$ *with monotone and contraction properties. Suppose that for any* $\psi \in C_{\mathrm{b}}^2(\mathbb{R}^d)$ *and* $\phi \in C_{\mathrm{bu}}(\mathbb{R}^d)$,

$$\lim_{t \to 0} \frac{1}{t}(\mathrm{S}_t(\psi + t\phi) - \psi(x)) = \mathrm{G}\psi(x) + \phi(x), \quad \forall x. \tag{4.119}$$

*Then, for* $g \in C_{\mathrm{bu}}(\mathbb{R}^d)$, $(\mathrm{S}_t g, t \in [0, T])$ *is the unique viscosity solution of the lower Isaacs equation* (4.118) *with the initial condition* $v(0, x) = g(x)$.

*Proof.* We can mimic the proof of Theorem 3.3. Put $U(t, x) = \mathrm{S}_t g(x)$. First let us show that

$$U \in C_{\mathrm{bu}}([0, T] \times \mathbb{R}^d). \tag{4.120}$$

Indeed, for $\varepsilon > 0$, there is $\Delta_\varepsilon > 0$ such that

$$\begin{aligned} \|U(t, \cdot) - U(\theta, \cdot)\|_C &= \|\mathrm{S}_\theta \mathrm{S}_{t-\theta} g - \mathrm{S}_\theta g\|_C \\ &\le \|\mathrm{S}_{t-\theta} g - g\|_C < \varepsilon \quad \text{for } |t - \theta| < \Delta_\varepsilon. \end{aligned} \tag{4.121}$$

Taking the $\frac{1}{2}\Delta_\varepsilon$-net $\{t_j, j = 1, \ldots, p\}$ of $[0, T]$ and noticing that $U(t_j, \cdot)$ is in $C_{\mathrm{bu}}(\mathbb{R}^d)$, we can choose a constant $\delta_\varepsilon > 0$, so that

$$|U(t_j, x) - U(t_j, \tilde{x})| < \varepsilon, \quad j = 1, \ldots, p \tag{4.122}$$

whenever $|x - \tilde{x}| < \delta_\varepsilon$. Thus, if $|x - \tilde{x}| < \delta_\varepsilon$ and $|t - \theta| < \frac{1}{2}\Delta_\varepsilon$, we take $t_j$, such that $|t - t_j|, |\theta - t_j| < \Delta_\varepsilon$, and obtain

$$\begin{aligned} &|U(t, x) - U(\theta, \tilde{x})| \\ &\le |U(t, x) - U(t_j, x)| + |U(t_j, x) - U(t_j, \tilde{x})| + |U(t_j, \tilde{x}) - U(\theta, \tilde{x})| < 3\varepsilon, \end{aligned}$$

which in turn yields (4.120).

Next we will show that $U(\cdot)$ is a viscosity subsolution. Let $\rho \in C^{12}((0, T] \times \mathbb{R}^d) \cap C([0, T] \times \mathbb{R}^d)$ be a test function. Since $U(\cdot)$ is bounded, we may assume that $\rho$ is constant outside some compact set. Thus, we have

$$\rho(t + \theta, x) = \rho(t, x) + \theta \partial_t \rho(t, x) + o(\theta) \tag{4.123}$$

where $o(\theta)$ is small uniformly in $x$.

Suppose that $U(t, x) - \rho(t, x)$ attains its global maximum at $(\hat{t}, \hat{x}) \in (0, T) \times \mathbb{R}^d$ and $U(\hat{t}, \hat{x}) = \rho(\hat{t}, \hat{x})$. Then, by the monotonicity of $\mathrm{S}_\theta$,

$$\begin{aligned} 0 &= U(\hat{t}, \hat{x}) - \rho(\hat{t}, \hat{x}) \\ &= \mathrm{S}_\theta U(\hat{t} - \theta, \cdot)(\hat{x}) - \rho(\hat{t}, \hat{x}) \\ &\le \mathrm{S}_\theta \rho(\hat{t} - \theta, \cdot)(\hat{x}) - \rho(\hat{t}, \hat{x}). \end{aligned} \tag{4.124}$$

On the other hand, the contractiveness of $\mathrm{S}_\theta$ together with (4.123) imply that

$$\begin{aligned}&\|\mathrm{S}_\theta\rho(\hat{t}-\theta,\cdot)-\mathrm{S}_\theta(\rho(\hat{t},\cdot)-\theta\partial_t\rho(\hat{t},\cdot))\|_C\\ &\le\|\rho(\hat{t}-\theta,\cdot)-(\rho(\hat{t},\cdot)-\theta\partial_t\rho(\hat{t},\cdot))\|_C\longrightarrow 0\quad\text{as }\theta\to 0.\end{aligned}\tag{4.125}$$

Thus, from (4.124) and (4.125) we deduce that

$$0\le\mathrm{S}_\theta(\rho(\hat{t},\cdot)-\theta\partial_t\rho(\hat{t},\cdot))(\hat{x})-\rho(\hat{t},\hat{x})+o(\theta).\tag{4.126}$$

Dividing both sides by $\theta$ and letting $\theta\to 0$, we have, by (4.119)

$$0\le\mathrm{G}\rho(\hat{t},\hat{x})-\partial_t\rho(\hat{t},\hat{x})$$

and so $U(\cdot)$ is a viscosity subsolution.

Since similar arguments show that $U(\cdot)$ is a viscosity supersolution, the uniqueness theorem (refer to Theorem 3.5 and Example 3.2) completes the proof. □

Next we will show that $(\mathrm{V}_t)$ satisfies (4.119), which in turn implies,

**Theorem 4.5.** *The lower value function is the unique bounded viscosity solution of the lower Isaacs equation.*

*Proof.* For $\psi\in C_{\mathrm{b}}^2(\mathbb{R}^d)$ and $\phi\in C_{\mathrm{bu}}(\mathbb{R}^d)$, Itô's formula, (4.38) and (4.75) lead to

$$\begin{aligned}&v(t,x;\psi+t\phi)-\psi(x)-t(\mathrm{G}\psi(x)+\phi(x))\\ &=\lim_{n\to\infty}\Big(\inf_{\zeta\in\mathscr{Z}_{\mathrm{EK}}}\sup_{\mathbf{y}\in\mathbb{Y}^{\mathcal{D}_n}}E_x\Big[\int_0^t f(X(s),\mathbf{y}(s),\zeta(\mathbf{y})(s))\,ds\\ &\quad+(\psi(X(t))-\psi(x))-t\mathrm{G}\psi(x)+t(\phi(X(t))-\phi(x))\Big]\Big)\end{aligned}\tag{4.127}$$

where $X:=X^{\mathbf{y}\zeta}$. Observing that, by (4.9),

$$tE_x[|\phi(X(t)-\phi(x)|]=o(t)\quad\text{uniformly in }\mathbf{y}\text{ and }\zeta,\tag{4.128}$$

we deduce (4.119) from (4.103) and (4.127). This completes the proof. □

Finally, we consider the upper value function, in the same way. Then the following two results are easy to verify.

**Proposition 4.8.** *For $\phi\in C_{\mathrm{bu}}(\mathbb{R}^d)$, the upper value function $u(t,x;\phi)$ is the unique bounded viscosity solution of the upper Isaacs equation*

$$\begin{cases}\partial_t u(t,x)-\inf_{z\in\mathrm{Z}}\sup_{y\in\mathrm{Y}}\mathrm{G}(y,z)u(t,x)=0,\quad t\in(0,T],x\in\mathbb{R}^d,\\ u(0,x)=\phi(x),\quad x\in\mathbb{R}^d.\end{cases}\tag{4.129}$$

**Proposition 4.9.** *Assume the Isaacs condition*

$$\inf_{z\in Z}\sup_{y\in Y}\Big(\frac{1}{2}\mathrm{tr}(a(x,y,z)A)+b(x,y,z)\cdot p+f(x,y,z)\Big)$$
$$=\sup_{y\in Y}\inf_{z\in Z}\Big(\frac{1}{2}\mathrm{tr}(a(x,y,z)A)+b(x,y,z)\cdot p+f(x,y,z)\Big) \tag{4.130}$$

*for* $x\in\mathbb{R}^d$, $A\in S^d$, $p\in\mathbb{R}^d$. *Then,*

$$v(t,x;\phi)=u(t,x;\phi). \tag{4.131}$$

## 4.4 Risk Sensitive Stochastic Controls and Differential Games

This section is concerned with a link between stochastic controls and differential games, via risk sensitive stochastic controls. Here we want to minimize the exponential-of-integral risk sensitive criterion

$$E\exp\Big(\frac{1}{\varepsilon}\int_0^T f(X^{\gamma(\cdot)}(t),\gamma(t))\,dt\Big),$$

where the response $X^{\gamma(\cdot)}$ evolves according to an SDE with small noise (see (4.132)). Applying the logarithmic transformation to the value function, we obtain the upper Isaacs equation, the corresponding stochastic differential games of which are considered in Sect. 4.4.1. Section 4.4.2 deals with differential games obtained as the limit of these stochastic differential games when $\varepsilon$ tends to 0. In Sect. 4.4.3, we mention a control problem with infinite time horizon.

### *4.4.1 Logarithmic Transformation*

Let $W$ be an $m$-dimensional Wiener process on $(\Omega,\mathcal{F},P)$. Let $\Gamma$ be convex and compact and put $\boldsymbol{\Gamma}^W=L^\infty([0,T]\times\Omega,(\mathcal{F}_t^W);\Gamma)$.

We assume the following condition: $(a_1)$ and $(a_2)$;

$(a_1)$

$$\alpha:\mathbb{R}^d\times\Gamma\longmapsto\mathbb{R}^d\otimes\mathbb{R}^d,$$
$$b:\mathbb{R}^d\times\Gamma\longmapsto\mathbb{R}^d,$$
$$f:\mathbb{R}^d\times\Gamma\longmapsto\mathbb{R}^1$$

are in $C^1_{\rm bu}(\mathbb{R}^d \times \Gamma)$ and $C^2_{\rm bu}(\mathbb{R}^d)$ uniformly on $\Gamma$, say

$$\|z\|_{C^1} + \sup_{\gamma} \|z(\cdot, \boldsymbol{\gamma})\|_{C^2} \le K_0 \quad \text{for } z = \alpha, b, f.$$

(a$_2$) ***Uniform positive definiteness***
There exists $\lambda_0 > 0$, such that

$$y^\top a(x,\gamma)y \ge \lambda_0 |y^2|, \quad \forall y \in \mathbb{R}^d, \quad \forall (x,\gamma),$$

where $a = \alpha\alpha^\top$.

Let $\varepsilon \in (0,1)$ be given. For a control process $\gamma(\cdot) \in \boldsymbol{\Gamma}^W$, its response evolves according to the SDE

$$dX(t) = b(X(t), \gamma(t))\,dt + \sqrt{\varepsilon}\alpha(X(t), \gamma(t))\,dW(t) \tag{4.132}$$

and the payoff is given by the exponential-of-integral risk sensitive criterion

$$J^\varepsilon(t,x;\gamma(\cdot)) = E_x \exp\Big(\frac{1}{\varepsilon}\int_0^t f(X(s), \gamma(s))\,ds\Big). \tag{4.133}$$

Since a controller wants to minimize the payoff, the value function is defined by

$$v^\varepsilon(t,x) = \inf_{\gamma \in \boldsymbol{\Gamma}^W} J^\varepsilon(t,x;\gamma(\cdot)). \tag{4.134}$$

Regarding the regularity of value function, we recall;

**Proposition 4.10 (cf. [FS06], IV Th. 4.2).**

$$v^\varepsilon \in C^{12}_{\rm b}([0,T] \times \mathbb{R}^d) \tag{4.135}$$

*and the Feynman–Kac formula yields the equation*

$$\partial_t v^\varepsilon - \inf_{\gamma \in \Gamma}\Big(\frac{\varepsilon}{2}\mathrm{tr}(a(x,\gamma)\partial_{xx}v^\varepsilon) + b(x,\gamma)\cdot\partial_x v^\varepsilon + \frac{1}{\varepsilon}f(x,\gamma)v^\varepsilon\Big) = 0 \tag{4.136}$$

*on* $(0,T] \times \mathbb{R}^d$, *with the initial condition*

$$v^\varepsilon(0,x) = 1 \quad \text{on } \mathbb{R}^d. \tag{4.137}$$

Next we explain how $v^\varepsilon$ is related to the upper value function of stochastic differential game, by performing the logarithmic transformation of $v^\varepsilon$.

Since the boundedness of $f$ implies $v^\varepsilon(t,x) > 0$, we can put

$$V^\varepsilon(t,x) = \varepsilon \log v^\varepsilon(t,x). \tag{4.138}$$

Then (4.135) implies

$$V^\varepsilon \in C_b^{12}([0,T] \times \mathbb{R}^d). \tag{4.139}$$

Further, a direct computation together with the equality

$$\frac{1}{2} p^\top a p = \sup_{y \in \mathbb{R}^d} \left( p^\top \alpha y - \frac{1}{2}|y|^2 \right) \tag{4.140}$$

yields the upper Isaacs equation

$$\partial_t V^\varepsilon - \inf_{\gamma \in \Gamma} \sup_{y \in \mathbb{R}^d} \mathrm{G}^\varepsilon(y,\gamma) V^\varepsilon = 0 \quad \text{on } (0,T] \times \mathbb{R}^d, \tag{4.141}$$

with

$$V^\varepsilon(0,x) = 0 \quad \text{on } \mathbb{R}^d, \tag{4.142}$$

where

$$\begin{aligned} \mathrm{G}^\varepsilon(y,\gamma)\psi(x) &= \frac{\varepsilon}{2} \mathrm{tr}(a(x,\gamma)\partial_{xx}\psi(x)) \\ &\quad + \partial_x \psi(x) \cdot (b(x,\gamma) + \alpha(x,\gamma)y) + f(x,\gamma) - \frac{1}{2}|y|^2. \end{aligned} \tag{4.143}$$

Since, by (4.139), $|\partial_x V^\varepsilon(t,x)| \le M$ with some constant $M = M_\varepsilon$, the function $\partial_x V^\varepsilon(t,x)^\top \alpha(x,r) y - \frac{1}{2}|y|^2$ attains its maximum at $\hat{\mathbf{y}}_{t,x,\gamma} \in \mathrm{S}_{MK_0}$. Consequently (4.141) is recast as

$$\partial_t V^\varepsilon - \inf_{\gamma \in \Gamma} \sup_{|y| \le N} \mathrm{G}^\varepsilon(y,\gamma) V^\varepsilon = 0, \quad \text{on} \quad (0,T] \times \mathbb{R}^d, \tag{4.141$_N$}$$

for $N \ge MK_0$.

**Proposition 4.11.** *Let $U \in C_b([0,T] \times \mathbb{R}^d)$ be a viscosity solution of* (4.141$_N$)–(4.142) *for some large $N$. Then $U$ coincides with the upper value function of the* stochastic differential game *and $U = V^\varepsilon$ holds.*

*Proof.* Since (4.141$_N$) is the upper Isaacs equation, $U$ is equal to the upper value function of an stochastic differential game between player I and the controller (player II) with control regions $\mathrm{S}_N$ and $\Gamma$, respectively. Now the proposition follows from the uniqueness of the viscosity solution. □

### 4.4.2 Small Noise Limit

Let us consider the upper Isaacs equation when $\varepsilon = 0$,

$$\partial_t V - \inf_{\gamma \in \Gamma} \sup_{y \in \mathbb{R}^d} \mathrm{G}^0(y, \gamma) V = 0 \quad \text{on } (0, T] \times \mathbb{R}^d, \tag{4.144}$$

with

$$V(0, x) = 0 \quad \text{on } \mathbb{R}^d. \tag{4.145}$$

Now we recall the following;

**Proposition 4.12 ([FS06], XI Th 7.1).** *Equations* (4.144)–(4.145) *admits a unique viscosity solution. This solution is bounded and Lipschitz continuous.*

*Proof.* We sketch the proof in four steps.

**Step 1.** ***Differential games***
Let us consider the auxiliary equation

$$\partial_t V - \inf_{\gamma \in \Gamma} \sup_{|y| \le R} \mathrm{G}^0(y, \gamma) V = 0 \quad on\ (0, T] \times \mathbb{R}^d. \tag{4.144$_R$}$$

Put $\mathbb{Y}_R = L^\infty([0, T]; \mathrm{S}_R)$, $\mathbb{Y}_0 = \bigcup_{R>0} \mathbb{Y}_R$ and $\boldsymbol{\Gamma} = L^\infty([0, T]; \Gamma)$. For $\mathbf{y} \in \mathbb{Y}_R$ and $\gamma(\cdot) \in \boldsymbol{\Gamma}$, the response is given by the solution of the differential equation

$$\frac{dX}{dt}(t) = b(X(t), \gamma(t)) + \alpha(X(t), \gamma(t))\mathbf{y}(t), \quad t \in (0, T], \tag{4.146}$$

with initial condition

$$X(0) = x \ (\in \mathbb{R}^d).$$

The payoff and the value function are respectively

$$J(t, x, \mathbf{y}, \gamma(\cdot)) = \int_0^t \left( f(X(s), \gamma(s)) - \frac{1}{2} |\mathbf{y}(s)|^2 \right) ds \tag{4.147}$$

and

$$v_R(t, x) = \sup_{\eta \in \mathscr{Y}_{\mathrm{REK}}} \inf_{\gamma(\cdot) \in \boldsymbol{\Gamma}} J(t, x, \eta(\gamma(\cdot)), \gamma(\cdot)), \tag{4.148}$$

where $\mathscr{Y}_{\mathrm{REK}}$ is the set of strategies from $\boldsymbol{\Gamma}$ onto $\mathbb{Y}_R$.
**Step 2.** There is a Lipschitz constant $M_1$, independent of $t, x$ and $R$, such that

$$|v_R(t, x) - v_R(t, \tilde{x})| \le M_1 |x - \tilde{x}|, \quad \forall t, R. \tag{4.149}$$

Indeed, the estimates

$$J(t,x;0,\gamma(\cdot)) \geq -\|f\|_C t, \quad \forall\gamma(\cdot) \tag{4.150}$$

and

$$J(t,x;\mathbf{y},\gamma(\cdot)) \leq \|f\|_C t - \frac{1}{2}\int_0^t |\mathbf{y}(s)|^2\, ds, \tag{4.151}$$

imply that the maximizer $\hat{\mathbf{y}}$ of $J(t,x,\cdot,\gamma(\cdot))$ satisfies

$$\int_0^t |\hat{\mathbf{y}}(s)|^2\, ds \leq 4\|f\|_C T =: K_1. \tag{4.152}$$

Hence it suffices to consider $\mathbf{y}(\cdot)$ satisfying (4.154).
Let $\tilde{X}$ denote the solution of (4.146) with $\tilde{X}(0) = \tilde{x}$. Then $(a_1)$ yields

$$|X(t) - \tilde{X}(t)|^2 \leq |x - \tilde{x}|^2 \exp 2(K_0(T + \sqrt{TK_1})), \tag{4.153}$$

from which (4.149) follows.
**Step 3.** Noticing that

$$|v_R(t,x) - v_R(\hat{t},x)| \leq \|f\|_C |t - \tilde{t}|, \tag{4.154}$$

we conclude that $v_R$ is bounded and Lipschitz continuous, and is the unique viscosity solution of $(4.144_R)$–(4.145) (cf. [FS06], II Th 9.1).
**Step 4.** Let $R > M_1\|\alpha\|_C$. Then

$$v_R(t,x) = \lim_{R\to\infty} v_R(t,x) =: V^0(t,x). \tag{4.155}$$

Indeed, (4.149) shows that the sub- and super differential of $v_R(t,\cdot)$ are in $S_{M_1}$. Hence for $|p| \leq M_1$ and $R > M_1\|\alpha\|_C$ it holds that

$$\sup_{y\in\mathbb{R}^d}\left(p^\top\alpha(x,\gamma)y - \frac{1}{2}|y|^2\right) = \sup_{|y|\leq R}\left(p^\top\alpha(x,\gamma)y - \frac{1}{2}|y|^2\right), \tag{4.156}$$

which in turn yields (4.155).

Thus, $v_R(\cdot)$ and $V^0(\cdot)$ satisfy (4.144)–(4.145), whenever $R > M_1\|\alpha\|_C$. This concludes the proof of the proposition. □

Now the stability of viscosity solution yields;

**Theorem 4.6 ([FS06] XI, Th 7.2).** *Under conditions* $(a_1)$ *and* $(a_2)$, $V^\varepsilon(\cdot)$ *converges to* $V^0(\cdot)$ *uniformly on any compact set of* $[0,T]\times\mathbb{R}^d$.

*Example 4.3.* ***Investment model with small noise***
We shall consider a small noise limit of the investment model given in Sect. 2.4.

**1. Formulation**

Let $B$ and $W$ be mutually independent $d$- and $m$-dimensional Wiener processes. $S^0$ and $S^i$ $(i = l, \dots, m)$ denote the price processes of bond and $i$-th asset, respectively, and evolve according to the SDE

$$\begin{cases} dS^0(t) = rS^0(t)\,dt, & t \in (0,T], \\ S^0(0) = s^0 \ (> 0), \end{cases} \tag{4.157}$$

and

$$\begin{cases} dS^i(t) = S^i(t)\Big(g^i(X(t))\,dt + \sqrt{\dfrac{\varepsilon}{1+\varepsilon}} \displaystyle\sum_{k=1}^{m} \sigma_k^i(X(t))\,dW^k(t)\Big), & t \in (0,T], \\ S^i(0) = s^i > 0, \end{cases} \tag{4.158}$$

where $X$ is the $d$-dimensional factor process, described by the SDE

$$dX(t) = b(X(t))\,dt + \sqrt{\varepsilon}\,dB(t). \tag{4.159}$$

We assume that $b : \mathbb{R}^d \mapsto \mathbb{R}^d, \sigma : \mathbb{R}^d \mapsto \mathbb{R}^m \otimes \mathbb{R}^m$ and $g : \mathbb{R}^d \mapsto \mathbb{R}^m$ are in $C^1_{\mathrm{bu}}(\mathbb{R}^d)$ and $\sigma\sigma^\top$ is uniformly positive definite. Further, $r$ and $\varepsilon$ are positive constants. We use the negative power utility $U(z) = -\varepsilon z^{-\frac{1}{\varepsilon}}$.

Using data of $X$ and $S^i, i = 1, \dots, m$, an agent invests at $t \in (0,T)$ a proportion $\pi^i(t)$ of his/her wealth in the $i$-th asset and $\pi^0(t) = 1 - \sum_{i=1}^m \pi^i(t)$ in the bond. $\pi(\cdot) = (\pi^1(\cdot), \dots, \pi^m(\cdot)) \in L^\infty([0,T], (\mathcal{F}_t^{BW}); \mathbb{R}^m)$ is called a control process, where we admit selling $(\pi^i(t) < 0)$ and borrowing $(\pi^i(t) > 1)$. Accordingly, the wealth process $Z^{\pi(\cdot)}$ is given by

$$\begin{aligned} Z^{\pi(\cdot)}(t) = \exp\Big\{ & \int_0^t (r + \pi(s) \cdot \mu(X(s))\,ds \\ & + \lambda \int_0^t \pi(s)^\top \sigma(X(s))\,dW(s) - \frac{\lambda^2}{2} \int_0^t |\pi(s)^\top \sigma(X(s))|^2\,ds \Big\} \end{aligned} \tag{4.160}$$

where $\lambda := \sqrt{\frac{\varepsilon}{1+\varepsilon}}$, $\mu^i(x) = g^i(x) - r$, $i = 1, \dots, m$, and we suppose $Z^{\pi(\cdot)}(0) = 1$ (see (2.187)).

The agent wants to maximize the expected utility from the terminal wealth, $E[U(Z^{\pi(\cdot)}(T))]$, by choosing an appropriate control process $\pi(\cdot)$, where

$$\begin{aligned} & U(Z^{\pi(\cdot)}(t)) \\ & = -\varepsilon \exp\Big\{ -\frac{1}{\varepsilon} \int_0^t \Big( r + \pi \cdot \mu - \frac{\lambda^2}{2} |\pi^\top \sigma|^2 \Big)\,ds - \frac{\lambda}{\varepsilon} \int_0^t \pi^\top \sigma\,dW \Big\}. \end{aligned} \tag{4.161}$$

**2. Value function and its HJB equation**

Since $\pi(\cdot)$ is bounded,

$$M^{\pi(\cdot)}(t) := \exp\Big(-\frac{\lambda}{\varepsilon}\int_0^t \pi^\top\sigma\, dW(s) - \frac{\lambda^2}{2\varepsilon^2}\int_0^t |\pi^\top\sigma|^2\, ds\Big)$$

is an exponential martingale. Under the new probability $P^{\pi(\cdot)} := M^{\pi(\cdot)}(T)\circ P$, $B$ and $\hat{W}(:= W(t) + \frac{\lambda}{\varepsilon}\int_0^t \pi^\top\sigma\, ds)$ are mutually independent Wiener processes. Hence, the value function $u^\varepsilon$ and its HJB equation are as follows (cf. (2.199) and (2.202)):

$$\begin{aligned} u^\varepsilon(t,x) &= \sup_{\pi(\cdot)} E_x[U(Z^{\pi(\cdot)}(T))] \\ &= \sup_{\pi(\cdot)} E_x^{\pi(\cdot)}\Big[-\varepsilon\exp\Big\{-\frac{1}{\varepsilon}\int_0^t\Big(r+\pi\cdot\mu-\frac{1}{2}|\pi^\top\sigma|^2\Big)\,ds\Big\}\Big] \end{aligned} \tag{4.162}$$

and

$$\begin{aligned} 0 = &-\frac{\partial u^\varepsilon}{\partial t} + \frac{\varepsilon}{2}\Delta u^\varepsilon + b(x)\cdot\partial_x u^\varepsilon - \frac{r}{\varepsilon}u^\varepsilon \\ &+ \sup_{\pi\in\mathbb{R}^m}\Big\{-\frac{1}{\varepsilon}\Big(\pi\cdot\mu(x) - \frac{1}{2}|\pi^\top\sigma(x)|^2\Big)u^\varepsilon\Big\}, \quad t\in(0,T], \quad x\in\mathbb{R}^d, \end{aligned} \tag{4.163}$$

with $u^\varepsilon(0,x) = -\varepsilon$.

Noting that $u^\varepsilon < 0$, $\sigma\sigma^\top$ is uniformly positive definite and that coefficients are bounded, we see that

$$\text{the 5th term of RHS} = \sup_{|\pi|\le c}\Big(\pi\cdot\mu(x) - \frac{1}{2}|\pi^\top\sigma(x)|^2\Big)\Big(-\frac{1}{\varepsilon}u^\varepsilon\Big),$$

with a positive constant $c$, independent of $t, x$ and $\varepsilon$. Thus, (4.163) admits a classical solution with bounded $\partial_x u^\varepsilon$, where the bound depends on $\varepsilon$.

**3. Logarithmic transformation and small noise limit**

Take the logarithmic transformation, $v^\varepsilon = -\varepsilon\log(-\frac{1}{\varepsilon}u^\varepsilon)$. Then

$$rt \le v^\varepsilon(t,x) \le t\sup_{\pi,x}\Big(r + \pi\cdot\mu(x) - \frac{1}{2}|\pi^\top\sigma(x)|^2\Big). \tag{4.164}$$

By the boundedness of $\partial_x u^\varepsilon$, there is a constant $d_\varepsilon > 0$, such that

$$\inf_y\Big(\frac{1}{2}|y|^2 - y\cdot\partial_x v^\varepsilon(t,x)\Big) = \inf_{|y|\le d_\varepsilon}\Big(\frac{1}{2}|y|^2 - y\cdot\partial_x v^\varepsilon(t,x)\Big).$$

Hence, $v^\varepsilon$ is the unique viscosity solution of the Isaacs equation

$$\begin{cases} 0 = -\dfrac{\partial v^\varepsilon}{\partial t} + \dfrac{\varepsilon}{2}\Delta v^\varepsilon + b(x)\cdot\partial_x v^\varepsilon + r \\ \qquad + \inf\limits_{|y|\le d_\varepsilon}\Big(\dfrac{1}{2}|y|^2 - y\cdot\partial_x v^\varepsilon\Big) + \sup\limits_{|\pi|\le c}\Big(\pi\cdot\mu(x) - \dfrac{1}{2}|\pi^\top\sigma(x)|^2\Big), \\ \qquad\qquad t\in(0,T], \quad x\in\mathbb{R}^d, \\ v^\varepsilon(0,x) = 0, \quad x\in\mathbb{R}^d. \end{cases} \tag{4.165}$$

On the other hand, the Isaacs equation associated with the differential game,

$$\begin{cases} 0 = -\dfrac{\partial v}{\partial t} + b(x)\cdot\partial_x v + r \\ \qquad + \inf\limits_{y}\Big(\dfrac{1}{2}|y|^2 - y\cdot\partial_x v\Big) + \sup\limits_{|\pi|\le c}\Big(\pi\cdot\mu(x) - \dfrac{1}{2}|\pi^\top\sigma(x)|^2\Big), \\ \qquad\qquad t\in(0,T], \quad x\in\mathbb{R}^d, \\ v(0,x) = 0, \quad x\in\mathbb{R}^d \end{cases} \tag{4.166}$$

admits a unique viscosity solution, which is bounded and Lipschitz continuous.

Consequently, $v^\varepsilon$ converges to $v$ uniformly on any compact subset of $[0,T]\times\mathbb{R}^d$ (cf. Theorem 4.6).

### 4.4.3 Note on Control with Infinite Time Horizon

Finally we sketch a model problem with infinite time horizon, given in [FMcE95], Sect. 7.

For a control process $\gamma(\cdot)\in\boldsymbol{\Gamma}^W$, its response $X\in X^{\gamma(\cdot)}$ evolves according to the SDE

$$dX(t) = b(X(t),\gamma(t))\,dt + \sqrt{\frac{\varepsilon}{2}\frac{1}{\mu}}\,dW(t), \tag{4.167}$$

with a positive constant $\mu$. The payoff is given by the long-run expected rate

$$J^\varepsilon(\gamma(\cdot)) = \varepsilon\liminf_{T\to\infty}\frac{1}{T}\log E_x\exp\Big(\frac{1}{\varepsilon}\int_0^T f(X^{\gamma(\cdot)}(s),\gamma(s))\,ds\Big), \tag{4.168}$$

where $\varepsilon\in(0,1)$ is called the noise intensity and the control region $\Gamma$ is convex and compact.

Now we assume that the following conditions are satisfied:

$(\mathrm{a}_1)'$

$$f \geq 0,\ f \in C_{\mathrm{b}}^1(\mathbb{R}^d) \quad \text{and} \quad b(\cdot, \gamma) \in C^1(\mathbb{R}^d)$$

with $\sup_{\gamma x} |\partial_x b(x, \gamma)| < \infty$.

$(\mathrm{a}_2)'$

$$(x - y) \cdot (b(x, \gamma) - b(y, \gamma)) \leq -c_0 |x - y|^2, \quad \forall x, y, \gamma$$

with a positive constant $c_0$.

Hence the payoff $J^\varepsilon$ is independent of the initial state $x$. The problem is to minimize $J^\varepsilon$ over $\boldsymbol{\Gamma}^W$.

Following [FMcE95], we will outline a link between ergodic controls and ergodic Isaacs equations.

1. By (4.167)–(4.168), we have the following ergodic dynamic programming equation

$$\begin{aligned} \Lambda^\varepsilon = & \frac{\varepsilon}{4\mu^2} \Delta W^\varepsilon(x) + \frac{1}{4\mu^2} |\partial_x W^\varepsilon(x)|^2 \\ & + \min_{\gamma \in \Gamma} (b(x, \gamma) \cdot \partial_x W^\varepsilon(x) + f(x, \gamma)), \quad x \in \mathbb{R}^d. \end{aligned} \tag{4.169}$$

2. Under $(\mathrm{a}_1)'$ and $(\mathrm{a}_2)'$, (4.169) has a solution $(\Lambda^\varepsilon \in \mathbb{R}^1, W^\varepsilon \in C^2(\mathbb{R}^d))$, satisfying

$$|\partial_x W^\varepsilon(x)| \leq B, \quad \forall x, \quad \forall \varepsilon \in (0, 1), \tag{4.170}$$

for some positive constant $B$.

Further, a minimum selector $\hat{\gamma}_\varepsilon : \mathbb{R}^d \mapsto \Gamma$, of $(b(x, \gamma) \cdot \partial_x W^\varepsilon(x) + f(x, \gamma))$ provides an optimal Markovian policy, and

$$\Lambda^\varepsilon = \inf_{\gamma(\cdot) \in \boldsymbol{\Gamma}^W} J^\varepsilon(\gamma(\cdot)) = J^\varepsilon(\hat{\gamma}(X^{\hat{\gamma}}(\cdot))) \tag{4.171}$$

holds.

3. Using (4.140), Eq. (4.169) is recast of an ergodic Isaacs equation:

$$\begin{aligned} \Lambda^\varepsilon = & \frac{\varepsilon}{4\mu^2} \Delta W^\varepsilon(x) + \max_{y \in \mathbb{R}^d} (y \cdot \partial_x W^\varepsilon(x) - \mu^2 |y|^2) \\ & + \min_{\gamma \in \Gamma} (b(x, \gamma) \cdot \partial_x W^\varepsilon(x) + f(x, \gamma)). \end{aligned} \tag{4.172}$$

4. In order to obtain a stochastic differential game associated with (4.172), we consider the following discounted cost stochastic differential game. For $\gamma(\cdot) \in \boldsymbol{\Gamma}^W$ and $\mathbf{y} \in L^\infty([0,\infty), (\mathcal{F}_t^W); \mathbb{R}^d)$, the response $\xi = \xi^{\gamma(\cdot),\mathbf{y}}$ is given by the SDE

$$d\xi(t) = (b(\xi(t), \gamma(t)) + \mathbf{y}(t))\,dt + \sqrt{\frac{\varepsilon}{2}\frac{1}{\mu}}\,dW(t), \tag{4.173}$$

while the payoff with the discount rate $\rho$ is given by

$$J_\rho^\varepsilon(t, x, \gamma(\cdot), \mathbf{y}) = E_x \int_0^t e^{-\rho s}(f(\xi(s), \gamma(s)) - \mu^2 |\mathbf{y}(s)|^2)\,ds. \tag{4.174}$$

Let us take the upper value function, $v := v_\rho^\varepsilon$, in much the same way as in Sect. 4.1.2. Then

$$\begin{aligned} &\partial_t v + \rho v - \frac{\varepsilon}{4\mu^2}\Delta v \\ &+ \min_{\gamma \in \Gamma}(b(x,\gamma)\cdot\partial_x v + f(x,\gamma)) \\ &+ \max_{y \in \mathbb{R}^d}(y \cdot \partial_t v - \mu^2|y|^2) = 0 \quad \text{on } (0,\infty)\times\mathbb{R}^d, \end{aligned} \tag{4.175}$$

with the initial condition

$$v(0,x) = 0, \quad x \in \mathbb{R}^d. \tag{4.176}$$

Equation (4.175)–(4.176) has a classical solution. By using various a priori estimates, we see that

$$W_\rho^\varepsilon(x) := \lim_{t\to\infty} v_\rho^\varepsilon(t,x) \quad \text{exists in } C^2(\mathbb{R}^d)$$

and $W_\rho^\varepsilon$ satisfies the stationary form of (4.175).

5. For a suitable sequence $\rho_n \to 0$, the constant $\Lambda^\varepsilon := \lim_{n\to\infty} \rho_n W_{\rho_n}^\varepsilon(x)$ exists and for a fixed $x_0$, the difference $W_{\rho_n}^\varepsilon(x) - W_{\rho_n}^\varepsilon(x_0)$ converges to some $W^\varepsilon$ $(\in C^2(\mathbb{R}^d))$ uniformly on any compact set. Hence $(\Lambda^\varepsilon, W^\varepsilon)$ satisfies (4.172) in the sense of viscosity. Observing that $W^\varepsilon \in C^2(\mathbb{R}^d)$, we see that $(\Lambda^\varepsilon, W^\varepsilon)$ is the solution.

   This fact shows that (4.172) is associated with the ergodic stochastic differential game.

6. Since $\Lambda^\varepsilon \in [0, \|f\|_\infty]$ and $|\partial_x W^\varepsilon| \le B$, we can take $\varepsilon_n \to 0$, so that $\Lambda^{\varepsilon_n} \to \Lambda^0$ and $W^{\varepsilon_n} \to W^0$ uniformly on any compact set. Therefore, $(\Lambda^0, W^0)$ satisfies (4.172) with $\varepsilon = 0$, in the sense of viscosity, i.e., the small noise limit is related to the ergodic differential game.

7. Let us consider the lower value function $u_\rho^\varepsilon$ for the stochastic differential game given by (4.173)–(4.174). Then $u_\rho^\varepsilon$ satisfies (4.175)–(4.176), by changing $v$ to $u_\rho^\varepsilon$.
8. Combining 5, 6, and 7, we see that the ergodic differential game related to the small noise limit has the value.

For various applications of ergodic type stochastic controls, we can refer to [Br06, BP99, FSh99, FSh00, FSh02, HNSh10] and [HS10].

# Chapter 5
# Stochastic Parabolic Equations

**Abstract** This chapter is devoted to stochastic evolution equations in Hilbert spaces, in particular stochastic parabolic type equations of the form

$$du(t,x,\omega) = \mathcal{L}(t,\omega)u(t,x,\omega)\,dt + \mathcal{M}(t,\omega)u(t,x,\omega)\,dW_Q(t),$$

where $\mathcal{L}$ and $\mathcal{M}$ are second-order elliptic and first-order differential operators and $W_Q$ is a colored Wiener process (see Example 5.1).

These equations are generalization of finite-dimensional SDEs and appear in the study of random phenomena in natural sciences and the unnormalized conditional probability of finite-dimensional diffusion processes (see Sect. 5.5), related to filtering equations derived in Fujisaki et al. (Osaka J Math 9:19–40, 1972) and Kushner (J Differ Equ 3:179–190, 1967).

In Sect. 5.1 we collect basic definitions and results for Hilbert space-valued processes; in particular, for continuous martingales, quadratic variations and correlation operators are treated. Stochastic integrals are introduced in Sect. 5.2. Section 5.3 is devoted to the study of stochastic parabolic equations from the viewpoint of Hilbert space-valued SDEs, following Rozovskii (Stochastic evolution systems. Kluwer Academic, Dordrecht/Boston, 1990). By using the results presented, we also consider a semilinear stochastic parabolic equation with Lipschitz nonlinearity in Sect. 5.3.4. Section 5.4 deals with Itô's formula and in Sect. 5.5 Zakai equations related to filtering problems are given.

## 5.1 Preliminaries

This section gives preliminaries for SDEs in Hilbert spaces. Here Hilbert space always means separable real one. We first collect basic definitions and results for stochastic processes taking values in Hilbert spaces. Proofs can be founded in the accessible books (cf. [DaPZ92, M88, R90]).

For a Hilbert space $\mathbb{H}$, $\| * \| := \| * \|_{\mathbb{H}}$ and $( \ , \ ) := ( \ , \ )_{\mathbb{H}}$ denote its norm and inner product, respectively. $\mathbb{H}$ is equipped with a usual Borel field $\mathcal{B}(\mathbb{H})$, generated by the open sets. $\mathbb{H}^*$ denotes the conjugate space, namely the set of all continuous linear functional on $\mathbb{H}$. $h^* \in \mathbb{H}^*$ can be identified with $h \in \mathbb{H}$, such that

© Springer Japan 2015

M. Nisio, *Stochastic Control Theory*, Probability Theory and Stochastic Modelling 72, DOI 10.1007/978-4-431-55123-2_5

$$h^*(z) = (h, z), \quad \forall z \in \mathbb{H}. \tag{5.1}$$

$\mathcal{B}(\mathbb{H})$ is nothing but the smallest $\sigma$-field which makes $h^*$ measurable for any $h^* \in \mathbb{H}^*$.

### 5.1.1 $\mathbb{H}$-Random Variables

Let $\Omega$ be a Polish space and $(\Omega, \mathcal{F}, P)$ be a complete probability space, where $\mathcal{F}$ denotes the completion of $\mathcal{B}(\Omega)$ w.r.t. $P$.

**Definition 5.1.** $X : \Omega \mapsto \mathbb{H}$ is called an $\mathbb{H}$-random variable, if $X$ is $\mathcal{F}/\mathcal{B}(\mathbb{H})$-measurable. $X = \tilde{X}$ means $P(X = \tilde{X}) = 1$.

Let $(e_i, i = 1, 2, \dots)$ be an ONB (orthonormal basis) of $\mathbb{H}$. $\Pi_N$ denotes projection onto the linear space spanned by $(e_1, \dots, e_N)$ and $\Pi_N^{\perp} = \mathbf{1} - \Pi_N$, where $\mathbf{1}$ is the identity map.

For an $\mathbb{H}$-random variable $X$, $X^i := (X, e_i)$ is a real random variable and

$$\lim_{n\to\infty} \left\| X - \sum_{i=1}^{n} X^i e_i \right\| = 0 \quad P\text{-a.e.} \tag{5.2}$$

$P_X$ defined by $P_X(A) = P(X^{-1}(A))$, $A \in \mathcal{B}(\mathbb{H})$, is called the distribution (or law) of $X$.

#### Expectation of $X$

When $\|X\|$ is integrable, we write $X \in L^1(\Omega, \mathcal{F}; \mathbb{H})$ and the expectation of $X$, denoted by $EX$, is defined as the integral $EX = \int_\Omega X(\omega)\, dP(\omega)$, as follows:

1. Let $X$ be simple, say $X = \sum_{i=1}^{n} X^i e_i$, with $X^i \in L^1(\Omega, \mathcal{F}; \mathbb{R}^1)$. Then set

$$EX = \sum_{i=1}^{n} EX^i e_i. \tag{5.3}$$

Hence, $EX$ is an element of $\mathbb{H}$ satisfying

$$(EX, h) = E(X, h), \quad \forall h \in \mathbb{H}. \tag{5.4}$$

Taking $h = EX$, we see that

$$\|EX\| \le E\|X\|. \tag{5.5}$$

2. For $X \in L^1(\Omega, \mathcal{F}; \mathbb{H})$, (5.5) implies that $(E[\Pi_n X], n = 1, 2, \dots)$ is a Cauchy sequence in $\mathbb{H}$. Now define

$$EX = \lim_{n \to \infty} E[\Pi_n X]. \tag{5.6}$$

Then (5.4) characterizes $EX$. Moreover $EX$ does not depend on the choice of the basis $(e_i, i = 1, 2, \dots)$.

**Proposition 5.1.** ***Properties of*** $EX$

(i) $E[aX + bY] = aEX + bEY,\ \forall a, b \in \mathbb{R}^1$;
(ii) $(EX, h) = E(X, h),\ \forall h \in \mathbb{H}$;
(iii) ***Inequality***

$$\|EX\|^p \le E\|X\|^p, \quad \forall p \ge 1; \tag{5.7}$$

(iv) ***Convergence theorem***
*If* $\|X_n - X\|$ *converges to 0* $P$*-a.s. as* $n \to \infty$, *and* $E[\sup_n \|X_n\|] < \infty$, *then*

$$\lim_{n \to \infty} \|EX_n - EX\| = 0; \tag{5.8}$$

(v) *Let* $f : \mathbb{H} \mapsto \mathbb{R}^1$ *be* $\mathcal{B}(\mathbb{H})/\mathcal{B}(\mathbb{R}^1)$*-measurable. Then* $f(X)$ *is a real random variable. If* $f(X)$ *is integrable, then*

$$Ef(X) = \int_{\mathbb{H}} f(h)\, dP_X(h), \tag{5.9}$$

*where the integral in the RHS is the Bochner integral.*

### Conditional Expectation

Let $X \in L^1(\Omega, \mathcal{F}; \mathbb{H})$ and let a $\sigma$-subfield $\mathcal{G} \subset \mathcal{F}$ be given.

**Definition 5.2.** A $\mathcal{G}/\mathcal{B}(\mathbb{H})$-measurable $\mathbb{H}$-random variable $Y$ is called the conditional expectation of $X$ given $\mathcal{G}$, if for any $h \in \mathbb{H}$

$$(Y, h) = E((X, h)|\mathcal{G}) \quad P\text{-a.s.}. \tag{5.10}$$

$Y$ exists and is unique $P$-a.s., and is denoted by $E(X|\mathcal{G})$.

Regarding the regular conditional probability, we refer to Theorem 1.1. Since $\mathcal{F}$ is the completion of $\mathcal{B}(\Omega)$ by $P$, the regular conditional probability given $\mathcal{G}$ exists and is unique.

The following result is immediate.

**Proposition 5.2.** ***Properties of the conditional expectations***

$$\|E(X|\mathcal{G})\| \le E(\|X\||\mathcal{G}) \quad P\text{-}a.s., \tag{5.11}$$

$$\|E(X|\mathcal{G})\|^2 \le E(\|X\|^2|\mathcal{G}) \quad P\text{-}a.s., \tag{5.12}$$

$$E(X|\tilde{\mathcal{G}}) = E(E(X|\mathcal{G})|\tilde{\mathcal{G}}) \quad P\text{-}a.s. \tag{5.13}$$

*for* $\tilde{\mathcal{G}} \subset \mathcal{G} \subset \mathcal{F}$.

### *5.1.2 Continuous Martingales*

Let $(\Omega, \mathcal{F}, (\mathcal{F}_t), P)$ be a filtered probability space.

**Definition 5.3.** Let $X; (T_0, T) \times \Omega \mapsto \mathbb{H}$ be given. $X$ is called

1. An $\mathbb{H}$-process, if it is $\mathcal{B}((T_0, T)) \times \mathcal{F}/\mathcal{B}(\mathbb{H})$-measurable;
2. $(\mathcal{F}_t)$-progressively measurable, if $X_{/(T_0,t]\times\Omega}$ is $\mathcal{B}((T_0, t]) \times \mathcal{F}_t/\mathcal{B}(\mathbb{H})$-measurable for any $t \in (T_0, T)$;
3. $(\mathcal{F}_t)$-adapted, if $X(t)$ is $\mathcal{F}_t/\mathcal{B}(\mathbb{H})$-measurable for any $t \in (T_0, T)$;
4. Continuous, if $X$ has continuous paths, that is there exist $X(T_0)$ and $X(T)$ and

$$\lim_{s\to t} \|X(s) - X(t)\| = 0, \quad \forall t \in [T_0, T], \quad P\text{-a.s.}$$

If $X$ is $(\mathcal{F}_t)$-progressively measurable, then it is $(\mathcal{F}_t)$-adapted. If $X$ is continuous, then the converse is true.

$C([0, T] \times \Omega, (\mathcal{F}_t); \mathbb{H})$ denotes the set of all continuous and $(\mathcal{F}_t)$-adapted $\mathbb{H}$-processes on $[0, T]$.

**Definition 5.4.** $M \in C([0, T] \times \Omega, (\mathcal{F}_t); \mathbb{H})$ is called

1. An $\mathbb{H}$-valued continuous $(\mathcal{F}_t)$-martingale (more precisely, an $(\mathcal{F}_t, P)$-martingale), if
   (a) $M(0) = 0$ $P$-a.s.;
   (b) $E\|M(t)\| < \infty, \forall t$;
   (c) ***Martingale property***
   For any $s \le t$,

$$E(M(t)|\mathcal{F}_s) = M(s) \quad P\text{-a.s.};$$

   equivalently, $E((M(t), h)|\mathcal{F}_s) = (M(s), h)$ $P$-a.s., for any $h \in \mathbb{H}$;
2. An $\mathbb{H}$-valued continuous local $(\mathcal{F}_t)$-martingale, if there is a sequence of $(\mathcal{F}_t)$-stopping times, $\tau_n, n = 1, 2, \ldots$, such that $\tau_n$ is increasing to $T$, $P$-a.s., as $n \to \infty$, and $M(\cdot \wedge \tau_n)$ is a continuous $(\mathcal{F}_t)$-martingale;

3. An $\mathbb{H}$-valued continuous $(\mathcal{F}_t)$-martingale is called square integrable, if $E\|M(t)\|^2 < \infty,\ \forall t$.

Similar notions are introduced for local martingales.

Further, we introduce the following notations:
$\mathbb{M}^{c}([0,T],(\mathcal{F}_t);\mathbb{H})$ = set of $\mathbb{H}$-valued continuous $(\mathcal{F}_t)$-martingales,
$\mathbb{M}^{2c}([0,T],(\mathcal{F}_t);\mathbb{H})$ = set of $\mathbb{H}$-valued square integrable continuous $(\mathcal{F}_t)$-martingales.

For local martingales, we use similar notations: $\mathbb{M}^{c}_{\mathrm{loc}}$ and $\mathbb{M}^{2c}_{\mathrm{loc}}$.

For $M \in \mathbb{M}^{2c}([0,T],(\mathcal{F}_t);\mathbb{H})$, $M^i(t) := (M(t),e_i)$ is in $\mathbb{M}^{2c}([0,T],(\mathcal{F}_t);\mathbb{R}^1)$ and

$$\|M(t)\|^2 = \lim_{n\to\infty}\sum_{i=1}^{n} M^i(t)^2, \tag{5.14}$$

uniformly in $t$, $P$-a.s.

Hence $\|M(\cdot)\|^2$ is a continuous $(\mathcal{F}_t)$-submartingale. Now the Doob–Meyer decomposition asserts that there exists a unique continuous increasing $(\mathcal{F}_t)$-adapted process $\langle M\rangle(\cdot)$, such that

$$\|M(\cdot)\|^2 - \langle M\rangle(\cdot) \in M^{c}([0,T],(\mathcal{F}_t);\mathbb{R}^1). \tag{5.15}$$

**Definition 5.5.** A continuous increasing $(\mathcal{F}_t)$-adapted process $N(\cdot)$, satisfying $\|M(\cdot)\|^2 - N(\cdot) \in M^{c}((\mathcal{F}_t);\mathbb{R}^1)$, is called the quadratic variational process of $M$ and is denoted by $\langle M\rangle(\cdot)$.

The following theorem justifies the following formal notation:

$$\|dM(t)\|^2 = d\langle M\rangle(t).$$

**Theorem 5.1.**

$$\langle M\rangle(t) = \lim_{n\to\infty}\sum_{i=1}^{n}\langle M^i\rangle(t), \quad \textit{uniformly in } t,\ P\textit{-a.s.}, \tag{5.16}$$

$$E\|M(t)-M(\theta)\|^2 = E[\langle M\rangle(t) - \langle M\rangle(\theta)], \tag{5.17}$$

$$\lim_{n\to\infty} E\left|\sum_{j=k_n}^{l_n}\left\|M(t^n_{j+1}) - M(t^n_j)\right\|^2 - (\langle M\rangle(t) - \langle M\rangle(\theta))\right| = 0, \tag{5.18}$$

*where* $\mathcal{D}_n = (t^n_j, j = 1,2,\dots,)$ *is a division of* $[0,T]$ *satisfying* $\mathcal{D}^n \subset \mathcal{D}^{n+1}$ *and* $\max_j |t^n_{j+1} - t^n_j| < 2^{-n}$, *and* $l_n = \min\{l; t < t^n_l\}$ *and* $k_n = \max\{k; t^n_k \le \theta\}$.

The following statements are also useful.

**Theorem 5.2.** *Let $M \in \mathbb{M}^c([0,T],(\mathcal{F}_t);\mathbb{H})$.*

(i) $\|M(\cdot)\|$ *is a real* $(\mathcal{F}_t)$*-submartingale.*
(ii) *If* $E\|M(T)\|^p < \infty$*, for* $p > 1$*, then*

$$E[\sup_{0\le t\le T} \|M(t)\|^p] \le \left(\frac{p}{p-1}\right)^p E\|M(T)\|^p.$$

(refer to [DaPZ92], p. 78).

For $M, N \in \mathbb{M}^{2c}([0,T],(\mathcal{F}_t);\mathbb{H})$, we define $\langle M, N\rangle(\cdot)$ by

$$\langle M, N\rangle(t) = \frac{1}{4}(\langle M+N\rangle(t) - \langle M-N\rangle(t)), \quad t \in [0,T]. \tag{5.19}$$

Then $\langle M, N\rangle(\cdot)$ is a continuous $(\mathcal{F}_t)$-adapted process which has bounded variation and one has

$$(M(\cdot), N(\cdot)) - \langle M, N\rangle(\cdot) \in \mathbb{M}^c([0,T],(\mathcal{F}_t);\mathbb{R}^1). \tag{5.20}$$

Conversely suppose that $\xi(\cdot)$ is continuous, bounded variation and $(\mathcal{F}_t)$-adapted, and moreover $(M(\cdot), N(\cdot)) - \xi(\cdot) \in \mathbb{M}^c([0,T],(\mathcal{F}_t);\mathbb{R}^1)$.

Then

$$\xi(t) = \langle M, N\rangle(t), \quad \forall t \in [0,T], \quad P\text{-a.s.},$$

because a continuous martingale with bounded variation is 0 process.

**Definition 5.6.** $\langle M, N\rangle(\cdot)$ is called the quadratic variational process corresponding to $M$ and $N$.

The following corollary is immediate from Theorem 5.1.

**Corollary 5.1.**

$$\langle M, N\rangle(t) = \langle N, M\rangle(t), \quad \forall t, \quad P\text{-a.s.}, \tag{5.21}$$

$$\langle M, N\rangle(t) = \sum_i \langle M^i, N^i\rangle(t), \quad \forall t, \quad P\text{-a.s.}, \tag{5.22}$$

$$\lim_{n\to\infty} E\Big|\langle M, N\rangle(t) - \langle M, N\rangle(\theta) - \sum_j (M(t^n_{j+1}) - M(t^n_j), N(t^n_{j+1}) - N(t^n_j))\Big| = 0, \tag{5.23}$$

*where* $\{t^n_j\}$ *is the same as in* (5.18)

$$|\langle M, N\rangle(t) - \langle M, N\rangle(\theta)|^2 \le (\langle M\rangle(t) - \langle M\rangle(\theta))(\langle N\rangle(t) - \langle N\rangle(\theta)), \quad \forall \theta < t, \quad P\text{-a.s.} \tag{5.24}$$

### 5.1.3 Correlation Operators

By (5.16) and (5.24) $\langle M^i, M^j\rangle(dt)$ is absolutely continuous w.r.t. $\langle M\rangle(dt)$. Hence there is a unique $(\mathcal{F}_t)$-progressively measurable process $q^{ij}$ such that

$$\langle M^i, M^j\rangle(t) = \int_0^t q^{ij}(s)\, d\langle M\rangle(s), \quad \forall t, \quad P\text{-a.s.} \tag{5.25}$$

Let us define $q_M(t,\omega) \in L(\mathbb{H};\mathbb{H})$ by

$$q_M(t,\omega)e_i = \sum_j q^{ij}(t,\omega)e_j, \quad i = 1,2,\ldots \tag{5.26}$$

i.e., for any $h, g \in \mathbb{H}$,

$$\langle M^h, M^g\rangle(t) = \int_0^t (q_M(s)h, g)\, d\langle M\rangle(s), \quad \forall t, \quad P\text{-a.s.} \tag{5.27}$$

where $M^h(t) = (M(t), h)$ and $M^g(t) = (M(t), g)$. Hence, $q_M(t)$ becomes a symmetric, nonnegative nuclear (trace class) operator on $\mathbb{H}$ with $\operatorname{tr} q_M(t) = 1$.

**Definition 5.7.** $q_M$ is called the correlation operator of $M$ (refer to [R90], Ch. 2.1.13 for details). Generally, in our context a symmetric nonnegative nuclear operator on $\mathbb{H}$ is called a covariance operator.

*Example 5.1 (Colored Wiener process).* $M \in \mathbb{M}^{2c}([0,T], (\mathcal{F}_t); \mathbb{H})$ is called a colored $(\mathcal{F}_t)$-Wiener process with the covariance operator $Q$, if the following two conditions are satisfied.

(a) $\langle M\rangle(t) = t \operatorname{tr} Q$,
(b) $q_M(t) = (\operatorname{tr} Q)^{-1} Q$.

$M$ can be expressed in terms of real Wiener processes in the following way. Since we can take the ONB $(\tilde{e}_i, i = 1,2,\ldots)$ so that

$$Q\tilde{e}_i = \lambda_i \tilde{e}_i, \quad i = 1,2,\ldots \tag{5.28}$$

with $\lambda_1 \geq \lambda_2 \geq \cdots \geq 0$, and $\sum_i \lambda_i = \operatorname{tr} Q (\in (0,\infty))$, we put $M^i(t) = (M(t), \tilde{e}_i)$. Then by (5.27) and (5.28)

$$\langle M^i, M^j\rangle(t) = t\lambda_i \delta_{ij}. \tag{5.29}$$

Thus, $\beta^i(t) = \frac{1}{\sqrt{\lambda_i}} M^i(t), i = 1,\ldots,l$ are mutually independent $(\mathcal{F}_t)$-Wiener processes, if $\lambda_l > 0$. Consequently, if $\lambda_i > 0,\ i = 1,2,\ldots$, then

$$M(t) = \sum_{i=1}^{\infty} \sqrt{\lambda_i}\, \beta^i(t) \tilde{e}_i \quad \text{in } L^2(\Omega, C([0,T])). \tag{5.30}$$

If $\lambda_l > 0$ and $\lambda_{l+1} = 0$, then

$$M(t) = \sum_{i=l}^{l} \sqrt{\lambda_i}\beta^i(t)\tilde{e}_i.$$

$W_Q$ denotes the colored Wiener process with covariance operator $Q$.

## 5.2 Stochastic Integrals

This section deals with Hilbert space valued stochastic integrals. We are mainly concerned with stochastic integrals driven by colored Wiener processes. First we treat a broad class. For an $\mathbb{H}$-valued martingale $M$ and a progressively measurable integrand $\Phi$ whose values are Hilbert–Schmidt operators from $\mathbb{H}$ into $\mathbb{Y}$, we define the stochastic integral, $\int \Phi(s)\,dM(s)$, and list some of its basic properties in Sect. 5.2.1, following [R90]. Section 5.2.2 is devoted to the Burkholder–Davis–Gundy inequality for $\mathbb{H}$-valued martingales.

### *5.2.1 Definitions and Basic Properties*

First we define stochastic integral in the case $\mathbb{H} = \mathbb{R}^1$. Let $\xi \in \mathbb{M}^{2c}([0,T],(\mathcal{F}_t);\mathbb{R}^1)$ be given. By $L^2(\langle\xi\rangle \cdot P,(\mathcal{F}_t);\mathbb{Y})$, we denote the set of all $(\mathcal{F}_t)$-progressively measurable $\mathbb{Y}$-process $\phi$, satisfying

$$E\int_0^T \|\phi(s)\|_{\mathbb{Y}}^2\, d\langle\xi\rangle(s) < \infty. \tag{5.31}$$

Let $(y_i, i = 1,2,\dots)$ be an ONB of $\mathbb{Y}$ and $\phi \in L^2(\langle\xi\rangle \cdot P,(\mathcal{F}_t);\mathbb{Y})$. Put

$$I_n(t) = \sum_{i=1}^{n} \int_0^T (\phi(s), y_i)_{\mathbb{Y}}\, d\xi(s) y_i, \quad n = 1,2,\dots. \tag{5.32}$$

Then there is $\eta \in \mathbb{M}^{2c}([0,T],(\mathcal{F}_t);\mathbb{Y})$ such that

$$\lim_{n\to\infty} E[\sup_t \|\eta(t) - I_n(t)\|_{\mathbb{Y}}^2] = 0, \tag{5.33}$$

by the Burkholder–Davis–Gundy inequality for real martingales.

The stochastic integral $I_\phi(t) := \int_0^t \phi(s)\,d\xi(s)$ is defined by

$$I_\phi(t) = \eta(t).$$

It is clear that

$$\langle I_\phi \rangle(t) = \int_0^t \|\phi(s)\|_{\mathbb{Y}}^2 \, d\langle \xi \rangle(s). \tag{5.34}$$

Next we suppose that $M \in \mathbb{M}^{2c}([0,T], (\mathcal{F}_t); \mathbb{H})$ is given. $\mathcal{L}^2(\mathbb{H};\mathbb{Y})$ denotes the Hilbert space of Hilbert–Schmidt operators, equipped with the norm

$$\|\phi\|_{\mathcal{L}^2(\mathbb{H};\mathbb{Y})} = \Big(\sum_i \|\phi e_i\|_{\mathbb{Y}}^2\Big)^{\frac{1}{2}}. \tag{5.35}$$

By $L^2(\langle M \rangle \cdot P, (\mathcal{F}_t); \mathcal{L}^2(\mathbb{H};\mathbb{Y}))$ we denote the set of $\mathcal{L}^2(\mathbb{H};\mathbb{Y})$-valued processes $\Phi$, such that $(\Phi(\cdot)h, y)_{\mathbb{Y}}$ is $(\mathcal{F}_t)$-progressively measurable for any $(h, y) \in \mathbb{H} \times \mathbb{Y}$, and

$$E \int_0^T \|\Phi(s)\|_{\mathcal{L}^2(\mathbb{H};\mathbb{Y})}^2 \, d\langle M \rangle(s) < \infty. \tag{5.36}$$

We define the stochastic integral $I_\Phi(t) := \int_0^t \Phi(s)\, dM(s)$ as an element of $\mathbb{M}^{2c}([0,T], (\mathcal{F}_t); \mathbb{Y})$, following [R90], Ch. 2.2.2. We still omit the subscript $\mathbb{H}$ in $()_{\mathbb{H}}$ and $\| \, \|_{\mathbb{H}}$. For $M^i(t) = (M(t), e_i)$, the stochastic integral $J_i(t) := \int_0^t \Phi(s)e_i \, dM^i(s)$ is given as an element of $\mathbb{M}^{2c}([0,T], (\mathcal{F}_t); \mathbb{Y})$ by (5.33). Define $\Pi_{np}\Phi \in L^2(\langle M \rangle \cdot P, (\mathcal{F}_t); \mathcal{L}^2(\mathbb{H};\mathbb{Y}))$ by

$$\Pi_{np}\Phi(s)e_i = \begin{cases} \Phi(s)e_i, \ i = n, n+1, \ldots, p, \\ 0, \qquad \text{otherwise}. \end{cases}$$

Since $\| \sum_{i=n}^p J_i(t)\|_{\mathbb{Y}}^2$ is a real submartingale, we have

$$\begin{aligned} E\Big[\sup_{0 \le t \le T} \Big\|\sum_{i=n}^p J_i(t)\Big\|_{\mathbb{Y}}^2\Big] &\le 4E\Big[\Big\|\sum_{i=n}^p J_i(T)\Big\|_{\mathbb{Y}}^2\Big] \\ &= 4E \int_0^T \|\Pi_{np}\Phi(s) q_M^{\frac{1}{2}}(s)\|_{\mathcal{L}^2(\mathbb{H};\mathbb{Y})}^2 \, d\langle M \rangle(s) \\ &\le 4E \int_0^T \|\Pi_{np}\Phi(s)\|_{\mathcal{L}^2(\mathbb{H};\mathbb{Y})}^2 \, d\langle M \rangle(s) \longrightarrow 0 \text{ as } n, p \to \infty. \end{aligned}$$

Now we define the stochastic integral.

**Definition 5.8.**

$$I_\Phi(t) := \int_0^t \Phi(s)\, dM(s) = \lim_{n\to\infty} \sum_{i=1}^n \int_0^t \Phi(s)e_i \, dM^i(s) \quad \text{in } L^2(\Omega; C([0,T];\mathbb{Y})). \tag{5.37}$$

The following properties are easy to verify.

**Proposition 5.3.** *For* $I_\Psi, I_\Phi \in \mathbb{M}^{2c}([0,T],(\mathcal{F}_t);\mathbb{Y})$,

(i) $I_{a\Phi+b\Psi} = aI_\Phi + bI_\Psi$, *P-a.s.*, $\forall a,b \in \mathbb{R}^1$,
(ii) $\int_0^t \Phi(s)\,dM(s) = \int_0^T \chi_{[0,t]}(s)\Phi(s)\,dM(s)$, $\forall t$, *P-a.s.*,
(iii) $\langle I_\Phi\rangle(t) = \int_0^t \|\Phi(s)q_M^{\frac{1}{2}}(s)\|^2_{\mathcal{L}^2(\mathbb{H};\mathbb{Y})}\,d\langle M\rangle(s)$, *P-a.s.*,
(iv) $\langle I_\Phi, I_\Psi\rangle(t) = \int_0^t (\Phi(s)q_M(s),\Psi(s))_{\mathcal{L}^2(\mathbb{H};\mathbb{Y})}\,d\langle M\rangle(s)$, *P-a.s.*,
(v) $E[\sup_{0\le t\le T}\|I_\Phi(t)\|^2_{\mathbb{Y}}] \le 4E\int_0^T \|\Phi(s)\|^2_{\mathcal{L}^2(\mathbb{H};\mathbb{Y})}\,d\langle M\rangle(s)$.

$I_\Phi$ does not depend on the choice of ONB of $\mathbb{H}$ and $\mathbb{Y}$.

Finally, we introduce a weaker condition for $\Phi$:

$$\int_0^T \|\Phi(s)\|^2_{\mathcal{L}^2(\mathbb{H};\mathbb{Y})}\,d\langle M\rangle(s) < \infty \quad P\text{-a.s.}. \tag{5.38}$$

We will define the stochastic integral of $\Phi$ as a local martingale, in the same way as in Step 3 of Sect. 1.1.3. Namely, put

$$\tau_N = \begin{cases} \inf\Big\{t < T;\ \int_0^t \|\Phi(s)\|^2_{\mathcal{L}^2(\mathbb{H};\mathbb{Y})}\,d\langle M\rangle(s) \ge N\Big\}, \\ T, \quad \text{if } \{\cdots\} = \text{empty} \end{cases} \tag{5.39}$$

and

$$\Phi_N(s) = \chi_{[0,\tau_N]}(s)\Phi(s). \tag{5.40}$$

Then there is $I_\Phi(t)$, such that

$$I_\Phi(t) = I_{\Phi_N}(t) \quad \text{on } (\tau_N \ge t), \tag{5.41}$$

which implies that $I_\Phi \in \mathbb{M}^{2c}_{\text{loc}}([0,T],(\mathcal{F}_t);\mathbb{Y})$. We define $\int_0^t \Phi(s)\,dM(s)$ as the integral $I_\Phi(t)$ appearing in (5.41).

*Example 5.2 (Stochastic integral driven by a colored Wiener process).* Let $W_Q$ be an $(\mathcal{F}_t)$-Wiener process with covariance operator $Q$, say

$$W_Q(t) = \sum_i \sqrt{\lambda_i}\beta^i(t)e_i.$$

For $\Phi \in L^2([0,T]\times\Omega,(\mathcal{F}_t);\mathcal{L}^2(\mathbb{H};\mathbb{Y}))$, we introduce the stochastic integral $I_\Phi$ by

$$I_\Phi(t) = \int_0^t \Phi(s)\,dW_Q(s) = \sum_i \sqrt{\lambda_i}\int_0^t \Phi(s)e_i\,d\beta^i(s) \tag{5.42}$$

as an element of $\mathbb{M}^{2c}([0,T],(\mathcal{F}_t);\mathbb{Y})$.

The following properties hold:

$$\langle I_\Phi\rangle(t) = \sum_i \lambda_i \int_0^t \|\Phi(s)e_i\|_{\mathbb{Y}}^2\, ds$$

$$= \int_0^t (\Phi(s)Q, \Phi(s))_{\mathcal{L}^2(\mathbb{H};\mathbb{Y})}\, ds, \quad \forall t, \quad P\text{-a.s.}, \tag{5.43}$$

$$\langle I_\Phi, I_\Psi\rangle(t) = \int_0^t (\Phi(s)Q, \Psi(s))_{\mathcal{L}^2(\mathbb{H};\mathbb{Y})}\, ds, \quad \forall t, \quad P\text{-a.s..} \tag{5.44}$$

*Example 5.3 (Stochastic integral for $\mathbb{H}$-processes).* Let $\phi \in L^2(\langle M\rangle \cdot P, (\mathcal{F}_t); \mathbb{H})$ be given. Under $\mathbb{H} = \mathbb{H}^*$, we can regard $\phi$ as an element of $L^2(\langle M\rangle \cdot P, (\mathcal{F}_t); \mathcal{L}^2(\mathbb{H}; \mathbb{R}^1))$ and define the stochastic integral $I_\phi(t) := \int_0^t (\phi(s), dM(s))$ by (5.37), namely

$$\int_0^t (\phi(s), dM(s)) = \sum_i \int_0^t (\phi(s), e_i)\, dM^i(s). \tag{5.45}$$

Thus, $I_\phi$ is in $\mathbb{M}^{2c}([0,T], (\mathcal{F}_t); \mathbb{R}^1)$ and satisfies

$$\langle I_\phi, I_\psi\rangle(t) = \int_0^t (\phi(s)q_M(s), \psi(s))\, d\langle M\rangle(s) \tag{5.46}$$

and

$$E\Big[\sup_{0\le t\le T}\Big|\int_0^t (\phi(s), dM(s))\Big|^2\Big] \le 4E\int_0^T \|\phi(s)\|^2\, d\langle M\rangle(s). \tag{5.47}$$

*Example 5.4 (Martingale part of $\|M(t)\|^2$).* Put $M^\#(t) = \int_0^t (M(s), dM(s))$. Then $M^\#$ is in $\mathbb{M}^c([0,T], (\mathcal{F}_t); \mathbb{R}^1)$ and $2M^\#(t)$ gives the martingale part of $\|M(t)\|^2$, namely

$$\|M(t)\|^2 = 2M^\#(t) + \langle M\rangle(t). \tag{5.48}$$

Indeed, for the exit time $\tau_N$ of $\|M(t)\|$ from $[0,N]$, $M^i(t) := (M(t), e_i)$ satisfies

$$M^i(t\wedge\tau_N)^2 = 2\int_0^{t\wedge\tau_N} M^i(s)\, dM^i(s) + \langle M^i\rangle(t\wedge\tau_N), \quad \forall t, \quad P\text{-a.s.} \tag{5.49}$$

Summing w.r.t. $i$ leads to

$$\|M(t\wedge\tau_N)\|^2 = 2M^\#(t\wedge\tau_N) + \langle M\rangle(t\wedge\tau_N). \tag{5.50}$$

Hence $M^\#$ is in $\mathbb{M}^c_{\rm loc}([0,T], (\mathcal{F}_t); \mathbb{R}^1)$.

Further, (5.50) yields

$$E\Big[\sup_{0\le\theta\le\tau_N}\Big|\int_0^\theta (M(s), dM(s))\Big|\Big] \le \frac{1}{2}E\Big[\sup_{s\le T}\|M(s)\|^2 + \langle M\rangle(T)\Big]$$
$$\le \frac{5}{2}E\langle M\rangle(T). \tag{5.51}$$

Thus the monotone convergence theorem for (5.50) implies that $E|M^{\#}(t)| < \infty$ and (5.48).

Note that if $M^{\#} \in \mathbb{M}^{2c}([0,T],(\mathcal{F}_t);\mathbb{R}^1)$, then

$$\langle M^{\#}\rangle(t) = \int_0^t (M(s)q_M(s), M(s))\, d\langle M\rangle(s). \tag{5.52}$$

Therefore,

$$d\langle M^{\#}\rangle(t) \le \|M(t)\|^2\, d\langle M\rangle(s). \tag{5.53}$$

### 5.2.2 Martingale Inequalities

We have already stated a martingale inequality in Theorem 5.2. Now we are going to prove the Burkholder–Davis–Gundy inequality for $\mathbb{H}$-valued martingales.

**Theorem 5.3.** *Let $M \in \mathbb{M}^{2c}([0,T],(\mathcal{F}_t);\mathbb{H})$ and put*

$$M^*(t) = \sup_{0\le s\le t}\|M(s)\|. \tag{5.54}$$

*Then we have*

$$\frac{1}{4}E[M^*(t)^2] \le E\langle M\rangle(t) \le E[M^*(t)^2], \tag{5.55}$$
$$c_pE[M^*(t)^{2p}] \le E\langle M\rangle(t)^p \le C_pE[M^*(t)^{2p}] \tag{5.56}$$

*where*

$$\begin{cases} c_p = \left(\left(\frac{2p}{2p-1}\right)^{2p} p(2p-1)\right)^{-p}, & C_p = (4p)^p, \text{ if } p > 1, \\ c_p = \left(\frac{p}{16}\right)^p, \quad C_p = p^{-2p}, & \text{if } p \in (0,1). \end{cases} \tag{5.57}$$

*Proof.* Since (5.55) is clear from Theorem 5.2, we will prove (5.56), by using arguments similar to those in the real martingale cases.

Let $\tau_N$ be the exit time of $\|M(t)\|$ from $[0, N]$. By considering $M(t \wedge \tau_N)$ instead of $M(t)$, we assume that $M$ is bounded. We divide the proof into four steps.

**Step 1.** ***Compute*** $c_p$ ***for*** $p > 1$

Let $\varepsilon > 0$ be given. Put $X(t) = \|M(t)\|^2$. Applying Itô's formula to (5.48) and using (5.52), we obtain

$$\begin{aligned} d(X(t)+\varepsilon)^p =& p(X(t)+\varepsilon)^{p-1}(2dM^{\#}(t) + d\langle M\rangle(t)) \\ &+ 2p(p-1)(X(t)+\varepsilon)^{p-2}(M(t)q_M(t), M(t))\, d\langle M\rangle(t) \\ &\le 2p(X(t)+\varepsilon)^{p-1}\, dM^{\#}(t) + p(2p-1)(X(t)+\varepsilon)^{p-1}\, d\langle M\rangle(t) \end{aligned} \tag{5.58}$$

since $X(t) \ge 0$. Letting $\varepsilon \to 0$ leads to

$$\begin{aligned} EX(t)^p &\le p(2p-1)E[M^*(t)^{2p-2}\langle M\rangle(t)] \\ &\le p(2p-1)(EM^*(t)^{2p})^{\frac{p-1}{p}}(E\langle M\rangle(t)^p)^{\frac{1}{p}}, \end{aligned} \tag{5.59}$$

which in conjunction with Theorem 5.2 (ii) yields the value of $c_p$.

**Step 2.** ***Compute*** $C_p$ ***for*** $p > 1$

Put

$$N^i(t) = \int_0^t \langle M\rangle(s)^{\frac{p-1}{2}}\, dM^i(s) \tag{5.60}$$

and

$$N(t) = \sum_i N^i(t)e_i = \int_0^t \langle M\rangle(s)^{\frac{p-1}{2}}\, dM(s). \tag{5.61}$$

Then $N \in \mathbb{M}^{2C}([0,T], (\mathcal{F}_t); \mathbb{H})$ and

$$\langle N\rangle(t) = \int_0^t \langle M\rangle(s)^{p-1}\, d\langle M\rangle(s) = \frac{1}{p}\langle M\rangle(t)^p. \tag{5.62}$$

From (5.61) it follows that

$$N(t) = \langle M\rangle(t)^{\frac{p-1}{2}} M(t) - \int_0^t M(s)\, d\langle M\rangle^{\frac{p-1}{2}}(s) \tag{5.63}$$

and

$$\|N(t)\|^2 \le 4M^*(t)^2\langle M\rangle(t)^{p-1}. \tag{5.64}$$

Thus

$$E\langle N\rangle(t) = E\|N(t)\|^2 \le 4(EM^*(t)^{2p})^{\frac{1}{p}}(E\langle M\rangle(t)^p)^{\frac{p-1}{p}} \tag{5.65}$$

gives $C_p$ by using (5.62).

**Step 3.** ***Compute*** $c_p$ ***for*** $p \in (0,1)$

Noting that

$$N(t)\langle M\rangle(t)^{\frac{1-p}{2}} = M(t) + \int_0^t N(s)\,d\langle M\rangle^{\frac{1-p}{2}}(s) \tag{5.66}$$

by (5.61), we get

$$\|M(t)\| \le 2N^*(t)\langle M\rangle(t)^{\frac{1-p}{2}}, \tag{5.67}$$

where $N^*(t) = \sup_{s\le t}\|N(s)\|$.

Since the RHS of (5.67) is increasing in $t$,

$$M^*(t) \le 2N^*(t)\langle M\rangle(t)^{\frac{1-p}{2}}. \tag{5.68}$$

Applying Hölder's inequality, we have

$$EM^*(t)^{2p} \le 4^p(EN^*(t)^2)^{\frac{1}{p}}(E\langle M\rangle(t)^p)^{1-p} \quad (p\in(0,1)) \tag{5.69}$$

and now using Theorem 5.2 (ii) for $N(t)$, we obtain $c_p$.

**Step 4.** ***Compute*** $C_p$ ***for*** $p \in (0,1)$

Since, for $\varepsilon > 0$,

$$\langle M\rangle(t)^p = \langle M\rangle(t)^p(\varepsilon + M^*(t))^{-2p(1-p)}(\varepsilon + M^*(t))^{2p(1-p)}, \tag{5.70}$$

Hölder inequality yields

$$E\langle M\rangle(t)^p \le (E\langle M\rangle(t)(\varepsilon+M^*(t))^{-2(1-p)})^p(E(\varepsilon+M^*(t))^{2p})^{1-p}. \tag{5.71}$$

Let us put

$$N^i(t) = \int_0^t (\varepsilon + M^*(s))^{p-1}\,dM^i(s). \tag{5.72}$$

Then, since $p-1<0$, it holds

$$\langle N^i\rangle(t) \ge (\varepsilon + M^*(t))^{2(p-1)}\langle M^i\rangle(t) \tag{5.73}$$

From (5.72) it follows that

$$N(t) := \sum_i N^i(t)e_i$$
$$= M(t)(\varepsilon + M^*(t))^{p-1} + (1-p)\int_0^t M(s)(\varepsilon + M^*(s))^{p-2}dM^*(s) \tag{5.74}$$

and

$$\|N(t)\| \leq \frac{1}{p}(\varepsilon + M^*(t))^p. \tag{5.75}$$

Now (5.71), (5.73), and (5.75) yield

$$E\langle M\rangle(t)^p \leq p^{-2p}E(\varepsilon + M^*(t)^{2p}). \tag{5.76}$$

Letting $\varepsilon \to 0$, we obtain $C_p$.

This completes the proof. □

For example, we have

$$EM^*(t) \leq \sqrt{32}E\sqrt{\langle M\rangle(t)} \leq 6E\sqrt{\langle M\rangle(t)}. \tag{5.77}$$

## 5.3 Stochastic Parabolic Equations with Colored Wiener Noises

This section is mainly concerned with the stochastic parabolic equation

$$du(t,x,\omega) = \Big\{\sum_{i=1}^d \partial_i\Big(\sum_{j=1}^d a^{ij}(t,x,\omega)\partial_j u(t,x,\omega) + b^i(t,x,\omega)u(t,x,\omega)\Big)$$
$$+ \rho(t,x,\omega)u(t,x,\omega) + \mathfrak{z}(t,x,\omega)\Big\}\,dt$$
$$+ \sum_k \sqrt{\lambda_k}\Big\{\sum_{j=1}^d g_k^j(t,x,\omega)\partial_j u(t,x,\omega)$$
$$+ \hbar_k(t,x,\omega)u(t,x,\omega) + f_k(t,x,\omega)\Big\}\,dW^k(t). \tag{5.78}$$

We will consider (5.78) from the viewpoint of Hilbert space-valued SDE, following [P79], [P93], [R90], Chs. 3 and 4. In Sect. 5.3.1, we introduce concepts, notations, and basic results needed for later sections. We study the Cauchy problem for (5.78) and properties of solutions in Sects. 5.3.2 and 5.3.3. By changing $\mathfrak{z}(t,x,\omega)$ to $\mathfrak{z}(t,x,u(t,x),\partial_x u(t,x),\omega)$, we will consider a semilinear stochastic parabolic equation with Wiener noise in Sect. 5.3.4, based on results obtained in Sect. 5.3.3.

### 5.3.1 Preliminaries

In this subsection, we list some notations and basic results we need later.

Let $\mathbb{X}$ and $\mathbb{H}$ be Hilbert spaces. By $\mathbb{X} \curvearrowright \mathbb{H}$, we mean that the embedding; $\mathbb{X} \mapsto \mathbb{H}$ is continuous and dense. We use the identification $\mathbb{H} = \mathbb{H}^*$ (= conjugate space of $\mathbb{H}$). The triple $(\mathbb{X}, \mathbb{H}, \tilde{\mathbb{X}})$ is called normal, if $\mathbb{X} \curvearrowright \mathbb{H} = \mathbb{H}^* \curvearrowright \tilde{\mathbb{X}}$ and there is a constant $N$ such that

$$|(y,x)_{\mathbb{H}}| \le N\|y\|_{\tilde{\mathbb{X}}}\|x\|_{\mathbb{X}}, \quad \forall x \in \mathbb{X}, \quad y \in \mathbb{H}. \tag{5.79}$$

Let $(\mathbb{X}, \mathbb{H}, \tilde{\mathbb{X}})$ be normal and $x \in \mathbb{X}$ and $y \in \tilde{\mathbb{X}}$ be given. Since there is a sequence $y_n \in \mathbb{H}$ $(n = 1, 2, \ldots)$ such that $\lim_{n\to\infty} \|y_n - y\|_{\tilde{\mathbb{X}}} = 0$, we can define the dual product between $y$ and $x$ by

$$\langle y, x\rangle_{(\tilde{\mathbb{X}},\mathbb{X})} = \lim_{n\to\infty} (y_n, x)_{\mathbb{H}}. \tag{5.80}$$

The RHS of (5.80) does not depend on the choice of $(y_n)$, and

$$|\langle y, x\rangle_{(\tilde{\mathbb{X}},\mathbb{X})}| \le N\|y\|_{\tilde{\mathbb{X}}}\|x\|_{\mathbb{X}}, \quad \forall y \in \tilde{\mathbb{X}}, \quad x \in \mathbb{X}. \tag{5.81}$$

Let $W_Q$ be an $\mathbb{H}$-valued Wiener process with covariance operator $Q$. Let $(e_k, k = 1, 2, \ldots)$ be an ONB of $\mathbb{H}$ with

$$Qe_k = \lambda_k e_k, \quad k = 1, 2, \ldots. \tag{5.82}$$

Put

$$\mathcal{L}_Q^2(\mathbb{H};\mathbb{H}) = \{\psi \in L(\mathbb{H};\mathbb{H});\ \|\psi\|_Q^2 := \sum_k \lambda_k \|\psi e_k\|_{\mathbb{H}}^2 < \infty,\} \tag{5.83}$$

i.e., $\psi \in \mathcal{L}_Q^2(\mathbb{H};\mathbb{H})$ means that $\psi Q^{\frac{1}{2}}$ is a Hilbert-Schmidt operator.

For given $A(t,\omega) \in L(\mathbb{X};\tilde{\mathbb{X}})$ and $G(t,\omega) \in L(\mathbb{X};\mathcal{L}_Q^2(\mathbb{H};\mathbb{H}))$, consider the SDE

$$\begin{aligned} du(t,\omega) =& (A(t,\omega)u(t,\omega) + Z(t,\omega))\,dt \\ &+ (G(t,\omega)u(t,\omega) + \Psi(t,\omega))\,dW_Q(t,\omega), \quad t \in (T_0, T] \end{aligned} \tag{5.84}$$

with the initial condition

$$u(T_0, \omega) = u_0(\omega). \tag{5.85}$$

Assume the following conditions are satisfied:

$(A_0)$ For any $\phi \in \mathbb{X}$, $A(t)\phi$ and $G(t)\phi$ are $(\mathcal{F}_t)$-progressively measurable $\tilde{\mathbb{X}}$- and $\mathcal{L}_Q^2(\mathbb{H};\mathbb{H})$-valued processes, respectively,

$(A_1)$ There is a constant $K$, such that

$$\|A(t)\phi\|_{\tilde{\mathbb{X}}}^2 + \|G(t)\phi\|_Q^2 \le K\|\phi\|_{\mathbb{X}}^2, \quad \forall \phi \in \mathbb{X} \quad \text{and} \quad \forall t, \quad P\text{-a.s..}$$

$(A_2)$ ***Coerciveness***
There are constants $\bar{\lambda} > 0$ and $\bar{\mu} \in \mathbb{R}^1$, such that

$$2\langle A(t)\phi, \phi\rangle_{(\tilde{\mathbb{X}},\mathbb{X})} + \|G(t)\phi\|_Q^2 \le -\bar{\lambda}\|\phi\|_{\tilde{\mathbb{X}}}^2 + \bar{\mu}\|\phi\|_{\mathbb{H}}, \quad \forall \phi \in \mathbb{X}, \quad \forall t, \quad P\text{-a.s..}$$

$(A_3)$ $Z \in L^2([T_0, T] \times \Omega, (\mathcal{F}_t); \tilde{\mathbb{X}})$.

$(A_4)$ $\Psi \in L^2([T_0, T] \times \Omega, (\mathcal{F}_t); \mathcal{L}_Q^2(\mathbb{H};\mathbb{H}))$.

$(A_5)$ $u_0 \in L^2(\Omega, \mathcal{F}_{T_0}, \mathbb{H})$.

Considering (5.84) on $(\mathbb{X}, \mathbb{H}, \tilde{\mathbb{X}})$, we define the solution as follows;

**Definition 5.9.** The $\mathbb{X}$-valued $(\mathcal{F}_t)$-progressively measurable process $u(\cdot)$ is called a solution of (5.84)–(5.85) if

$$u \in L^2([T_0, T]; \mathbb{X}) \cap C([T_0, T]; \mathbb{H}) \quad P\text{-a.s.}$$

and, for any $\psi$ from some countable dense subset $\mathbb{D}$ of $\mathbb{X}$,

$$\begin{aligned}(u(t), \psi)_{\mathbb{H}} =& (u_0, \psi)_{\mathbb{H}} \\ &+ \int_{T_0}^t \langle A(s)u(s) + Z(s), \psi\rangle_{(\tilde{\mathbb{X}},\mathbb{X})}\, ds \\ &+ \int_{T_0}^t ((G(s)u(s) + \Psi(s))\, dW_Q(s), \psi)_{\mathbb{H}}, \quad \forall t, \quad P\text{-a.s.}\end{aligned} \tag{5.86}$$

We identify two solutions $u_1(\cdot)$ and $u_2(\cdot)$, if

$$P(\|u_1(t) - u_2(t)\|_{\mathbb{H}} = 0, \quad \forall t \in [T_0, T]) = 1.$$

*Remark 5.1.* Noticing that

$$\int_{T_0}^T \|G(s)u(s)\|_Q^2\, ds \le K \int_{T_0}^T \|u(s)\|_{\mathbb{X}}^2\, ds < \infty, \quad P\text{-a.s.}$$

we confirm that $M(t) := \int_{T_0}^t G(s)u(s)\,dW_Q(s), t \in [T_0, T]$ is a continuous local martingale with

$$\langle M\rangle(t) = \int_{T_0}^t \|G(s)u(s)\|_Q^2\,ds.$$

**Theorem 5.4 (Theorem 3.1.4 in [R90], p. 90).** *Under conditions* $(A_0)$*–*$(A_5)$*, the* SDE *(5.84)–(5.85) has the unique solution* $u(\cdot)$*, such that*

$$u \in L^2([T_0, T] \times \Omega, (\mathcal{F}_t); \mathbb{X}) \cap L^2(\Omega; C([T_0, T]; \mathbb{H}))$$

*and, for any* $\psi \in \mathbb{X}$*,* (5.86) *holds.*

Further, (i) and (ii) below are valid.

(i) ***Energy equality***
For any $(\mathcal{F}_t)$-stopping time $\tau$ taking values in $[T_0, T]$,

$$\begin{aligned}
\|u(t\wedge\tau)\|_{\mathbb{H}}^2 &= \|u_0\|_{\mathbb{H}}^2 + \\
&\int_{T_0}^{t\wedge\tau} \left(2\langle A(s)u(s) + Z(s), u(s)\rangle_{(\tilde{\mathbb{X}},\mathbb{X})} + \|G(s)u(s) + \Phi(s)\|_Q^2\right) ds \\
&+ 2\int_{T_0}^{t\wedge\tau} \Big((G(s)u(s) + \Psi(s))\,dW_Q(s), u(s)\Big)_{\mathbb{H}}, \\
&\qquad \forall t \in [T_0, T], \quad P\text{-a.s.},
\end{aligned} \tag{5.87}$$

(ii) ***Estimate***

$$\begin{aligned}
&E\Big[\sup_{T_0\le t\le T} \|u(t)\|_{\mathbb{H}}^2 + \int_{T_0}^T \|u(t)\|_{\mathbb{X}}^2\,dt\Big] \\
&\le NE\Big[\|u_0\|_{\mathbb{H}}^2 + \int_{T_0}^T (\|Z(s)\|_{\tilde{\mathbb{X}}}^2 + \|\Psi(s)\|_Q^2)\,ds\Big],
\end{aligned} \tag{5.88}$$

where $N$ is a constant depending only on $K, \bar{\lambda}$, and $\bar{\mu}$.

From now on, we are mainly concerned with $L^2(\mathbb{R}^d)$ as $\mathbb{H}$ and introduce the following notations:

$H_0 := L^2(\mathbb{R}^d)$ with the usual inner product $(\ ,\ )$ and the norm $\|\quad\|$.

$H_p := \{\phi \in H_0;$ the generalized derivatives of order $\le p$ belong to $H_0\}$, endowed with the inner product

$$(\phi, \psi)_{H_p} = ((\mathbf{1} - \Delta)^{\frac{p}{2}}\phi, (\mathbf{1} - \Delta)^{\frac{p}{2}}\psi) \tag{5.89}$$

and the norm $\|\phi\|_{H_p} = \|(\mathbf{1} - \Delta)^{\frac{p}{2}}\phi\|$, where $\mathbf{1}$ is the identity map and $\Delta$ is the $d$-dimensional Laplacian operator.

For example, the triple $(H_2, H_1, H_0)$ is normal and is used later in Proposition 5.4.

$$H_{-1} := \{\phi; \phi \text{ is a Borel function such that } (\mathbf{1}-\Delta)^{-\frac{1}{2}}\phi \in H_0\},$$

endowed with the inner product and the norm such that

$$(\phi, \psi)_{H_{-1}} = ((\mathbf{1}-\Delta)^{-\frac{1}{2}}\phi, (\mathbf{1}-\Delta)^{-\frac{1}{2}}\psi) \tag{5.90}$$

and

$$\begin{aligned}\|\phi\|_{H_{-1}} &= \|(\mathbf{1}-\Delta)^{-\frac{1}{2}}\phi\| \\ &= \sup(|(\phi, \psi)|; \psi \in H_1 \text{ with } \|\psi\|_{H_1} = 1).\end{aligned} \tag{5.91}$$

The triple $(H_1, H_0, H_{-1})$ is also normal. For simplicity, we put

$$\begin{aligned}&|||\ \ ||| = \|\ \ \|_{H_1}, \quad \|\ \ \|_* = \|\ \ \|_{H_{-1}}, \quad \mathbb{I}\ \ \mathbb{I} = \|\ \ \|_{H_2}, \\ &\langle\ ,\ \rangle = \langle\ ,\ \rangle_{(H_{-1}, H_1)}, \quad (\!(\ ,\ )\!) = (\ ,\ )_{H_1}.\end{aligned}$$

Then it is easy to see that

$$\begin{aligned}&|\langle y, x\rangle| \le \|y\|_* |||x||| \quad \text{for } y \in H_{-1}, \quad x \in H_1, \\ &\langle y, x\rangle = (y, x) \quad \text{for } y \in H_0, \quad x \in H_1.\end{aligned} \tag{5.92}$$

(Refer to [A03] for $H_p$).

### 5.3.2 Linear Stochastic Parabolic Equations

Let us come back to (5.78). Here we will study equation (5.78) in the Hilbert space framework. But we firstly introduce the notion of generalized solution from the viewpoint of parabolic equations (cf. [R90], p. 130).

Let

$$u_0 \in L^2(\Omega, \mathcal{F}_{T_0}; H_0). \tag{5.93}$$

The $(\mathcal{F}_t)$-progressively measurable $H_1$-process $u(\cdot)$ is called a generalized solution of (5.78) with $u(T_0) = u_0$, if $u \in L^2([T_0, T]; H_1) \cap C([T_0, T]; H_0)$, $P$-a.s., and for any $\phi \in C_K^\infty(\mathbb{R}^d)$,

$$\begin{aligned}&(u(t), \phi) = (u_0, \phi) \\ &\quad - \sum_{i=1}^d \int_{T_0}^t \Big(\sum_{j=1}^d a^{ij}(s, \cdot)\partial_j u(s) + b^i(s, \cdot)u(s), \partial_i \phi\Big)\, ds\end{aligned}$$

$$+\int_{T_0}^{t}((\rho(s,\cdot)u(s),\phi)+\langle\mathfrak{z}(s,\cdot),\phi\rangle)\,ds$$

$$+\sum_k\sqrt{\lambda_k}\int_{T_0}^{t}\Big(\sum_{j=1}^{d}g_k^j(s,\cdot)\partial_j u(s)+\hbar_k(s,\cdot)u(s)+f_k(s,\cdot),\phi\Big)\,dW^k(s),$$

$$\forall t\in[T_0,T],\quad P\text{-a.s.,} \tag{5.94}$$

where $\lambda_1\geq\lambda_2\geq\cdots\geq 0$ and $\lambda^*=\sum_k\lambda_k\in(0,\infty)$. Two generalized solutions $u$ and $\tilde{u}$ are identified, if

$$P(\|u(t)-\tilde{u}(t)\|=0,\quad\forall t\in[T_0,T])=1. \tag{5.95}$$

We assume the following three conditions;

$(B_1)$ $z=a^{ij},b^i,g_k^i,\hbar_k,\rho,i,j=1,\ldots,d,k=1,2,\ldots$ satisfy

- (a) $\mathcal{B}([0,T]\times\mathbb{R}^d)\times\mathcal{F}/\mathcal{B}(\mathbb{R}^1)$-measurable.
- (b) For any $x\in\mathbb{R}^d$, $z(\cdot,x)$ is$(\mathcal{F}_t)$-progressively measurable.
- (c) ***Boundedness***

  There is a constant $K$, such that

$$\sup_{t,x}|z(t,x)|\leq K,\quad P\text{-a.s.}. \tag{5.96}$$

$(B_2)$ ***Superparabolicity***

$$\wedge(t,x):=\Big(2a^{ij}(t,x)-\sum_k\lambda_k g_k^i(t,x)g_k^j(t,x)\Big)_{ij=1,\ldots,d}$$

is symmetric and uniformly positive definite:

$$y^\top\wedge(t,x)y\geq\lambda_0(y)^2,\quad\forall y\in\mathbb{R}^d,\quad\forall t,x,\quad P\text{-a.s.,} \tag{5.97}$$

with a positive constant $\lambda_0$.

$(B_3)$ $\mathfrak{z}$ and $f_k$ $(k=1,2,\ldots)$ satisfy

$$\mathfrak{z}\in L^2([0,T]\times\Omega,(\mathcal{F}_t);H_{-1}),$$

$$f_k\in L^2([0,T]\times\Omega,(\mathcal{F}_t);H_0)\text{ with }\sum_k\lambda_k E\int_0^T\|f_k(s)\|^2\,ds<\infty.$$

We will consider (5.78) on $(H_1,H_0,H_{-1})$. Let $e_i, i=1,2,\ldots$ be an ONB in $H_0$, and put

$$W_Q(t)=\sum_k\sqrt{\lambda_k}W^k(t)e_k.$$

Next let us define $A(t,\omega)\in L(H_1;H_{-1}), G_k(t,\omega)\in L(H_1;H_0)$,

$G(t,\omega) \in L(H_1; \mathcal{L}_Q^2(H_0; H_0))$ and $\Psi \in L^2([0,T] \times \Omega, (\mathcal{F}_t); \mathcal{L}_Q^2(H_0; H_0))$ by

$$A(t,\omega)\phi = \sum_{i=1}^{d} \partial_i \Big( \sum_{i=1}^{d} a^{ij}(t,x,\omega)\partial_j \phi + b^i(t,x,\omega)\phi \Big) + \rho(t,x,\omega)\phi, \tag{5.98}$$

$$G_k(t,\omega)\phi = \sum_{j=1}^{d} g_k^i(t,x,\omega)\partial_j \phi + \hbar_k(t,x,\omega)\phi, \tag{5.99}$$

$$G(t,\omega)\phi e_k = G_k(t,\omega)\phi, \tag{5.100}$$

and

$$\Psi(t,\omega)e_k = f_k(t,x,\omega) \tag{5.101}$$

for $\phi \in H_1$.

Thus, (5.78) can be recast as the following SDE on $(H_1, H_0, H_{-1})$:

$$du(t) = (A(t)u(t) + \mathfrak{z}(t))\,dt + (G(t)u(t) + \Psi(t))\,dW_Q. \tag{5.102}$$

Since $C_K^\infty(\mathbb{R}^d)$ is dense in $H_1$ and since (5.96) and $(B_2)$ imply $(A_1)$ and $(A_2)$, Theorem 5.4 leads to the following

**Theorem 5.5.** *Under conditions* $(B_1)$–$(B_3)$, *Eq.* (5.78) *has a unique generalized solution* $u(\cdot) \in L^2([T_0,T]\times\Omega, (\mathcal{F}_t); H_1) \cap L^2(\Omega; C([T_0,T]; H_0))$ *with* $u(T_0) = u_0$. *Moreover,* (i) *and* (ii) *hold:*

(i) ***Energy equality***
*For any* $(\mathcal{F}_t)$*-stopping time* $\tau$ *taking values in* $[T_0,T]$,

$$\begin{aligned}\|u(t\wedge\tau)\|^2 =& \|u_0\|^2 \\ &+ \int_{T_0}^{t\wedge\tau} (2\langle A(s)u(s) + \mathfrak{z}(s), u(s)\rangle + \|G(s)u(s) + \Psi(s)\|_Q^2)\,ds \\ &+ 2\int_{T_0}^{t\wedge\tau} ((G(s)u(s) + \Psi(s))\,dW_Q(s), u(s)), \\ &\qquad \forall t \in [T_0,T], \quad P\text{-a.s.}.\end{aligned}$$

(ii) ***Estimate***

$$\begin{aligned}&E\Big[\sup_{T_0\le S\le T} \|u(s)\|^2 + \int_{T_0}^{T} |||u(s)|||^2\,ds\Big] \\ \le& NE\Big[\|u_0\|^2 + \int_{T_0}^{T} (\|\mathfrak{z}(s)\|_*^2 + \|\Psi(s)\|_Q^2)\,ds\Big],\end{aligned} \tag{5.103}$$

*where* $N$ *is a constant depending only on* $K$ *and* $\lambda_0$.

### 5.3.3 Regularities of Solutions

Let us introduce the following two conditions $(B_4)$ and $(B_5)$;

$(B_4)$ There is a constant $K_1$, such that

$$\begin{cases} \sup_t \|z(t,\cdot)\|_{C^2} \le K_1, & P\text{-a.s.} \quad \text{for } z = a^{ij}, b_i, \\ \sup_t \|z(t,\cdot)\|_{C^1} \le K_1, & P\text{-a.s.} \quad \text{for } z = g_k^i, \hbar_k. \end{cases} \tag{5.104}$$

$(B_5)$ (a) $\mathfrak{z} \in L^2([0,T] \times \Omega, (\mathcal{F}_t); H_0)$,
(b) $f_k \in L^2([0,T] \times \Omega, (\mathcal{F}_t); H_1), k = 1, 2, \ldots$
with $\sum \lambda_k E \int_0^T |||f_k(s)|||^2 \, ds < \infty$.

Let $\eta_k, k = 1, 2, \ldots$ be an ONB in $H_1$ and define $G$ and $\Psi$ by (5.100) and (5.101) with $\eta_k$ instead of $e_k$. Then (5.104) provides a constant $K_2$, such that for any $\phi \in H_2$,

$$\|A(t)\phi\|^2 + |||G(t)\phi|||_Q^2 \le K_2 \mathbb{I}\phi\mathbb{I}^2, \quad \forall t, \quad P\text{-a.s.} \tag{5.105}$$

Further $(B_2)$ and $(B_4)$ yield the coerciveness on $(H_2, H_1, H_0)$, namely there is a constant $\mu_0 (\in \mathbb{R}^1)$ depending only on $K_1, \lambda_0$, and $\lambda^*$, such that for any $\phi \in H_2$,

$$\begin{aligned} &2\langle A(t)\phi, \phi\rangle_{(H_0 H_2)} + |||G(t)\phi|||_Q^2 \\ &\le -\frac{\lambda_0}{2}\mathbb{I}\phi\mathbb{I} + \mu_0 |||\phi|||^2, \quad \forall t, \quad P\text{-a.s.} \end{aligned} \tag{5.106}$$

Indeed, when $\phi \in H_3$, standard computations lead to following relations:

$$\begin{aligned} \langle \partial_i (a^{ij} \partial_j \phi), \phi\rangle_{(H_0 H_2)} &= (\!(\partial_i (a^{ij} \partial_j \phi), \phi)\!) \\ &= (\partial_i (a^{ij} \partial_j \phi), (\mathbf{1} - \Delta)\phi) \\ &= -(a^{ij} \partial_j \phi, (\mathbf{1} - \Delta)\partial_i \phi), \\ \Big|\Big|\Big| \sum_i g_k^i \partial_i \phi \Big|\Big|\Big|^2 &= \Big(\sum_i g_k^i \partial_i \phi, (\mathbf{1} - \Delta) \sum_j g_k^j \partial_j \phi\Big). \end{aligned}$$

For any $\varepsilon > 0$, there is a constant $K_\varepsilon$, independent of $\phi$, such that

$$\begin{aligned} (a^{ij} \partial_j \phi, \partial_{ll} \partial_i \phi) &= -(a^{ij} \partial_j (\partial_l \phi), \partial_i (\partial_l \phi)) - (\partial_l a^{ij} \partial_j \phi, \partial_l \partial_i \phi) \\ &\le \text{1st term} + \varepsilon \mathbb{I}\phi\mathbb{I}^2 + K_\varepsilon |||\phi|||^2, \\ (g_k^i \partial_i \phi, \partial_{ll} (g_k^j \partial_j \phi)) &\le -(g_k^i \partial_i (\partial_l \phi), g_k^j \partial_j (\partial_l \phi)) + \varepsilon \mathbb{I}\phi\mathbb{I}^2 + K_\varepsilon |||\phi|||^2. \end{aligned}$$

We apply the same calculations to other terms in the LHS of (5.106). Then letting $\varepsilon$ small, depending only on $\lambda_0$ and $K_1$, and putting all estimates together, we conclude (5.106) for $\phi \in H_3$.

For $\phi \in H_2$, we can take $\psi_n \in H_3$, so that $\lim_{n\to\infty} \mathbb{I}\phi - \psi_n\mathbb{I} = 0$. Hence, $(B_4)$ and (5.105) yield

$$|\langle A(t)\psi_n, \psi_n\rangle_{(H_0H_2)} - \langle A(t)\phi, \phi\rangle_{(H_0H_2)}| + |||G(t)(\psi_n - \phi)|||_Q^2 \longrightarrow 0,\ \forall t,\ P\text{-a.s.}$$

This completes the proof of (5.106).

**Proposition 5.4.** *Suppose $u_0 \in L^2(\Omega, \mathcal{F}_{T_0}; H_1)$. Under conditions $(B_1)$, $(B_2)$, $(B_4)$ and $(B_5)$, the unique generalized solution $u(\cdot)$ of (5.78) with $u(T_0) = u_0$ is in $L^2([T_0, T] \times \Omega, (\mathcal{F}_t); H_2) \cap L^2(\Omega; C([T_0, T]; H_1))$.*

*Moreover, one has the estimate*

$$\begin{aligned} &E\Big[\sup_{T_0\le t\le T} |||u(t)|||^2 + \int_{T_0}^T \mathbb{I}u(t)\mathbb{I}^2\, dt\Big] \\ \le &NE\Big[|||u_0|||^2 + \int_{T_0}^T (\|\mathfrak{z}(s)\|^2 + |||\Psi(s)|||_Q^2)\, ds\Big] \end{aligned} \tag{5.107}$$

*with a constant $N$ depending only on $K_1$, $\lambda_0$, and $\lambda^*$.*

*Proof.* We use the arguments similar to those in the proof of Theorem 5.5.

Let $\eta_k, k = 1, 2, \ldots$ be ONB in $H_1$ and put

$$W_Q(t) = \sum_k \sqrt{\lambda} W^k(t)\eta_k.$$

We consider on $(H_2, H_1, H_0)$ the SDE

$$dv(t) = (A(t)v(t) + \mathfrak{z}(t))\, dt + (G(t)v(t) + \Psi(t))\, dW_Q(t), \quad t \in (T_0, T], \tag{5.108}$$

with the initial condition

$$v(T_0) = u_0. \tag{5.109}$$

Then Theorem 5.4 yields the unique solution of problem (5.108)–(5.109) in the space $L^2([T_0, T] \times \Omega, (\mathcal{F}_t); H_2) \cap L^2(\Omega; C([T_0, T]; H_1))$. Hence, for any $\psi \in H_2$,

$$\begin{aligned} ((v(t), \psi)) =&((u_0, \psi)) \\ &+ \int_{T_0}^t \langle A(s)v(s) + \mathfrak{z}(s), \psi\rangle_{(H_0H_2)}\, ds \\ &+ \sum_k \sqrt{\lambda_k} \int_{T_0}^t ((G_k(t)v(t) + f_k(t), \psi))\, dW^k(t), \quad t \in (T_0, T]. \end{aligned} \tag{5.110}$$

For $\phi \in C_k^\infty(\mathbb{R}^d)$, we take $\psi = (\mathbf{1} - \Delta)^{-1}\phi$. Then by (5.110),

$$(v(t), \phi) = (u_0, \phi) + \int_{T_0}^t \langle A(s)v(s) + \mathfrak{z}(s), \phi\rangle \, ds + \sum_k \sqrt{\lambda_k} \int_{T_0}^t (G_k(s)v(s) + f_k(s), \phi) \, dW^k(s).$$

Consequently, $v(\cdot)$ coincides with the unique generalized solution of (5.78) with $u(T_0) = u_0$.

This completes the proof. □

For the dependence on the time parameter, we have;

**Proposition 5.5.** *Suppose* $\mathfrak{z} = 0$ *and* $f_k = 0, k = 1, 2, \ldots$. *Let* $u(\cdot)$ *be the unique generalized solution of* (5.78) *with* $u(T_0) = u_0$.

(i) *Let* $u_0 \in L^2(\Omega, \mathcal{F}_{T_0}; H_1)$ *and* $\theta, t \in [T_0, T]$. *Under* $(B_1)$, $(B_2)$ *and* $(B_4)$,

$$E\Big[\sup_{\theta \le s \le t} \|u(s) - u(\theta)\|^2\Big] \le N_1 |t - \theta| E\,|||u_0|||^2, \tag{5.111}$$

*with a constant* $N_1$ *depending only on* $\lambda_0$, $\lambda^*$, *and* $K_1$.

(ii) *Let* $u_0 \in L^2(\Omega, \mathcal{F}_{T_0}; H_2)$ *and* $\theta, t \in [T_0, T]$. *Besides* $(B_1)$ *and* $(B_2)$, *we assume a stronger condition*

$(B_6)$ *There is a constant* $K_3$, *such that*

$$\sup_t \|z(t, \cdot)\|_{C^3} \le K_3, \quad P\text{-a.s.} \quad \text{for } z = a^{ij}, b^i,$$

$$\sup_t \|z(t, \cdot)\|_{C^2} \le K_3, \quad P\text{-a.s.} \quad \text{for } z = g_k^i, \hbar^k.$$

*Then*

$$E\Big[\sup_{\theta \le s \le t} |||u(s) - u(\theta)|||^2\Big] \le N_2 (t - \theta) E\,\mathbb{I}u_0\mathbb{I}^2, \tag{5.112}$$

*with a constant* $N_2$ *depending only on* $\lambda_0$, $\lambda^*$, *and* $K_3$.

*Proof.* (i) From Proposition 5.4, it follows that $u(\theta)$ is in $L^2(\Omega, \mathcal{F}_\theta; H_1)$ and $v(s) := u(s) - u(\theta), s \in [\theta, T]$, satisfies on $(H_1, H_0, H_{-1})$ the SDE

$$\begin{cases} dv(s) = (A(s)v(s) + A(s)u(\theta))\, ds \\ \qquad\qquad + (G(s)v(s) + G(s)u(\theta))\, dW_\theta(s), \quad s \in (\theta, T], \\ v(\theta) = 0. \end{cases} \tag{5.113}$$

Further, $\mathfrak{z}(\cdot) = A(\cdot)u(\theta)$ and $\Psi(\cdot) = G(\cdot)u(\theta)$ are in $L^2([\theta, T] \times \Omega, (\mathcal{F}_t); H_{-1})$ and $L^2([\theta, T] \times \Omega, (\mathcal{F}_t); \mathcal{L}_Q^2(H_0; H_0))$, respectively. $(B_1)$ yields

$$\|\mathfrak{z}(s)\|_*^2 + \|\Psi\|_Q^2 \leq K_0 |||u(\theta)|||^2, \quad \forall s \in [\theta, T], \quad P\text{-a.s.}, \tag{5.114}$$

with a constant $K_0$ depending only on $K$ of (5.96). Now (5.103) and (5.114) yield (5.111).

Since (ii) mimics (i), this completes the proof of Proposition 5.5. □

### 5.3.4 Semilinear Stochastic Parabolic Equations with Lipschitz Nonlinearity

Let us change $\mathfrak{z}(t, x, \omega)$ of (5.78) to $\mathfrak{z}(t, x, u, v, \omega)$ with a Lipschitz condition in the variables $u$ and $v$. Precisely, we assume

$(B_7)$ $\mathfrak{z} : [0, T] \times \mathbb{R}^d \times \mathbb{R}^1 \times \mathbb{R}^d \times \Omega \mapsto \mathbb{R}^1$ is a measurable map, satisfying

(a) $\mathfrak{z}(\cdot, x, u, v)$ is $(\mathcal{F}_t)$-progressively measurable for any $x, u, v$.

(b) Set $\mathfrak{z}_0(t, x, \omega) = \mathfrak{z}(t, x, 0, 0, \omega)$ and $\mathfrak{z}_1(t, x, u, v, \omega) = \mathfrak{z}(t, x, u, v, \omega) - \mathfrak{z}_0(t, x, \omega)$. There is a positive constant $l_0$, such that

$$\begin{aligned} &|\mathfrak{z}_1(t, x, u_1, v_1) - \mathfrak{z}_1(t, x, u_2, v_2)|^2 \\ \leq &\, l_0(|u_1 - u_2|^2 + |v_1 - v_2|^2), \quad \forall t, x, \quad P\text{-a.s.}. \end{aligned} \tag{5.115}$$

(c) $\mathfrak{z}_0$ is in $L^2([\theta, T] \times \Omega, (\mathcal{F}_t); H_0)$.

Now we consider the semilinear stochastic parabolic equation

$$\begin{aligned} du(t, x) =& (A(t)u(t, x) + \mathfrak{z}(t, x, u(t, x), \partial_x u(t, x)))\, dt \\ &+ \sum_k \sqrt{\lambda_k}(G_k(t)u(t, x) + f_k(t, x))\, dW^k(t), \end{aligned} \tag{5.116}$$

where $A(t)$ and $G_k(t)$ are given by (5.98) and (5.99), respectively.

The definition of generalized solution for (5.116) is given in the same way as (5.94), by replacing $\langle \mathfrak{z}(s, \cdot), \phi \rangle$ with $\langle \mathfrak{z}(s, \cdot, u(s), \partial_x u(s)), \phi \rangle$.

Let us introduce the SDE on $(H_1, H_0, H_{-1})$ associated with (5.116), whose solution provides the generalized solution of (5.116) (see (5.118)).

Using (b) and (c), we can define $Z : [0, T] \times H_1 \times \Omega \mapsto H_0$ by

$$\begin{aligned} Z(t, h, \omega)(x) &= \mathfrak{z}(t, x, h(x), \partial_x h(x), \omega) \\ &= \mathfrak{z}_1(t, x, h(x), \partial_x h(x), \omega) + \mathfrak{z}_0(t, x, \omega) \quad \text{for } h \in H_1. \end{aligned} \tag{5.117}$$

Hence, for $\xi \in L^2([0,T]\times\Omega,(\mathcal{F}_t);H_1)$, $\eta(t) := Z(t,\xi(t))$ is defined as an element of $L^2([0,T]\times\Omega,(\mathcal{F}_t);H_0)$ by (b) and (c), and

$$E\int_0^T \|Z(t,\xi_1(t)) - Z(t,\xi_2(t))\|^2\,dt \le l_0 E\int_0^T |||\xi_1(t)-\xi_2(t)|||^2\,dt$$

for $\xi_1,\xi_2 \in L^2([0,T]\times\Omega,(\mathcal{F}_t);H_1)$.

This subsection is devoted to studying the similinear SDE on $(H_2,H_1,H_0)$

$$\begin{aligned} du(t) &= (A(t)u(t) + Z(t,u(t)))\,dt \\ &\quad + (G(t)u(t)+\Psi(t))\,dW_Q(t), \quad t\in[T_0,T], \end{aligned} \tag{5.118}$$

that is

$$\begin{aligned} du(t) =&\Big\{\sum_{i=1}^d \partial_i\Big(\sum_{j=1}^d a^{ij}(t,x)\partial_j u(t) + b^i(t,x)u(t)\Big) \\ &+ \rho(t,x)u(t) + \mathfrak{z}(t,x,u(t,x),\partial_x u(t,x))\Big\}\,dt \\ &+\sum_k \sqrt{\lambda_k}\Big\{\sum_{i=1}^d g_k^j(t,x)\partial_j u(t) + \hbar_k(t,x)u(t) + f_k(t,x)\Big\}\,dW^k(t), \end{aligned}$$

with the initial condition

$$u(T_0) = u_0\ (\in L^2(\Omega,\mathcal{F}_{T_0};H_1)). \tag{5.119}$$

Now we aim to prove;

**Proposition 5.6.** *Under* $(B_1)$, $(B_2)$, $(B_4)$, $(B_7)$ *and* $(B_5)$(b), *problem* (5.71)–(5.72) *has a unique solution* $u(\cdot)$ *in* $L^2([T_0,T]\times\Omega,(\mathcal{F}_t);H_2)\cap L^2(\Omega;C([T_0,T];H_1))$. *Moreover, the following estimate holds*

$$\begin{aligned} &E\Big[\sup_{T_0\le t\le T}|||u(t)|||^2 + \int_{T_0}^T \mathbb{I}u(s)\mathbb{I}^2\,ds\Big] \\ \le&\tilde{N}\Big(E[|||u(0)|||^2] + E\int_{T_0}^T (\|\mathfrak{z}_0(t)\|^2 + |||\Psi(t)|||_Q^2)\,dt\Big), \end{aligned} \tag{5.120}$$

*where the constant* $\tilde{N}$ *depends only on* $\lambda_0$, $K_1$ *and* $l_0$.

*Proof.* We divide the proof into four steps.

**Step 1.** ***Approximate Solutions*** We construct a solution by using the successive approximation. Since the method is standard, we will only sketch the proof.

Put $u_0(t) = u_0$ and define $u_1(\cdot)$ as the solution of the SDE

$$\begin{cases} du_1(t) = (A(t)u_1(t) + Z(t, u_0(t)))\,dt \\ \qquad\qquad + (G(t)u_1(t) + \Psi(t))\,dW_Q(t), \quad t \in (T_0, T], \\ u_1(T_0) = u_0. \end{cases} \tag{5.121}$$

Since $Z(t, u_0(t))$ is an element of $L^2([T_0, T] \times \Omega, (\mathcal{F}_t); H_0)$ with

$$E\int_0^T \|Z(t, u_0(t))\|^2\,dt \le 2E\Big[\int_{T_0}^T \|\mathfrak{z}_0(t)\|^2\,dt + l_0|||u_0|||^2(T - T_0)\Big], \tag{5.122}$$

(5.121) admitsthe unique solution $u_1(\cdot)$ in $L^2([T_0, T] \times \Omega, (\mathcal{F}_t); H_2) \cap L^2(\Omega; C([T_0, T]; H_1)$, by Proposition 5.4.Now suppose $u_{n-1}(\cdot) \in L^2([T_0, T] \times \Omega, (\mathcal{F}_t); H_2) \cap L^2(\Omega; C([T_0, T]; H_1))$ is the solution of the approximate problem of index $n - 1$. We define $u_n(\cdot)$ as the solution of the SDE

$$\begin{cases} du_n(t) = (A(t)u_n(t) + Z(t, u_{n-1}(t)))\,dt \\ \qquad\qquad + (G(t)u_n(t) + \Psi(t))\,dW_Q(t), \quad t \in (T_0, T], \\ u_n(T_0) = u_0. \end{cases} \tag{5.123}$$

Again observing that

$$\begin{aligned} &E\int_{T_0}^T \|Z(t, u_{n-1}(t))\|^2\,dt \\ &\le 2E\Big[\int_{T_0}^T (\|\mathfrak{z}_0(t)\|^2 + l_0|||u_{n-1}(t)|||^2)\,dt\Big] < \infty, \end{aligned} \tag{5.124}$$

we obtain the unique solution of (5.123) in $L^2([T_0, T] \times \Omega, (\mathcal{F}_t); H_2) \cap L^2(\Omega; C([T_0, T]; H_1))$, satisfying

$$\begin{aligned} \mu_n(t) :=& E\Big[\sup_{T_0 \le s \le t} |||u_n(s)|||^2 + \int_{T_0}^t \mathbb{I}u_n(s)\mathbb{I}^2\,ds\Big] \\ \le& NE\Big[|||u_0|||^2 + \int_{T_0}^t (\|Z(s, u_{n-1}(s))\|^2 + |||\Psi(s)|||_Q^2)\,ds\Big] \\ \le& NE|||u_0|||^2 + NE\int_{T_0}^t (2\|\mathfrak{z}_0(s)\|^2 + |||\Psi(s)|||_Q^2)\,ds \\ &+ 2Nl_0\int_{T_0}^t \mu_{n-1}(s)\,ds, \end{aligned} \tag{5.125}$$

with a constant $N$ depending only on $K_1$, $\lambda_0$, and $\lambda^*$.

Applying Gronwall's inequality, we obtain

$$\mu_n(t) \le NE|||u_0|||^2 \sum_{j=1}^{n-1} \frac{(C_0(t-T_0))^{j-1}}{(j-1)!}$$
$$+ N \int_{T_0}^t E[2\|\mathfrak{z}_0(s)\|^2 + |||\Psi(s)|||_Q^2] \sum_{j=1}^{n-1} \frac{(C_0(t-s))^{j-1}}{(j-1)!} ds$$
$$+ \frac{(C_0(t-T_0))^n}{n!} E|||u_0|||^2, \tag{5.126}$$

where $C_0 = 2Nl_0$.

**Step 2.** ***Convergence of*** $u_n(\cdot)$
Put $\omega_1 = u_1$ and $\omega_{n+1} = u_{n+1} - u_n, n = 1, 2, \ldots$. Then $\omega_{n+1}$ satisfies

$$\begin{cases} d\omega_{n+1}(t) = (A(t)\omega_{n+1}(t) + Z(t, u_n(t)) - Z(t, u_{n-1}(t)))\, dt \\ \qquad\qquad + G(t)\omega_{n+1}(t)\, dW_Q(t), \quad t \in (T_0, T], \\ \omega_{n+1}(T_0) = 0. \end{cases} \tag{5.127}$$

Since (5.107) and (5.115) yield

$$\rho_{n+1}(t) := E\Big[\sup_{T_0 \le s \le t} |||\omega_{n+1}(s)|||^2 + \int_{T_0}^t \mathbb{I}\omega_{n+1}(s)\mathbb{I}^2\, ds\Big]$$
$$\le Nl_0 \int_{T_0}^t \rho_n(s)\, ds, \tag{5.128}$$

we have

$$\rho_{n+1}(t) \le \frac{(C_0(t-T_0))^{n+1}}{(n+1)!} \sup_{T_0 \le s \le t} \rho_1(s). \tag{5.129}$$

Therefore,

$$E\Big[\sum_n \sup_{T_0 \le s \le T} |||\omega_n(s)||| + \int_{T_0}^T \sum_n \mathbb{I}\omega_n(s)\mathbb{I}\, ds\Big] < \infty. \tag{5.130}$$

Since $\sup_{T_0 \le s \le T} \sum_n |||\omega_n(s)||| < \infty$, $P$-a.s., we deduce that $u_n(\cdot) = \sum_{k=1}^n \omega_k(\cdot)$ converges to $u(\cdot) := \sum_k \omega_k(\cdot)$ in $C([T_0, T]; H_1)$, $P$-a.s.

On the other hand, $u_n(\cdot)$ converges to some $\tilde{u}(\cdot)$ in $L^1([T_0, T]; H_2)$, $P$-a.s., by (5.130). Therefore, $u_n(\cdot)$ converges to $\tilde{u}(\cdot)$ in $L^1([T_0, T]; H_1)$, $P$-a.s., and so $u(t) = \tilde{u}(t)$, a.e. $t \in [T_0, T]$, $P$-a.s. By using (5.126) and the Fatou's lemma, we conclude that

$$u \in L^2([T_0, T] \times \Omega, (\mathcal{F}_t); H_2) \cap L^2(\Omega; C([T_0, T]; H_1)). \tag{5.131}$$

**Step 3.** ***$u(\cdot)$ satisfies (5.118)***

For $\psi \in H_2$, we have the following estimates:

$$E\Big[\sup_{T_0\le t\le T}\Big|\int_{T_0}^{t}\langle A(s)(u_n(s)-u(s)),\psi\rangle_{(H_0,H_2)}\,ds\Big|\Big]$$
$$\le K\|\psi\|E\int_{T_0}^{T}\|u_n(s)-u(s)\|\,ds, \tag{5.132}$$
$$E\Big[\sup_{T_0\le t\le T}\Big|\int_{T_0}^{t}\langle Z(s,u_n(s))-Z(s,u(s)),\psi)\rangle_{(H_0,H_2)}\,ds\Big|\Big]$$
$$\le\sqrt{l_0}\|\psi\|E\int_{T_0}^{T}|||u_n(s)-u(s)|||\,ds. \tag{5.133}$$

For the stochastic integral term, the Burkholder–Davis–Gundy inequality yields

$$E\Big[\sup_{T_0\le t\le T}\Big|\int_{T_0}^{t}((G(s)(u_n(s)-u(s))\,dW_Q,\psi))\Big|\Big]$$
$$\le 6E\Big(\int_{T_0}^{T}((G(s)(u_n(s)-u(s)),\psi))_Q^2\,ds\Big)^{\frac{1}{2}}$$
$$\le 6K|||\psi|||\sqrt{E\int_{T_0}^{T}\|u_n(s)-u(s)\|^2\,ds}. \tag{5.134}$$

Combining (5.132)–(5.134) and letting $n\to\infty$, we obtain (5.118).

**Step 4.** ***Uniqueness and (5.120)***

Let $v(\cdot)$ be a solution and put $\omega(t)=u(t)-v(t)$. Since $\omega(\cdot)$ satisfies

$$d\omega(t)=(A(t)\omega(t)+Z(t,u(t))-Z(t,v(t)))dt+G(t)\omega(t)dW_Q(t),\quad t\in(T_0,T] \tag{5.135}$$

with the initial condition

$$\omega(T_0)=0, \tag{5.136}$$

(5.107) shows that $\rho(t):=E[\sup_{T_0\le s\le t}|||\omega(s)|||^2]$ satisfies

$$\rho(t)\le NE\int_{T_0}^{t}\|Z(s,u(s))-Z(s,v(s))\|^2\le Nl_0\int_{T_0}^{t}\rho(s)ds. \tag{5.137}$$

Now Gronwall's inequality yields

$$\rho(t)=0,\quad t\in[T_0,T],$$

which establishes the uniqueness of solution.

Regarding (5.120) for $v(\cdot)$, we put

$$K(t) = E\Big[\sup_{T_0 \le s \le t} |||v(s)|||^2 + \int_{T_0}^{t} \mathbb{I}v(s)\mathbb{I}^2\, ds\Big].$$

Again (5.107) and $(B_7)$ show that

$$\begin{aligned} K(t) &\le NE[|||u_0|||^2 + \int_{T_0}^{t} (\|Z(s, v(s)\|^2 + |||\psi|||_Q^2)\, ds] \\ &\le c_1\Big(E[|||u_0|||^2 + \int_{T_0}^{t} \|\mathfrak{z}_0^2(s)\| + |||\psi|||_Q^2) ds] + \int_{T_0}^{t} K(s)\, ds\Big), \end{aligned} \tag{5.138}$$

with a constant $c_1$ independent of $t$ and $v(\cdot)$. Now Gronwall's inequality yields (5.120).

The proof is completed. □

If $\lambda_1 = \lambda_2 = \cdots = \lambda_m = 1$ and $\lambda_{m+1} = 0$, then $W_Q$ is an $m-$dimensional Wiener process and we write $\| \|_m$ instead of $\| \|_Q$.

*Example 5.5 (HJB equation with noise).* Let us consider a time-homogeneous HJB equation affected by a Wiener noise;

$$\begin{aligned} du(t,x) =&\Big\{\sum_{i=1}^{d} \partial_i\Big(\sum_{j=1}^{d} a^{ij}(x)\partial_j u(t,x)\Big) \\ &+ \sup_{\gamma\in\Gamma}\Big(\sum_{i=1}^{d} b^i(x,\gamma)\partial_i u(t,x) + \rho(x,\gamma)u(t,x) + \mathfrak{z}(x,\gamma)\Big)\Big\}\, dt \\ &+ \sum_{k=1}^{m}\Big(\sum_{i=1}^{d} g_k^i(x)\partial_i u(t,x) + \hbar_k(x)u(t,x) + f_k(x)\Big)\, dW^k(t), \\ & t \in (0,T], \quad x \in \mathbb{R}^d, \end{aligned} \tag{5.139}$$

with the initial condition

$$u(0,x) = \phi(x), \quad x \in \mathbb{R}^d, \tag{5.140}$$

where $\phi \in H_1$.

We assume that

(a) $\Gamma$ is convex and compact,
(b) All coefficients are time-homogeneous and bounded continuous non-random functions,
(c) $a^{ij} \in C_b^2(\mathbb{R}^d), g_k^i, \hbar_k \in C_b^1(\mathbb{R}^d), ij = 1,\ldots,d, k = 1,\ldots,m,$

(d) ***Superellipticity***

$$\Big(2a^{ij}(x) - \sum_{k=1}^{m} g_k^i(x) g_k^j(x)\Big)_{ij=1,\dots,d}$$

is uniformly positive definite,

(e) $f_k \in H_1, k = 1, \dots, m$, and there exists $\hat{\mathfrak{z}} \in H_0$, such that $|\mathfrak{z}(x,\gamma)| \leq \hat{\mathfrak{z}}(x), \forall x, \gamma$.

Put

$$\mathfrak{z}(x,u,v) = \sup_{\gamma \in \Gamma} \Big(\sum_{i=1}^{d} b^i(x,\gamma) v_i + \rho(x,\gamma) u + \mathfrak{z}(x,\gamma)\Big).$$

Then the SDE on $(H_2, H_1, H_0)$, associated with (5.139)–(5.140) has a unique solution, which provides the unique generalized solution of (5.139)–(5.140).

Stochastic HJB equations with lateral condition can be treated via backward SDE arguments. Refer to [Pe92, BM07] for details.

## 5.4 Itô's Formula

This section is devoted to Itô's formula for $\mathbb{H}$-valued semimartingale, in particular to the solution of linear stochastic parabolic equation (5.102). First we consider a continuous semimartingale $X$ given by

$$X(t) = X_0 + \int_0^t \xi(s)\, ds + M(t), \tag{5.141}$$

where $\xi \in L^2([0,T] \times \Omega, (\mathcal{F}_t); \mathbb{H})$, $X_0 \in L^2(\Omega, \mathcal{F}_0; \mathbb{H})$, $M \in \mathbb{M}^{2c}([0,T], (\mathcal{F}_t); \mathbb{H})$. In Sect. 5.4.2, we will show that, for $F \in C^{12}([0,T] \times \mathbb{H})$ satisfying local boundedness condition $(F_0)$, Itô's formula holds:

$$\begin{aligned} F(t\wedge\tau, X(t\wedge\tau)) &= F(0, X_0) \\ &+ \int_0^{t\wedge\tau} \Big\{\partial_s F(s, X(s)) + (\xi(s), DF(s, X(s)))_{\mathbb{H}}\Big\}\, ds \\ &+ \int_0^{t\wedge\tau} (DF(s, X(s)), dM(s))_{\mathbb{H}} \\ &+ \frac{1}{2}\int_0^{t\wedge\tau} \mathrm{tr}(D^2F(s, X(s)) q_M(s))\, d\langle M\rangle(s), \ \forall t, \quad P\text{-a.s.} \end{aligned} \tag{5.142}$$

for any $(\mathcal{F}_t)$-stopping time $\tau$ taking values in $[0, T]$. In the differential form,

$$dF(t, X(t)) = \{\partial_t F(t, X(t)) + (\xi(t), DF(t, X(t)))_{\mathbb{H}}\} \, dt \\ + (DF(t, X(t)), dM(t))_{\mathbb{H}} + \frac{1}{2}\mathrm{tr}(D^2 F(t, X(t)) q_M(t)) \, d\langle M\rangle(t).$$

In Sect. 5.4.3, we study Itô's formula for the solution of (5.102). In this case, we impose stronger conditions on $F$, to deal with the unbounded operator $A(t)$ (see $(F_1)$ in Sect. 5.4.3). When $F(t, \phi) = \|\phi\|^2$, Itô's formula leads to the energy equality. We also estimate the negative norm of the solution by using Itô's formula.

## 5.4.1 Preliminaries

This subsection lists basic definitions and results that will be needed in the sequel. We state linear- and bilinear maps, Hilbert–Schmidt operators, nuclear operators and first and second derivatives.

Proofs are given in accessible books, cf. [La83, Y80].

**1. Linear- and bilinear maps**

Let $\mathbb{X}, \mathbb{Y}$, and $\mathbb{Z}$ be Hilbert spaces. $L(\mathbb{Y}; \mathbb{Z})$ denotes the Banach space of continuous linear mappings from $\mathbb{Y}$ into $\mathbb{Z}$, endowed with the norm $\| \quad \|_{L(\mathbb{Y};\mathbb{Z})}$ given by

$$\|\Phi\|_{L(\mathbb{Y};\mathbb{Z})} = \sup_{\|y\|_{\mathbb{Y}}=1} \|\Phi y\|_{\mathbb{Z}}.$$

A map $\Psi : \mathbb{X} \times \mathbb{Y} \mapsto \mathbb{Z}$ is said to be bilinear, if $\Psi(x, \cdot) \in L(\mathbb{Y}; \mathbb{Z})$ for each fixed $x \in \mathbb{X}$, and $\Psi(\cdot, y) \in L(\mathbb{X}; \mathbb{Z})$ for each fixed $y \in \mathbb{Y}$. $L(\mathbb{X}, \mathbb{Y}; \mathbb{Z})$ denotes the set of all bilinear maps from $\mathbb{X} \times \mathbb{Y}$ into $\mathbb{Z}$, with the norm

$$\|\Psi\|_{L(\mathbb{X},\mathbb{Y};\mathbb{Z})} = \sup_{\|x\|_{\mathbb{X}}, \|y\|_{\mathbb{Y}}=1} \|\Psi(x, y)\|_{\mathbb{Z}}.$$

$\Psi \in L(\mathbb{X}, \mathbb{Y}; \mathbb{Z})$ can be identified with $\hat{\Psi} \in L(\mathbb{X}; L(\mathbb{Y}; \mathbb{Z}))$ by

$$\Psi(x, y) = \hat{\Psi}(x)y. \tag{5.143}$$

This identification gives an isometric isomorphism (norm preserving linear map) between $L(\mathbb{X}, \mathbb{Y}; \mathbb{Z})$ and $L(\mathbb{X}; L(\mathbb{Y}; \mathbb{Z}))$.

*Example 5.6.* Let $\mathbb{Z} = \mathbb{R}^1$ and $\mathbb{X} = \mathbb{Y}$. Then

$$L(\mathbb{X}, \mathbb{X}; \mathbb{R}^1) = L(\mathbb{X}; \mathbb{X}), \quad \text{under the identification } \mathbb{X} = \mathbb{X}^*, \tag{5.144}$$

i.e., bilinear map $\mathbb{X} \times \mathbb{X} \mapsto \mathbb{R}^1$ can be regarded as a continuous linear transformation on $\mathbb{X}$.

**2. Hilbert–Schmidt operators, nuclear operators**

Let $e_i, i = 1, 2, \ldots$ be an ONB of $\mathbb{H}$. $\Psi \in L(\mathbb{H}; \mathbb{H})$ is called a Hilbert–Schmidt operator, if $\|\Psi\|^2_{\mathcal{L}^2(\mathbb{H})} := \sum_i \|\Psi e_i\|^2_{\mathbb{H}} < \infty$. The set of Hilbert–Schmidt operators, denoted by $\mathcal{L}^2(\mathbb{H})$, becomes a Hilbert space with the norm $\|\cdot\|_{\mathcal{L}^2(\mathbb{H})}$. The norm does not depend on the choice of ONB.

$\Phi \in L(\mathbb{H}; \mathbb{H})$ is called a nuclear operator, or trace class, if there exist $N$ and $\Psi_j, \tilde{\Psi}_j \in \mathcal{L}^2(\mathbb{H}), j = 1, \ldots, N$, such that

$$\Phi = \sum_{j=1}^{N} \Psi_j^* \tilde{\Psi}_j \quad (\Psi_j^* = \text{adjoint of } \Psi_i). \tag{5.145}$$

$\Phi$ is called nonnegative, if $(\Phi h, h) \geq 0, \forall h \in \mathbb{H}$. The trace of $\Phi$, $\mathrm{tr}(\Phi)$, is given by

$$\mathrm{tr}(\Phi) := \sum_i (\Phi e_i, e_i) = \sum_{j=1}^{N} (\tilde{\Psi}_j, \Psi_j)_{\mathcal{L}^2(\mathbb{H})}. \tag{5.146}$$

For a nuclear operator $\Phi$, the following properties are valid:

(i) $\mathrm{tr}(\Phi)$ does not depend on the choice of ONB.
(ii) $\Phi$ is a compact operator.
(iii) There exists an ONB consisting of eigenvectors for $\Phi$.
(iv) For $A \in L(\mathbb{H}; \mathbb{H})$, $A\Phi$ and $\Phi A$ are also nuclear and

$$\mathrm{tr}(A\Phi) = \mathrm{tr}(\Phi A). \tag{5.147}$$

(v) Suppose that $A_n \in L(\mathbb{H}; \mathbb{H}), n = 1, 2, \ldots$ converge weakly to $A \in L(\mathbb{H}; \mathbb{H})$. Then

$$\lim_{n \to \infty} \mathrm{tr}(A_n \Phi) = \mathrm{tr}(A\Phi) \tag{5.148}$$

(refer to [La83], VII).

**3. First- and second derivatives**

$F : \mathbb{H} \mapsto \mathbb{Z}$ is said to be differentiable at $x \in \mathbb{H}$ if there exists an operator $DF(x) \in L(\mathbb{H}; \mathbb{Z})$, such that

$$\lim_{\|h\|_{\mathbb{H}} \to 0} \frac{\|F(x+h) - F(x) - DF(x)h\|_{\mathbb{Z}}}{\|h\|_{\mathbb{H}}} = 0. \tag{5.149}$$

Formally,

$$F(x+h) - F(x) = DF(x)h + o(h).$$

$DF(x)$ of (5.149) is unique and is called the first derivative of $F$ at $x$. When $F$ is differentiable at any point and the map $x \mapsto DF(x)$ is continuous, we say that $F \in C^1(\mathbb{H}; \mathbb{Z})$.

**Mean Value Theorem ([La83], p. 107)**

Let $F \in C^1(\mathbb{H}) := C^1(\mathbb{H}; \mathbb{R}^1)$. Under the identification $\mathbb{H}^* = \mathbb{H}$,

$$F(x+h) - F(x) = \int_0^1 (DF(x+th), h)_{\mathbb{H}}\, dt.$$

Suppose that $F \in C^1(\mathbb{H}; \mathbb{Z})$ and $DF$ is differentiable at $x$. Then we can define $D^2F(x) \in L(\mathbb{H}; L(\mathbb{H}; \mathbb{Z}))$ $(= L(\mathbb{H}, \mathbb{H}; \mathbb{Z}))$ in the same way as (5.149), namely

$$\lim_{\|h\|_{\mathbb{H}} \to 0} \frac{\|DF(x+h) - DF(x) - D^2F(x)h\|_{L(\mathbb{H};\mathbb{Z})}}{\|h\|_{\mathbb{H}}} = 0 \tag{5.150}$$

(The numerator $= \sup_{\|g\|_{\mathbb{H}}=1} \|DF(x+h)g - DF(x)g - D^2F(x)(h,g)\|_{\mathbb{Z}}$). When $D^2F$ is continuous, we write $F \in C^2(\mathbb{H}; \mathbb{Z})$. For $F \in C^2(\mathbb{H}; \mathbb{Z})$, $D^2F(x)$ is symmetric:

$$D^2F(x)(h, \tilde{h}) = D^2F(x)(\tilde{h}, h), \quad \forall h, \tilde{h} \in \mathbb{H}$$

(see [La83], p. 110).

**Taylor Formula**

Let $F \in C^2(\mathbb{H}) := C^2(\mathbb{H}; \mathbb{R}^1)$. Then

$$F(x+h) - F(x) = (DF(x), h)_{\mathbb{H}} + \int_0^1 (1-t) D^2F(x+th)(h,h)\, dt \tag{5.151}$$

(see [La83], V for details).

### 5.4.2 *Itô's Formula for $\mathbb{H}$-Valued Semimartingales*

We denote by $C^{12}([0,T]\times\mathbb{H})$ (or $C_{\mathrm{u}}^{12}([0,T]\times\mathbb{H})$) the set of $F \in C([0,T]\times\mathbb{H})$ such that $\partial_t F$, $DF$, and $D^2F$ are continuous (or uniformly continuous) on $[0,T] \times \mathbb{H}$.

For $F \in C^{12}([0,T] \times \mathbb{H})$, we assume the local boundedness condition;

$(F_0)$ $|F(\cdot)|$, $|\partial_t F(\cdot)|$, $\|DF(\cdot)\|_{\mathbb{H}}$ and $\|D^2F(\cdot)\|_{L(\mathbb{H};\mathbb{H})}$ are bounded on any bounded subset of $[0,T] \times \mathbb{H}$.

**Theorem 5.6.** *Assume* $(F_0)$. *For* $X$ *given by* (5.141), *Itô's formula* (5.142) *holds.*

*Proof.* Put

$$\tilde{\Omega} = \{\omega \in \Omega; X(\cdot, \omega) \in C([0,T]; \mathbb{H})\} \tag{5.152}$$

and, for $\omega \in \tilde{\Omega}$,

$$\mathbb{H}_\omega = \text{ closed convex hull of } \{X(t,\omega); t \in [0,T]\}. \tag{5.153}$$

Then $P(\tilde{\Omega}) = 1$ and $\mathbb{H}_\omega$ is compact in $\mathbb{H}$.

For the proof, we use a time discretization procedure, as in the finite-dimensional cases. So we only sketch the proof, assuming that $\tau = T$ and the following boundedness is satisfied:

$$\begin{aligned} &|F(t,\phi)| + |\partial_t F(t,\phi)| + \|DF(t,\phi)\|_{\mathbb{H}} + \|D^2 F(t,\phi)\|_{L(\mathbb{H},\mathbb{H})} \le K, \\ &\forall (t,\phi) \in [0,T] \times \mathbb{H} \end{aligned} \tag{5.154}$$

and $\|X_0\|_{\mathbb{H}} \le K$, $P$-a.s. During the proof, we omit the suffix $\mathbb{H}$ in $\| \quad \|_{\mathbb{H}}$ and $( \, , \, )_{\mathbb{H}}$, for simplicity.

From Taylor's formula (5.151) it follows that

$$\begin{aligned} &F(t, X(t)) - F(\theta, X(\theta)) \\ &= F(t, X(t)) - F(\theta, X(t)) + F(\theta, X(t)) - F(\theta, X(\theta)) \\ &= \int_\theta^t \partial_s F(s, X(t))\, ds + (DF(\theta, X(\theta)), \Delta(\theta, t)) \\ &\quad + \int_0^1 (1-\lambda) D^2 F(\theta, X(\theta) + \lambda \Delta(\theta,t))(\Delta(\theta,t), \Delta(\theta,t))\, d\lambda, \end{aligned} \tag{5.155}$$

where $\Delta(\theta, t) = X(t) - X(\theta)$.

We divide the proof into four steps.

**Step 1.** Let $\mathcal{D}_N = (t_j, j = 1, \dots, p)$ be a division of $[0,T]$, satisfying

$$\mathcal{D}_N \subset \mathcal{D}_{N+1}, \quad |\mathcal{D}_N| < 2^{-N}, \quad N = 1, 2, \dots .$$

Put

$$\theta_N(s) = t_j, \quad \bar{\theta}_N(s) = t_{j+1} \quad \text{on } s \in [t_j, t_{j+1}), \quad j = 0, 1, \dots, p. \tag{5.156}$$

From the continuity of $\partial_s F$ and the bounded convergence theorem, it is clear that

$$\lim_{N \to \infty} \int_0^T |\partial_s F(s, X(\bar{\theta}_N(s))) - \partial_s F(s, X(s))|\, ds = 0 \tag{5.157}$$

in $L^1(\Omega)$ and $P$-a.s. Similarly, observing that

$$\left(DF(\theta, X(\theta)), \int_\theta^t \xi(s)\,ds\right) = \int_\theta^t (DF(\theta, X(\theta)), \xi(s))\,ds, \tag{5.158}$$

we have

$$\lim_{N\to\infty} E\int_0^T |(DF(\theta_N(s), X(\theta_N(s))) - DF(s, X(s)), \xi(s))|\,ds = 0. \tag{5.159}$$

For the stochastic integral term, the Burkholder–Davis–Gundy inequality yields

$$\lim_{N\to\infty} E\Big[\sup_{0\le t\le T}\Big|\int_0^t (DF(\theta_N(s), X(\theta_N(s))) - DF(s, X(s)), dM(s))\Big|\Big] = 0. \tag{5.160}$$

**Step 2.** We divide the 3rd term in the RHS of (5.155) into four parts, $J_1$, $J_2$, $J_3$, and $J_4$. Let $\omega \in \tilde{\Omega}$ and $\varepsilon > 0$ be given. Put

$$J_1(\theta, t, \lambda, \omega) = D^2F(\theta, X(\theta,\omega) + \lambda(X(t,\omega) - X(\theta,\omega))) - D^2F(\theta, X(\theta,\omega)).$$

Since $\mathbb{H}_\omega$ is compact and $D^2F$ is continuous, there is $\delta_{\varepsilon,\omega} > 0$, such that

$$\sup_{\lambda\in[0,1]} \|J_1(\theta, t, \lambda, \omega)\|_{L(\mathbb{H};\mathbb{H})} < \varepsilon \quad \text{if } |t-\theta| < \delta_{\varepsilon,\omega}. \tag{5.161}$$

**Step 3.** Put

$$\triangle_M(\theta, t) = M(t) - M(\theta) \quad \text{and} \quad \mathcal{I}(\theta, t) = \int_\theta^t \xi(s)\,ds.$$

We will compute the following terms:

$$\begin{aligned} J_2(\theta, t) &= D^2F(\theta, X(\theta))(\mathcal{I}(\theta,t), \mathcal{I}(\theta,t)),\\ J_3(\theta, t) &= D^2F(\theta, X(\theta))(\mathcal{I}(\theta,t), \triangle_M(\theta,t)),\\ J_4(\theta, t) &= D^2F(\theta, X(\theta))(\triangle_M(\theta,t), \triangle_M(\theta,t)). \end{aligned} \tag{5.162}$$

From the Schwarz inequality and (5.17), it follows that

$$E\Big|\sum_{t_{j+1}\le t} J_3(t_j, t_{j+1})\Big| \le K2^{-\frac{N}{2}}\Big(E\int_0^t \|\xi(s)\|^2\,ds\Big)^{\frac12}(E\langle M\rangle(t))^{\frac12} \tag{5.163}$$

and

$$E\Big|\sum_{t_{j+1}\le t} J_2(t_j, t_{j+1})\Big| \le K2^{-N}E\int_0^t \|\xi(s)\|^2\,ds. \tag{5.164}$$

**Step 4.** For $J_4$, we have that

$$E\left|\sum_{t_{j+1}\leq t} D^2F(t_j,X(t_j))(\Pi_n^{\perp}\Delta_M(t_j,t_{j+1}),\Delta_M(t_j,t_{j+1}))\right|$$
$$\leq K(E\langle \Pi_n^{\perp}M\rangle(t))^{\frac{1}{2}}(E\langle M\rangle(t))^{\frac{1}{2}} \longrightarrow 0,$$
$$\text{as } n\to\infty, \quad \text{uniformly w.r.t. the divisions } \mathcal{D}_N. \tag{5.165}$$

On the other hand, (5.22) and (5.24) yield

$$\sum_{t_{j+1}\leq t} D^2F(t_j,X(t_j))(\Pi_n\Delta_M(t_j,t_{j+1}),\Pi_n\Delta_M(t_j,t_{j+1}))$$
$$=\sum_{k,p=1}^{n}\sum_{t_{j+1}\leq t} D^2F(t_j,X(t_j))(e_k,e_p)(\Delta_M(t_j,t_{j+1}),e_k)(\Delta_M(t_j,t_{j+1}),e_p)$$
$$\longrightarrow \sum_{k,p=1}^{n}\int_0^t D^2F(s,X(s))(e_k,e_p)q^{k,p}(s)\,d\langle M\rangle(s), \quad P\text{-a.s.} \quad \text{as } N\to\infty. \tag{5.166}$$

Now (5.165), (5.166), and (5.162) imply that

$$\lim_{N\to\infty}\sum_{t_{j+1}\leq t} J_4(t_j,t_{j+1}) = \int_0^t \mathrm{tr}(D^2F(s,X(s))q_M(s))\,d\langle M\rangle(s) \quad P\text{-a.s.} \tag{5.167}$$

Putting (5.157), (5.159)–(5.161), (5.163), (5.164), and (5.167) together, we complete the proof. □

*Example 5.7 (Energy equality).* For $F(\phi) = \|\phi\|_{\mathbb{H}}^2$, Itô's formula gives the energy equality;

$$d\|X(t)\|_{\mathbb{H}}^2 = 2(X(t),\xi(t))_{\mathbb{H}}\,dt + 2(X(t),dM(t))_{\mathbb{H}} + d\langle M\rangle(t). \tag{5.168}$$

*Example 5.8 (Stochastic integral driven by $W_Q$ on $H_0$).*
Suppose that

$$M(t) = \int_0^t \Psi(s)\,dW_Q(s),$$

with

$$W_Q(s) = \sum_k \sqrt{\lambda_k}\beta^k(s)e_k$$

and $\Psi \in L^2([0,T] \times \Omega, (\mathcal{F}_t); \mathcal{L}^2_Q(H_0; H_0))$. Putting $\Psi_k(t) = \Psi(t)e_k$, $(\in L^2([0,T] \times \Omega,\ (\mathcal{F}_t); H_0))$, we have

$$\begin{aligned} dF(t,X(t)) = \Big\{ & \partial_t F(t,X(t)) + (DF(t,X(t)), \xi(t)) \\ & + \frac{1}{2}\sum_k \lambda_k D^2F(t,X(t))(\Psi_k(t), \Psi_k(t)) \Big\} dt \\ & + \sum_k \sqrt{\lambda_k}(DF(t,X(t)), \Psi_k(t))\, d\beta^k(t). \end{aligned} \tag{5.169}$$

### 5.4.3 Itô's Formula for Linear Stochastic Parabolic Equations

In this subsection, we always assume the conditions $(B_1)$, $(B_2)$, $(B_4)$, and $(B_5)$ are satisfied. Since the solution $u(\cdot)$ of (5.102) with the initial state $u_0 \in L^2(\Omega, \mathcal{F}_{T_0}; H_1)$ belongs to $L^2([T_0,T]\times\Omega, (\mathcal{F}_t); H_2) \cap L^2(\Omega; C([T_0,T]; H_1))$, Itô's formula is valid for $F$ satisfying condition $(F_0)$, by using

$$\xi(s) := A(s)u(s) + \mathfrak{z}(s) \quad \text{and} \quad M(t) := \int_{T_0}^t (G(s)u(s) + \Psi(s))\, dW_Q(s).$$

However, if $u_0$ is an $H_0$-random variable, then $\xi$ may not be an $H_0$-process and the previous arguments do not work. So, we impose the following stronger condition $(F_1)$, besides $(F_0)$.

$(F_1)$ $DF(t,\cdot)$ maps $H_1$ into $H_1$, and there exists a modulus $m_{DF}$ such that

$$\begin{cases} |||DF(t_1,\phi_1) - DF(t_2,\phi_2)||| \le m_{DF}(|t_1 - t_2| + |||\phi_1 - \phi_2|||) \\ \qquad\qquad\qquad\qquad \text{for } \phi_1, \phi_2 \in H_1, \\ \|DF(t_1,\phi_1) - DF(t_2,\phi_2)\| \le m_{DF}(|t_1 - t_2| + \|\phi_1 - \phi_2\|) \\ \qquad\qquad\qquad\qquad \text{for } \phi_1, \phi_2 \in H_0. \end{cases} \tag{5.170}$$

*Remark 5.2.* The following facts are obvious: under $(F_0)$ and $(F_1)$,

(i) There is a constant $K_{DF}$, such that

$$\begin{cases} |||DF(t,\phi)||| \le K_{DF}(1 + |||\phi|||),\ \forall (t,\phi) \in [0,T] \times H_1, \\ \|DF(t,\phi)\| \le K_{DF}(1 + \|\phi\|), \quad \forall (t,\phi) \in [0,T] \times H_0. \end{cases} \tag{5.171}$$

(ii) Let $\eta$ and $\eta_n, n = 1, 2, \ldots$ be in $L^2([0,T] \times \Omega; H_1)$. If

$$\lim_{n\to\infty} E \int_0^T |||\eta_n(t) - \eta(t)|||^2\, dt = 0,$$

then

$$\lim_{n\to\infty} E\int_0^T |||DF(t,\eta_n(t)) - DF(t,\eta(t))|||^2\,dt = 0.$$

*Example 5.9.* By $F \in C_{\rm u}^{12}([0,T]\times H_{-1})$ we mean that there is $\tilde{F} \in C_{\rm u}^{12}([0,T]\times H_0)$, such that

$$F(t,\phi) = \tilde{F}(t,B_0\phi), \quad \phi \in H_{-1},$$

where $B_0 = (\mathbf{1}-\Delta)^{-\frac{1}{2}}$ with the $d$-dimensional Laplacian $\Delta$. The restriction of $F$ $(\in C_{\rm u}^{12}([0,T]\times H_{-1}))$ to $H_0$ satisfies $(F_0)$ and $(F_1)$.

Indeed, $DF(t,\phi) = B_0 D\hat{F}(t,B_0\phi)$ is in $H_1$. Since $\tilde{F}, D\tilde{F}$, and $D^2\tilde{F}$ are uniformly continuous, there is a modulus $\tilde{m}(\cdot)$, such that

$$\begin{aligned}&|\tilde{F}(t_1,\phi_1) - \tilde{F}(t_2,\phi_2)| + \|D\tilde{F}(t_1,\phi_1) - D\tilde{F}(t_2,\phi_2)\|\\ &+ \|D^2\tilde{F}(t_1,\phi_1) - D^2\tilde{F}(t_2,\phi_2)\|_{L(H_0;H_0)} \le \tilde{m}(|t_1-t_2| + \|\phi_1-\phi_2\|).\end{aligned}$$

Hence, it holds that

$$\begin{aligned}&|||DF(t_1,\phi_1) - DF(t_2,\phi_2)||| = \|D\tilde{F}(t_1,B_0\phi_1) - D\tilde{F}(t_2,B_0\phi_2)\|\\ &\le \tilde{m}(|t_1-t_2| + \|\phi_1-\phi_2\|_*).\end{aligned}$$

Since $D^2F(t,\phi) = B_0D^2\tilde{F}(t,B_0\phi)B_0$, we conclude that $(F_0)$ and $(F_1)$ hold.

**Theorem 5.7.** *Assume that conditions* $(B_1),(B_2),(B_4)$ *and* $(B_5)$ *are satisfied. Suppose that* $F \in C^{12}([T_0,T]\times H_0)$ *satisfies* $(F_0)$ *and* $(F_1)$. *Then Itô's formula holds for the solution* $u(\cdot)$ *of problem* (5.102)–(5.93).

$$\begin{aligned}&dF(t,u(t))\\ &=\Big\{\partial_t F(t,u(t)) + \langle A(t)u(t) + \mathfrak{z}(t), DF(t,u(t))\rangle\\ &+ \frac{1}{2}\mathrm{tr}\Big(((G(t)u(t) + \Psi(t))Q^{\frac{1}{2}})^* D^2F(t,u(t))(G(t)u(t) + \Psi(t))Q^{\frac{1}{2}})\Big)\Big\}\,dt\\ &+ ((G(t)u(t) + \Psi(t))\,dW_Q(t), DF(t,u(t)))\end{aligned} \tag{5.172}$$

*Proof.* By taking the exit time of $u(\cdot)$ from a ball of $H_0$, we may assume that the boundedness condition (5.154) holds for $\mathbb{H} = H_0$. We divide the proof into two cases.

**Case 1.** $\|u_0\| \le C_0$ (= constant), $P$-a.s.

**Step 1.** We approximate $u_0$ by $u_{n0} \in L^2(\Omega, \mathcal{F}_{T_0}; H_1)$ so that

$$E\|u_{n0} - u_0\|^2 < 2^{-n} \quad \text{and} \quad \|u_{n0}\| \le C_0 \quad P\text{-a.s.} \tag{5.173}$$

By $u_n(\cdot)$, we denote the solution of (5.102) with the initial state $u_{n0}$. Then $u_n \in L^2([T_0, T] \times \Omega, (\mathcal{F}_t); H_2) \cap L^2(\Omega; C([T_0, T]; H_1))$ and

$$\begin{aligned} &E\Big[\sup_t \|u_n(t) - u(t)\|^2 + \int_{T_0}^T |||u_n(t) - u(t)|||^2\, dt\Big] \\ \le &NE\|u_{n0} - u_0\|^2, \quad n = 1, 2, \ldots \end{aligned} \tag{5.174}$$

with a constant $N$ independent of $n$. Further, we may assume that

$$\lim_{n\to\infty}\Big(\sup_t \|u_n(t) - u(t)\|^2 + \int_{T_0}^T |||u_n(t) - u(t)|||^2\, dt\Big) = 0 \quad P\text{-a.s.} \tag{5.175}$$

and

$$\lim_{n\to\infty} |||u_n(t) - u(t)|||^2 = 0 \quad \text{a.e. on } [T_0, T] \times \Omega \tag{5.176}$$

by taking a subsequence $n'$, if necessary.

Since Itô's formula (5.172) is valid for $u_n(\cdot)$, we intend to show the convergence of each term in the RHS as $n \to \infty$.

From (5.175) it is immediate that

$$\lim_{n\to\infty} \sup_t |F(t, u_n(t)) - F(t, u(t))| = 0 \tag{5.177}$$

and

$$\lim_{n\to\infty} \int_{T_0}^T |\partial_s F(s, u_n(s)) - \partial_s F(s, u(s))|\, ds = 0, \quad P\text{-a.s. and in } L^1(\Omega). \tag{5.178}$$

**Step 2.** For DF terms, we have

$$\begin{aligned} &E\int_{T_0}^T |\langle A(s)(u_n(s) - u(s)), DF(s, u_n(s))\rangle|\, ds \\ &\le KK_{DF} E\int_{T_0}^T |||u_n(s) - u(s)|||(1 + |||u_n(s)|||)\, ds \\ &\le KK_{DF}\Big(E\int_{T_0}^T |||u_n(s) - u(s)|||^2\, ds\Big)^{\frac{1}{2}}\Big(E\int_{T_0}^T (1 + |||u_n(s)|||)^2\, ds\Big)^{\frac{1}{2}} \\ &\longrightarrow 0 \quad \text{as } n \to \infty, \end{aligned} \tag{5.179}$$

thanks to (5.174) and (5.103). Similarly, for any $\varepsilon > 0$, we have

$$
\begin{aligned}
&E \int_{T_0}^{T} |\langle A(s)u(s), DF(s, u_n(s)) - DF(s, u(s))\rangle| \, ds \\
&\le KE \int_{T_0}^{T} |||u(s)||| m_{DF}(|||u_n(s) - u(s)|||) \, ds \\
&\le KE \int_{T_0}^{T} |||u(s)|||(\varepsilon + C_\varepsilon |||u_n(s) - u(s)|||) \, ds, \qquad (5.180)
\end{aligned}
$$

with some constant $C_\varepsilon$ depending on $m_{DF}(\cdot)$.
Hence (5.179) and (5.180) together with (5.175) yield

$$
\begin{aligned}
&E\Big[\sup_t \Big| \int_{T_0}^{t} \langle A(s)u_n(s) + \mathfrak{z}(s), DF(s, u_n(s))\rangle \, ds \\
&- \int_{T_0}^{t} \langle A(s)u(s) + \mathfrak{z}(s), DF(s, u(s))\rangle \, ds \Big|\Big] \longrightarrow 0 \quad \text{as } n \to \infty. \qquad (5.181)
\end{aligned}
$$

**Step 3.** For the stochastic integral terms, (5.154) and the dominated convergence theorem yield

$$
\begin{aligned}
&E\Big[\int_{T_0}^{T} (G_k(s)(u_n(s) - u(s)), DF(s, u_n(s)))^2 \, ds \\
&+ \int_{T_0}^{T} (G_k(s)u(s), DF(s, u_n(s)) - DF(s, u(s)))^2 \, ds\Big] \longrightarrow 0 \quad \text{as } n \to \infty,
\end{aligned}
\qquad (5.182)
$$

whence, by the Burkholder–Davis–Gundy inequality,

$$
\begin{aligned}
&E\Big[\sup_t \Big| \int_{T_0}^{t} ((G(s)u_n(s) + \Psi(s)) \, dW_Q(s), DF(s, u_n(s))) \\
&- \int_{T_0}^{t} ((G(s)u(s) + \Psi(s)) \, dW_Q(s), DF(s, u(s)))\Big|^2\Big] \longrightarrow 0 \quad \text{as } n \to \infty.
\end{aligned}
\qquad (5.183)
$$

**Step 4.** For the second differential terms, we consider

$$
\begin{aligned}
&D^2F(s, u_n(s))(G_k(s)u_n(s), G_k(s)u_n(s)) \\
&- D^2F(s, u(s))(G_k(s)u(s), G_k(s)u(s)) \\
&= D^2F(s, u_n(s))(G_k(s)(u_n(s) - u(s)), G_k(s)(u_n(s) + u(s))) \\
&+ (D^2F(s, u_n(s)) - D^2F(s, u(s)))(G_k(s)u(s), G_k(s)u(s)) \\
&:= I_{kn}(s) + J_{kn}(s). \qquad (5.184)
\end{aligned}
$$

Using the same arguments, we have

$$\sum_k \lambda_k E\Big[\int_{T_0}^{T} |I_{kn}(s)| + |J_{kn}(s)|\, ds\Big] \longrightarrow 0 \quad \text{as } n \to \infty, \tag{5.185}$$

because by (5.154) $D^2F \in C_b([T_0, T] \times H_0; L(H_0; H_0))$.

Thus, putting the above results together, we obtain Itô's formula for $u(\cdot)$, since Itô's formula holds for $u_n(\cdot)$.

**Case 2.** $E\|u_0\|^2 < \infty$.

Let $C_n, n = 1, 2, \ldots$ be an increasing sequence of positive numbers with $\lim_{n\to\infty} C_n = \infty$. Put $u_{n0} = \chi_{[0,C_n]}(\|u_0\|)u_0$ and let $u_n(\cdot)$ and $u(\cdot)$ denote the solutions of (5.100) with the initial states $u_{n0}$ and $u_0$ respectively. Referring to (5.174)–(5.176), we may assume that, as $n \to \infty$

$$\sup_{T_0 \le t \le T} \|u_n(t) - u(t)\|^2 + \int_{T_0}^{T} |||u_n(t) - u(t)|||^2\, dt \longrightarrow 0 \quad P\text{-a.s. and in } L^1(\Omega). \tag{5.186}$$

Since Itô's formula (5.172) holds for $u_n(\cdot)$ and we can prove the convergence of each term of (5.172) for $u_n(\cdot)$ as $n \to \infty$, by using same arguments as in Case 1, we obtain Itô's formula for $u(\cdot)$. □

*Example 5.10 (Estimation of the $H_{-1}$-norm).*

We will evaluate $E\|u(t)\|_*^2$, by using Itô's formula. Let $u(\cdot)$ be the solution of (5.102) with the initial state $u_0 \in L^2(\Omega, \mathcal{F}_{T_0}; H_0)$. Put $F(\phi) = \|\phi\|_*^2 = (\phi, (\mathbf{1} - \Delta)^{-1}\phi)$. Since $F$ satisfies $(F_0)$ and $(F_1)$, Itô's formula gives the dynamics of $\|u(t)\|_*^2$:

$$\begin{aligned} d\|u(t)\|_*^2 =& 2\langle A(t)u(t) + \mathfrak{z}(t), (\mathbf{1} - \Delta)^{-1}u(t)\rangle\, dt \\ &+ \|(\mathbf{1} - \Delta)^{-\frac{1}{2}}(G(t)u(t) + \Psi(t))\|_Q^2\, dt \\ &+ 2((G(t)u(t) + \Psi(t))\, dW_Q(t), (\mathbf{1} - \Delta)^{-1}u(t)). \end{aligned} \tag{5.187}$$

From (5.187), we deduce

**Proposition 5.7.** *Under* $(B_1)$, $(B_2)$, $(B_4)$ *and* $(B_5)$, *one has*

(i) ***Estimate***

$$\begin{aligned} &E\Big[\sup_{T_0 \le s \le T} \|u(s)\|_*^2 + \int_{T_0}^{T} \|u(s)\|^2\, ds\Big] \\ \le & NE\Big[\|u_0\|_*^2 + \int_{T_0}^{T} (\|\mathfrak{z}(s)\|_{H_{-2}}^2 + \|\Psi(s)\|_{*Q}^2)\, ds\Big], \end{aligned} \tag{5.188}$$

(ii) *Let $u(\cdot)$ and $\tilde{u}(\cdot)$ be the solutions of* (5.100) *with initial states $u_0$ and $\tilde{u}_0$ respectively. Then*

$$E\Big[\sup_{T_0\le s\le T}\|u(s)-\tilde{u}(s)\|_*^2+\int_{T_0}^{T}\|u(s)-\tilde{u}(s)\|^2\,ds\Big]\le NE\|u_0-\tilde{u}_0\|_*^2 \tag{5.189}$$

*where the constant $N$ depends only on $\bar{\lambda}, \bar{\mu}$ and $K_1$ of* (5.104).

*Proof.* Put $A_0 = \mathbf{1} - \Delta$ and $B_0 = (\mathbf{1} - \Delta)^{-\frac{1}{2}}$ for simplicity. Referring to [GŚ00], Lemma 3.3, we have for $\psi \in H_3$

$$2\langle A(s)A_0\psi,\psi\rangle + \|B_0G(s)A_0\psi\|_Q^2 \le -\lambda\mathbb{I}\psi\mathbb{I}^2 + \mu|||\psi|||^2, \quad \forall s, \tag{5.190}$$

with constants $\lambda > 0$ and $\mu \in \mathbb{R}^1$.

First we assume $u_0 \in L^2(\Omega, \mathcal{F}_{T_0}; H_1)$. Let $u(\cdot) \in L^2(\Omega; C([T_0,T]; H_1))$ be a unique solution of (5.100) with initial state $u_0$. Consequently, for $v(t) := B_0^2u(t)$,

$$\begin{aligned}&2\langle A(s)A_0v(s)+\mathfrak{z}(s),v(s)\rangle + \|G(s)A_0v(s)+\Psi(s)\|_{*Q}^2\\ \le{}&2\langle A(s)A_0v(s),v(s)\rangle + \|G(s)A_0v(s)\|_{*Q}^2\\ &+2\langle\mathfrak{z}(s),v(s)\rangle + \varepsilon\|G(s)A_0v(s)\|_{*Q}^2 + \Big(1+\frac{1}{\varepsilon}\Big)\|\Psi(s)\|_{*Q}^2\end{aligned} \tag{5.191}$$

for any $\varepsilon > 0$. Since

$$\|G(s)A_0v(s)\|_{*Q}^2 \le c_1\|A_0v(s)\|^2 = c_1\mathbb{I}v(s)\mathbb{I}^2 \tag{5.192}$$

with a constant $c_1$ independent of $s$, relations (5.187), (5.191), and (5.192) yield

$$\begin{aligned}&|||v(t)|||^2 - \|u_0\|_*^2\\ \le{}&\int_{T_0}^{t}\Big\{-\frac{\lambda}{2}\mathbb{I}v(s)\mathbb{I}^2 + \tilde{\mu}|||u(s)|||^2 + c_2\Big(\|\mathfrak{z}(s)\|_{H_{-2}}^2 + \|\Psi(s)\|_{*Q}^2\Big)\Big\}\,ds\\ &+\text{ stochastic integral,}\end{aligned} \tag{5.193}$$

with $\tilde{\mu} = \mu + \frac{1}{2}$ and $c_2 = \max\{\frac{1}{2}, (1+\frac{2c_1}{\lambda})\frac{4}{\lambda}\}$. Taking the expectation and using Gronwall's inequality, we evaluate $E|||v(t)|||^2$ and obtain

$$\begin{aligned}&E\Big[|||v(t)|||^2 + \int_{T_0}^{t}\mathbb{I}v(s)\mathbb{I}^2\,ds\Big]\\ \le{}&c_3E\Big[\|u_0\|_*^2 + \int_{T_0}^{t}\|\mathfrak{z}(s)\|_{H_{-2}}^2 + \|\Psi(s)\|_{*Q}^2\,ds\Big],\end{aligned} \tag{5.194}$$

with a constant $c_3$.

Put

$$M(t) = \int_{T_0}^{t} ((G(s)A_0v(s) + \Psi(s))\, dW_Q(s), v(s)).$$

Then $M(t)$ is a continuous real martingale with

$$\langle M\rangle(t) = \int_{T_0}^{t} (G(s)A_0v(s) + \Psi(s), v(s))_Q^2\, ds.$$

Since the Burkholder–Davis–Gundy inequality yields

$$\begin{aligned} E\Big[\sup_{T_0\le s\le t} |M(s)|\Big] &\le 6E\sqrt{\langle M\rangle(t)} \\ &\le \frac{1}{2}E\sup_{s\le t}|||v(s)|||^2 + 18E\int_{T_0}^{t} \|G(s)A_0v(s) + \Psi(s)\|_{*Q}^2\, ds, \end{aligned} \tag{5.195}$$

(5.193)–(5.195) yield (i).

For $u_0 \in L^2(\Omega, \mathcal{F}_{T_0}; H_0)$, we take $u_{n,0} \in L^2(\Omega, \mathcal{F}_{T_0}; H_1)$, $n = 1, 2, \ldots$ such that $\lim_{n\to\infty} E[\|u_{n,0} - u_0\|^2] = 0$. Let $u_{n,0}(\cdot)$ and $u(\cdot)$ be solutions of (5.100) with initial states $u_{n,0}$ and $u_0$, respectively. Then,

$$E[\sup_{T_0\le t\le T} \|u_n(t) - u(t)\|^2] \le NE[\|u_{n,0} - u_0\|^2]$$

yields (i) for $u(\cdot)$.

(ii) is immediate from (i). Thus, we have completed the proof. □

*Example 5.11 (Estimate of $E[\sup_{\theta\le s\le t} \|u(s) - u(\theta)\|_*^2]$ when $\mathfrak{z} = 0$ and $\Psi = 0$).* Let $\theta, t \in [T_0, T)$ be given. By using Proposition 5.7, we have

$$E\Big[\sup_{\theta\le s\le t} \|u(s) - u(\theta)\|_*^2\Big] \le c_0(t-\theta)E\|u(\theta)\|^2 \le c_0N(t-\theta)E\|u_0\|^2 \tag{5.196}$$

with a constant $c_0$, independent of $t$ and $\theta$, and with $N$ the constant in (5.103).

Indeed, the right inequality is clear by (5.103). So, let us prove the left inequality. Take $u_n \in L^2(\Omega, \mathcal{F}_\theta; H_1)$ such that

$$\lim_{n\to\infty} E\|u(\theta) - u_n\|^2 = 0. \tag{5.197}$$

Put $v_n(s) = u(s) - u_n$, $s \in [\theta, T]$. Then $v_n(\cdot)$ satisfies

$$\begin{cases} dv_n(s) = (A(s)v_n(s)+A(s)u_n)\,ds+(G(s)v_n(s)+G(s)u_n)\,dW_Q(s), & s\in(\theta,T], \\ v_n(\theta) = u(\theta) - u_n \ (\in L^2(\Omega, \mathcal{F}_\theta; H_0)). \end{cases}$$

Further, there is a constant $c_1$, depending only on $K_1$ of $(B_4)$, such that

$$\|A(s)u_n\|_{H_{-2}}^2 + \|G(s)u_n\|_{*Q}^2 \le c_1\|u_n\|^2, \quad \forall s, \quad P\text{-a.s.} \tag{5.198}$$

Now from (5.188), it follows that

$$E\Big[\sup_{\theta\le s\le t}\|v_n(s)\|_*^2\Big] \le c_0 E[\|u(\theta) - u_n\|_*^2 + (t-\theta)\|u_n\|^2] \tag{5.199}$$

with a constant $c_0$ independent of $t, \theta$ and $n$. Since

$$u(s) - u(\theta) - v_n(s) = u_n - u(\theta), \quad \forall s \ge \theta, \tag{5.200}$$

(5.196) follows, from (5.197), (5.199), and (5.200). □

## 5.5 Zakai Equations

This section is devoted to certain stochastic parabolic equations, called the Zakai equations, related to filtering problems. Let $X$ be a diffusion described by an SDE. $X$ cannot be measured directly. Only a partial measurement of $X$ can be obtained by means of another process $Y$ affected by an observation noise. Hence, we need to estimate $X(t)$, by using the information of $Y$ up to time $t$. This is the filtering problem (cf. [BC09]). Hence the problem is how to compute the conditional probability of $X(t)$, given $\mathcal{F}_t^Y$, by using results obtained in Sects. 5.3 and 5.4.

In Sect. 5.5.1 we are concerned with partially observable controlled diffusions and in Sect. 5.5.2 we study an explicit solution of the Zakai equation for conditional Gaussion process.

### *5.5.1 Partially Observable Controlled Diffusion*

Let $X$ be a controlled diffusion, governed by the SDE

$$dX(t) = \alpha(t, X(t), \gamma(t))\,dB(t) + \sigma(t, X(t), \gamma(t))\,dW(t) + b(t, X(t), \gamma(t))\,dt,$$

where $B$ and $W$ are mutually independent Wiener processes and $\gamma(\cdot)$ is a control process. As mentioned above we suppose that $X$ cannot be measured directly. Only a partial measurement of $X$ can be obtained by means of another process $Y$, given by

$$Y(t) = \int_0^t h(s, X(s), \gamma(s))\, ds + W(t).$$

$Y$ is called the observation process and $W$ is called the observation noise. Accordingly, $\mathcal{F}_t^Y$ stands for the information field up to time $t$. In this model we assume that the control process $\gamma(\cdot)$ is $(\mathcal{F}_t^Y)$-progressively measurable.

The problem is how to determine the conditional probability of $X$, given the information of $Y$ up to $t$.

Let us formulate the problem precisely. Let $B$ and $W$ be $d_0$- and $m$-dimensional $(\mathcal{F}_t)$-Wiener processes defined on $(\Omega, \mathcal{F}, (\mathcal{F}_t), P)$. Suppose that the control region $\Gamma$ is convex and compact subset of $\mathbb{R}^q$. Suppose further that

$$\alpha : [0, T] \times \mathbb{R}^d \times \Gamma \mapsto \mathbb{R}^d \otimes \mathbb{R}^{d_0},$$

$$\sigma : [0, T] \times \mathbb{R}^d \times \Gamma \mapsto \mathbb{R}^d \otimes \mathbb{R}^m,$$

$$b : [0, T] \times \mathbb{R}^d \times \Gamma \mapsto \mathbb{R}^d,$$

and

$$h : [0, T] \times \mathbb{R}^d \times \Gamma \mapsto \mathbb{R}^m$$

are continuous and satisfy the following conditions:

$(d_1^*)$ $z = \alpha$, $\sigma$ and $h$ are Lipschitz continuous w.r.t. $x$ uniformly on $[0, T] \times \Gamma$, say

$$|z(t, x, \gamma) - z(t, \tilde{x}, \gamma)| \le l|x - \tilde{x}|, \quad \forall t, \gamma \tag{5.201}$$

and

$$|z(t, x, \gamma)| \le K(1 + |x|), \quad \forall t, x, \gamma; \tag{5.202}$$

$(d_2^*)$ $b$ satisfies (5.202) and is locally Lipschitz continuous w.r.t. $x$ uniformly on $[0, T] \times \Gamma$, say, for any $R > 0$,

$$|b(t, x, \gamma) - b(t, \tilde{x}, \gamma)| \le l_R|x - \tilde{x}| \quad \text{for } |x|, |\tilde{x}| \le R, \quad \forall t, \gamma;$$

$(d_3^*)$ $\alpha$ and $\sigma$ are bounded.

**Definition 5.10.** $\zeta : [0, T] \times C([0, T]; \mathbb{R}^m) \mapsto \Gamma$ is called a policy, if it is non-anticipative, namely $\zeta_{/[0,t] \times C([0,T];\mathbb{R}^m)}$ is $\mathcal{B}([0, t]) \times \mathcal{B}(C[0, t]; \mathbb{R}^m)$-measurable for any $t \in [0, T]$.

For a given policy $\zeta$, we consider the following SDEs:

### System Equation

$$
\begin{cases}
dX(t) = \alpha(t, X(t), \zeta(t, Y))\, dB(t) + \sigma(t, X(t), \zeta(t, Y))\, dW(t) \\
\qquad\qquad + b(t, X(t), \zeta(t, Y))\, dt, \quad t \in (0, T], \\
X(0) = X_0 \ (\in L^2(\Omega, \mathcal{F}_0; \mathbb{R}^d)),
\end{cases}
\tag{5.203}
$$

and

### Observation Equation

$$
\begin{cases}
dY(t) = h(t, X(t), \zeta(t, Y))\, dt + dW(t), \quad t \in (0, T], \\
Y(0) = 0.
\end{cases}
\tag{5.204}
$$

Now we seek a solution $(X, Y)$ and then compute $P(X(t) \in \cdot | \mathcal{F}_t^Y)$. So the next step is to introduce the dynamics of the conditional probability density process, according to [R90], V.

Let $\beta$ and $\tilde{\beta}$ be $d_0$- and $m$-dimensional Wiener processes, defined on $(\Omega, \mathcal{F}, P)$, and $\xi_0$ be a random variable with the same probability distribution as $X_0$. We assume that $\beta, \tilde{\beta}$, and $\xi_0$ are mutually independent. Set

$$
\begin{aligned}
\mathcal{F}_t = \mathcal{F}_t^{\xi_0, \beta, \tilde{\beta}} &= \text{right continuous and complete } \sigma\text{-field} \\
&\qquad \text{generated by } \{\xi_0, \beta(s), \tilde{\beta}(s), s \le t\}.
\end{aligned}
$$

Hence $\beta$ and $\tilde{\beta}$ are mutually independent $(\mathcal{F}_t)$-Wiener processes and $\xi_0$ is in $L^2(\Omega, \mathcal{F}_0; \mathbb{R}^d)$.

Let us consider the auxiliary SDEs

$$
\begin{cases}
d\xi(t) = \alpha(t, \xi(t), \zeta(t, \eta))\, d\beta(t) + \sigma(t, \xi(t), \zeta(t, \eta))\, d\tilde{\beta}(t) \\
\qquad\qquad + \tilde{b}(t, \xi(t), \zeta(t, \eta))\, dt, \quad t \in (0, T], \\
\xi(0) = \xi_0,
\end{cases}
\tag{5.205}
$$

where

$$
\tilde{b}(t, x, \gamma) = b(t, x, \gamma) - \sigma(t, x, \gamma) h(t, x, \gamma),
$$

and

$$\begin{cases} d\eta(t) = d\tilde{\beta}(t), & t \in (0, T], \\ \eta(0) = 0. \end{cases} \tag{5.206}$$

By conditions ($\mathrm{d}_1^*$)–($\mathrm{d}_3^*$), the auxiliary SDEs (5.205)–(5.206) admit a unique strong solution $(\xi, \eta)$.

Applying Girsanov's transformation, we obtain

**Proposition 5.8.** *The* SDE *(5.203)–(5.204) has a weak solution. Moreover, the probability distribution of a weak solution is unique.*

*Proof.* Put

$$M(t) = \exp\Big(\int_0^t h(s, \xi(s), \zeta(s, \eta))\, d\tilde{\beta} - \frac{1}{2}\int_0^t |h(s, \xi(s), \zeta(s, \eta))|^2\, ds\Big). \tag{5.207}$$

Then by Proposition 1.5 $M$ is an exponential martingale and $\tilde{P} := M(T) \circ P$ gives a probability on $\mathcal{F}_T$. It is easy that $\tilde{P}(\Lambda) = \int_\Lambda M(t)\, dP$ for $\Lambda \in \mathcal{F}_t$. Put

$$\tilde{W}(t) = \tilde{\beta}(t) - \int_0^t h(s, \xi(s), \zeta(s, \tilde{\beta}))\, ds, \quad t \in [0, T]. \tag{5.208}$$

Then Girsanov's theorem shows that $\beta$ and $\tilde{W}$ are mutually independent $(\mathcal{F}_t)$-Wiener processes under $\tilde{P}$. Since $(\xi, \eta)$ satisfies (5.203)–(5.204) with $\beta$ and $\tilde{W}$ on $(\Omega, \mathcal{F}, (\mathcal{F}_t), \tilde{P})$, $(\xi, \eta)$ gives a weak solution of (5.203)–(5.204). Moreover, the distribution of the weak solution is unique, because (5.205)–(5.206) has a unique strong solution. □

Next we study $\tilde{E}(F(\xi(t))|\mathcal{F}_t^\eta)$, where $\tilde{E}$ = the expectation w.r.t. $\tilde{P}$. Since $E(M(t)|\mathcal{F}_t^\eta) \in (0, \infty)$ $P$-a.s., we can use Bayes formula (Proposition 1.3)

$$\tilde{E}(F(\xi(t))|\mathcal{F}_t^\eta) = \frac{E(M(t)F(\xi(t))|\mathcal{F}_t^\eta)}{E(M(t)|\mathcal{F}_t^\eta)}, \quad P\text{-a.s.}, \tag{5.209}$$

which is called the Kallianpur–Stiebel formula. We compute the RHS of (5.209). Since $\xi(t)$ is $\sigma(\xi_0, \beta(s), \eta(s), s \le t)$-measurable and $\beta$, $\eta$, and $\xi_0$ are mutually independent, we can compute $E(M(t)F(\xi(t))|\mathcal{F}_t^\eta)$, by taking the expectation w.r.t. $(\xi_0, \beta)$, freezing $\eta$ as if it were deterministic.

**Definition 5.11.** $E(M(t)F(\xi(t))|\mathcal{F}_t^\eta)$ is called the unnormalized conditional expectation of $F(\xi(t))$ given $\mathcal{F}_t^\eta$. An $(\mathcal{F}_t^\eta)$-progressively measurable $H_0$-process $\tilde{q}$ is called a density-valued process, if

$$\tilde{q}(t, x) \ge 0, \quad \forall (t, x), \quad P\text{-a.s.} \tag{5.210}$$

and for any $F \in C_K(\mathbb{R}^d)$,

$$E(M(t)F(\xi(t)))|\mathcal{F}_t^\eta) = \int_{\mathbb{R}^d} F(x)\tilde{q}(t,x)\,dx, \quad \forall t, \quad P\text{-a.s.} \tag{5.211}$$

**Proposition 5.9.**

(i) *Besides* $(\mathrm{d}_1^*)$–$(\mathrm{d}_3^*)$, *we assume that* $\alpha$ *and* $\sigma$ *are in* $C^1(\mathbb{R}^d)$ *w.r.t.* $x$. *Suppose that* $\xi_0$ *has the probability density* $p_0 \in H_0$. *If the density-valued process* $\tilde{q}$ *is in* $L^2([0,T];H_1) \cap C([0,T];H_0)$, $P$*-a.s., then* $\tilde{q}$ *is a generalized solution of the following stochastic linear parabolic equation* (5.212)–(5.213).

(ii) *If* $\alpha\alpha^\top$ *is uniformly positive definite, then the generalized solution is unique.*

$$d\tilde{q}(t) = \mathcal{L}^{\zeta(t,\eta)}(t)^*\tilde{q}(t)\,dt + \sum_{k=1}^m l_k^{\zeta(t,\eta)}(t)^*\tilde{q}(t)\,d\eta^k(t), \quad t \in (0,T], \tag{5.212}$$

*with the initial condition*

$$\tilde{q}(0) = p_0, \tag{5.213}$$

*where*

$$\mathcal{L}^\gamma(t)\phi(x) = \frac{1}{2}\mathrm{tr}((\alpha\alpha^\top + \sigma\sigma^\top)(t,x,\gamma)\partial_{xx}\phi(x)) + b(t,x,\gamma)\cdot\partial_x\phi(x), \tag{5.214}$$

$$l_k^\gamma(t)\phi(x) = \sum_{i=1}^d \sigma_k^i(t,x,\gamma)\partial_i\phi(x) + h^k(t,x,\gamma)\phi(x). \tag{5.215}$$

*Proof.* (i) Let $f \in C_b(\mathbb{R}^d \times \Gamma)$ and $F \in C_K^\infty(\mathbb{R}^d)$. Since

$$\begin{aligned} E(M(t)f(\xi(t),\zeta(t,\eta))|\mathcal{F}_T^\eta) &= E(M(t)f(\xi(t),\zeta(t,\eta))|\mathcal{F}_t^\eta) \\ &= (f(\cdot,\zeta(t,\eta)),\tilde{q}(t)), \quad \forall t, \quad P\text{-a.s.}, \end{aligned}$$

Itô's formula for $M(t)F(\xi(t))$ gives

$$\begin{aligned} (F,\tilde{q}(t)) - (F,p_0) &= \int_0^t E\Big(M(s)\mathcal{L}^{\zeta(s,\eta)}(s)F(\xi(s))|\mathcal{F}_T^\eta\Big)\,ds \\ &\quad + \sum_{k=1}^m \int_0^t E\Big(M(s)l_k^{\zeta(s,\eta)}(s)F(\xi(s))|\mathcal{F}_T^\eta\Big)\,d\eta^k(s) \\ &= \int_0^t \Big(\mathcal{L}^{\zeta(s,\eta)}(s)F,\tilde{q}(s)\Big)\,ds + \sum_{k=1}^m \int_0^t \Big(l_k^{\zeta(s,\eta)}(s)F,\tilde{q}(s)\Big)\,d\eta^k(s). \end{aligned} \tag{5.216}$$

This yields (i).

Since the superparabolicity condition ensures the uniqueness of the generalized solution [R90], Ch. 4, Th. 4.1.1, we obtain (ii).

This completes the proof. □

Regarding the relation between $\tilde{q}$ and (5.212), one can find details in [R90], Ch. 5.5.3.

*Example 5.12 (Filtering of factor process).* We consider a factor market model (refer to Sect. 2.4). Let $X$ be a $d$-dimensional factor process determining the performance of market. Suppose that $X$ evolves according to the SDE;

$$dX(t) = b(X(t))\,dt + \alpha(X(t))\,dB(t), \quad t > 0,$$

and $X(0)$ has probability density $p_0 \in H_0$. The price of the $k$-th asset is given by

$$\begin{cases} dS^k(t) = S^k(t)(h^k(X(t))\,dt + dW^k(t)), \quad t > 0, \\ S^k(0) = s_0^k > 0, \quad k = 1, \dots, m, \end{cases} \tag{5.217}$$

where $B$ and $W = (W^1, \dots, W^m)$ are mutually independent $d$- and $m$-dimensional Wiener processes.

We assume that $\alpha_j^i, b^i \in C_b^3(\mathbb{R}^d)$, $h^k \in C_b^2(\mathbb{R}^d)$, and $a(x) = \alpha(x)\alpha(x)^\top$ is uniformly positive definite.

We seek the unnormalized conditional probability of $X(t)$ given $\mathcal{F}_t^{(S^1,\dots,S^m)}$. From (5.217) we deduce that

$$\begin{aligned} Y^k(t) &:= \log S^k(t) \\ &= \log s_0^k + \int_0^t \left(h^k(X(s)) - \frac{1}{2}\right) ds + W^k(t) \end{aligned}$$

and $\mathcal{F}_t^{(S^1,\dots,S^m)} = \mathcal{F}_t^{(Y^1,\dots,Y^m)}$. Hence Proposition 5.9 yields the following stochastic linear parabolic equation for its density-valued process $q(t)$:

$$\begin{cases} dq(t) = \displaystyle\sum_{i=1}^d \partial_i \Big(\sum_{j=1}^d a^{ij}(x)\partial_j q(t) - \Big(b^i(x) - \sum_{j=1}^d \partial_j a^{ij}(x)\Big) q(t)\,dt\Big) \\ \qquad\qquad + \displaystyle\sum_{k=1}^m \Big(h^k(x) - \frac{1}{2}\Big) q(t)\,dW^k(t), \quad t > 0, \\ q(0) = p_0. \end{cases}$$

Hence (5.209) and (5.211) lead to

$$E(F(X(t))|\mathcal{F}_t^{(S^1,\dots,S^m)}) = \frac{\int_{\mathbb{R}^d} F(x)q(t,x)dx}{\int_{\mathbb{R}^d} q(t,x)dx}.$$

Now we will give the definition of the (controlled) Zakai equation.

Let $W$ be an $m$-dimensional Wiener process and $\gamma(t)$ be a $W$-adapted control process. The following stochastic linear parabolic equation is called the (controlled) Zakai equation:

$$du(t,x) = A^{\gamma(t)}(t)u(t,x)\,dt + \sum_{k=1}^{m} G_k^{\gamma(t)}(t)u(t,x)\,dW^k(t), \tag{5.218}$$

where

$$A^{\gamma}(t)\phi(x) = \sum_{i=1}^{d}\partial_i\Big(\sum_{j=1}^{d} a^{ij}(t,x,\gamma)\partial_j\phi(x) + b^i(t,x,\gamma)\phi(x)\Big) + \rho(t,x,\gamma)\phi(x),$$

$$G_k^{\gamma}(t)\phi(x) = \sum_{j=1}^{d} g_k^j(t,x,\gamma)\partial_j\phi(x) + \hbar_k(t,x,\gamma)\phi(x).$$

Hence (5.212) is the Zakai equation, because $\eta(t)$ is a Wiener process on $(\Omega, \mathcal{F}, P)$.

Let us return to (5.209). For the normalization factor, $N(t) := E(M(t)|\mathcal{F}_t^{\eta})$, observing that

$$dN(t) = \sum_{k=1}^{m} E\Big(M(t)h^k(t,\xi(t),\zeta(t,\eta))|\mathcal{F}_t^{\eta}\Big)d\eta^k(t),$$

we can easily deduce

*Remark 5.3.* $N(\cdot)$ is an $(\mathcal{F}_t^{\eta})$-exponential martingale under $P$, satisfying

$$N(t) = \exp\Big(\int_0^t \tilde{h}(s)\,d\eta(s) - \frac{1}{2}\int_0^t |\tilde{h}(s)|^2\,ds\Big), \tag{5.219}$$

where

$$\tilde{h}(t) = \tilde{E}(h(t,\xi(t),\zeta(t,\eta))|\mathcal{F}_t^{\eta}).$$

because (5.209) yields $dN(t) = N(t)\tilde{h}(t)d\eta(t)$.

### 5.5.2 Zakai Equation for a Conditional Gaussian Process

Let us consider the following system-observation SDEs:

### $d$-Dimensional System Equation

$$
\begin{cases}
dX(t) = (r_0(t,Y) + r_1(t,Y)X(t))\,dt \\
\qquad\qquad + \alpha(t,Y)\,dB + \sigma(t,Y)\,dW, \quad t \in (0,T], \\
X(0) = X_0 \quad (N(0,\mathrm{v}_0)\text{-distributed random variable}).
\end{cases}
\tag{5.220}
$$

### $m$-Dimensional Observation Equation

$$
\begin{cases}
dY(t) = h(t,Y)X(t)\,dt + \lambda\,dW(t), \quad t \in (0,T], \\
Y(0) = 0.
\end{cases}
\tag{5.221}
$$

where $\lambda$ is a positive constant.
Starting from (5.212)–(5.213), formal computations yield the Zakai equation

$$
\begin{cases}
dq(t) = \mathcal{L}^Y(t)^* q(t)\,dt + \sum_{k=1}^{m} l_k^Y(t)^* q(t)\,dY^k(t), \quad t \in (0,T], \\
q(0) = ((2\pi)^d \det \mathrm{v}_0)^{-\frac{1}{2}} \exp\Big(-\frac{1}{2} x^\top \mathrm{v}_0^{-1} x\Big),
\end{cases}
\tag{5.222}
$$

where

$$
\begin{cases}
\mathcal{L}^Y(t)\phi(x) = \frac{1}{2}\mathrm{tr}\{(\alpha\alpha^\top(t,Y) + \sigma\sigma^\top(t,Y))\partial_{xx}\phi(x)\} \\
\qquad\qquad + (r_0(t,Y) + r_1(t,Y)x)\cdot\partial_x\phi(x), \\
l_k^Y(t)\phi(x) = \lambda^{-1}\{\sigma_k(t,Y)\cdot\partial_x\phi(x) + h^k(t,Y)x\phi(x)\}, \quad k = 1,\dots,m.
\end{cases}
\tag{5.223}
$$

Here we will give an explicit formula for density valued processes (see (5.235)) and show that it gives the unique generalized solution of (5.222).

Let $B$ and $W$ be mutually independent $d_0$- and $m$-dimensional Wiener processes. Suppose that

$$
\begin{aligned}
&\alpha : [0,T] \times C([0,T];\mathbb{R}^m) \mapsto \mathbb{R}^d \otimes \mathbb{R}^{d_0}, \\
&\sigma : [0,T] \times C([0,T];\mathbb{R}^m) \mapsto \mathbb{R}^d \otimes \mathbb{R}^{m}, \\
&r_0 : [0,T] \times C([0,T];\mathbb{R}^m) \mapsto \mathbb{R}^d, \\
&r_1 : [0,T] \times C([0,T];\mathbb{R}^m) \mapsto \mathbb{R}^d \otimes \mathbb{R}^d, \\
&h : [0,T] \times C([0,T];\mathbb{R}^m) \mapsto \mathbb{R}^m \otimes \mathbb{R}^d
\end{aligned}
$$

are bounded, continuous, and non-anticipative. We assume the superparabolicity condition: $\alpha\alpha^\top + \sigma\sigma^\top(1-\lambda^{-2})$ is uniformly positive definite.

Referring to [LS01], Ch. 11, 12, 13, we study equations (5.220)–(5.221).

**1. Weak solution**

Let $\beta$ and $\tilde{\beta}$ be $d_0$- and $m$-dimensional Wiener processes and $\xi_0$ be a Gaussian random variable with $N(0, \mathrm{v}_0)$-distribution, defined on $(\Omega, \mathcal{F}, P)$. We assume that $\beta$, $\tilde{\beta}$, and $\xi_0$ are mutually independent. Put

$$\mathcal{F}_t = \mathcal{F}_t^{\xi_0 \beta \tilde{\beta}}.$$

We consider the following auxiliary SDEs

$$\begin{cases} d\xi(t) = \{r_0(t,\eta) + (r_1(t,\eta) - \lambda^{-1}\sigma(t,\eta)h(t,\eta))\xi(t)\}\,dt \\ \qquad\qquad + \alpha(t,\eta)\,d\beta(t) + \sigma(t,\eta)\,d\tilde{\beta}(t), \quad t \in (0,T], \\ \xi(0) = \xi_0 \end{cases} \tag{5.224}$$

and

$$\begin{cases} d\eta(t) = \lambda\,d\tilde{\beta}(t), \quad t \in (0,T], \\ \eta(0) = 0. \end{cases} \tag{5.225}$$

Since the system (5.224)–(5.225) has the unique strong solution $(\xi, \eta)$ and $(\xi_0, \beta, \eta)$ are mutually independent, (5.224) shows that the conditional probability of $(\xi(t_i), i = 1, \dots, n)$ given $\mathcal{F}_t^\eta$, is Gaussian for any $t$ and $0 \le t_1 < t_2 < \cdots < t_n \le t$.

*Note.* This property is called the conditional Gaussian property.

Now we define a new probability $\tilde{P}$ by

$$\tilde{P} = M(T) \circ P \quad \text{on } \mathcal{F}_T, \tag{5.226}$$

where $M(\cdot)$ is the exponential martingale given by

$$M(t) = \exp\Big(\lambda^{-1}\int_0^t h(s,\eta)\xi(s)\,d\tilde{\beta}(s) - \frac{\lambda^{-2}}{2}\int_0^t |h(s,\eta)\xi(s)|^2\,ds\Big). \tag{5.227}$$

From Girsanov's theorem it follows that

$$\tilde{W}(t) = \tilde{\beta}(t) - \tilde{\lambda}^{-1}\int_0^t h(s,\eta)\xi(s)\,ds, \quad t \in [0,T], \tag{5.228}$$

is an $m$-dimensional Wiener process independent of $(\beta, \xi_0)$ under $\tilde{P}$. Thus $(\xi, \eta)$ satisfies (5.220)–(5.221) with $\beta$ and $\tilde{W}$ on $(\Omega, \mathcal{F}, (\mathcal{F}_t), \tilde{P})$.

Hence (5.220)–(5.221) has one and only one weak solution.

**2. Conditional probability $\tilde{P}(\xi(t)|\mathcal{F}_t^\eta)$**

According to [LS01], Th. 12.6, the conditional Gaussian property is also valid under $\tilde{P}$.

We see that $\tilde{P}(\xi(t) \in \cdot|\mathcal{F}_t^\eta)$ is Gaussian distribution with mean $m(t) = \tilde{E}(\xi(t)|\mathcal{F}_t^\eta)$ and variance $V(t) = \tilde{E}((\xi(t)-m(t))(\xi(t)-m(t))^\top|\mathcal{F}_t^\eta)$. Since $(\xi,\eta)$ satisfies (5.220)–(5.221) with $\beta$ and $\tilde{W}$ on $(\Omega,\mathcal{F},(\mathcal{F}_t),\tilde{P})$, filtering theory asserts that

$$\bar{W}(t) := \lambda^{-1}\Big\{\eta(t) - \int_0^t h(s,\eta)\tilde{E}(\xi(s)|\mathcal{F}_s^\eta)\,ds\Big\} \tag{5.229}$$

is an $(\mathcal{F}_t^\eta)$-Wiener process and $(V(t),m(t))$ formally satisfies the following equations:

**Riccati Equation**

$$\begin{cases} \dfrac{dV}{dt}(t) = r_1(t,\eta)V(t) + V(t)r_1(t,\eta)^\top \\ \qquad\qquad + (\alpha(t,\eta)\alpha(t,\eta)^\top + \sigma(t,\eta)\sigma(t,\eta)^\top) \\ \qquad\qquad - \lambda^{-2}(\lambda\sigma(t,\eta) + V(t)h(t,\eta)^\top)(\lambda\sigma(t,\eta) + V(t)h(t,\eta)^\top)^\top, \\ \qquad\qquad t \in (0,T], \\ V(0) = v_0. \end{cases} \tag{5.230}$$

**Filtering Equation**

$$\begin{cases} dm(t) = (r_0(t,\eta) + r_1(t,\eta)m(t))\,dt \\ \qquad\qquad + (\lambda\sigma(t,\eta) + V(t)h(t,\eta)^\top)\,d\bar{W}(t), \quad t \in (0,T], \\ m(0) = 0. \end{cases} \tag{5.231}$$

When (5.230) admits a unique global solution $V(\cdot)$ in $C([0,T];S_{++}^d)$ for any path of $\eta$, (5.231) yields that for $F \in C_b(\mathbb{R}^d)$,

$$\tilde{E}(F(\xi(t))|\mathcal{F}_t^\eta) = \int_{\mathbb{R}^d} F(x)g(t,x)\,ds, \tag{5.232}$$

where $g$ is an $(\mathcal{F}_t^\eta)$-progressively measurable process given by the Gaussian density

$$g(t,x) = ((2\pi)^d \det V(t))^{-\frac{1}{2}} \exp\Big(-\frac{1}{2}(x-m(t))V^{-1}(t)(x-m(t))^\top\Big). \tag{5.233}$$

**3. Density-valued process $q(\cdot)$**

From the Bayes formula and (5.232) it follows that

$$\begin{aligned}\int_{\mathbb{R}^d} F(x)q(t,x)\,dx &= E(M(t)F(\xi(t))|\mathcal{F}_t^\eta)\\ &= \tilde{E}(F(\xi(t))|\mathcal{F}_t^\eta)E(M(t)|\mathcal{F}_t^\eta) \quad P\text{-a.s.}\end{aligned} \tag{5.234}$$

Thus (5.232) and Remark 5.3 yield

$$q(t,x) = g(t,x)\exp\Big(\lambda^{-1}\int_0^t h(s,\eta)m(s)\,d\eta - \frac{\lambda^{-2}}{2}\int_0^t |h(s,\eta)m(s)|^2\,ds\Big), \quad \forall t, \quad P\text{-a.s.} \tag{5.235}$$

Since $V(t)$ is positive definite and continuous in $t$, and $g(\cdot)$ is in $C([0,T];H_1)$, $P$-a.s., we obtain

$$q(\cdot) \in L^2([0,T];H_1) \cap C([0,T];H_0) \quad P\text{-a.s.} \tag{5.236}$$

Now Proposition 5.9 shows that $q(\cdot)$ is the unique generalized solution of (5.222).

Refer to [Be92], 4 and 6 for the solution of Zakai equations.

# Chapter 6
# Optimal Controls for Zakai Equations

**Abstract** This chapter is an application of previous one. In Sect. 5.5, we have introduced the (controlled) Zakai equation, which is a stochastic linear parabolic equation with a Brownian adapted control process. By using the results in Chap. 5, we will study control problems for Zakai equations related to partially observable diffusions. The control problem for partially observable diffusions turns out to be a completely observable control problem on a Hilbert space, by using the unnormalized conditional probability density given by the Zakai equation (cf. Bensoussan A, Stochastic control of partially observable systems. Cambridge University Press, Cambridge/New York, 1992; Lions, J Commun PDE 8:1101–1134, 1983, I, II; Gozzi and Świech, J Funct Analy 172:466–510, 2000). Section 6.1 is devoted to the analysis of controlled Zakai equations. In Sect. 6.2, we formulate control problems for a system governed by Zakai equations, in the same way as in Chap. 2. When a control process $\gamma(\cdot)$ is chosen, the cost on a time internal $[T_0, t]$ is given by $\int_{T_0}^{t} r(u^{\gamma(\cdot)}(s), \gamma(s))\, ds + F(u^{\gamma(\cdot)}(t))$, where $u^{\gamma(\cdot)}(\cdot)$ is the response of $\gamma(\cdot)$. By taking a suitable control process, we want to minimize (or maximize) the expectation of the cost. In Sect. 6.3 we formulate the DPP via the semigroup constructed from the value function, whose generator is related to the HJB equation on a Hilbert space. The viscosity solution of HJB equation is introduced following Gozzi and Świech (J Funct Analy 172:466–510, 2000) in Sect. 6.4. Example 6.1 makes explicitly the connection between controlled Zakai equations and control of partially observable diffusions.

## 6.1 Controlled Zakai Equations

We assume that the control region $\Gamma$ $(\subset \mathbb{R}^q)$ is convex and compact. Let $W$ be an $m$-dimensional Wiener process, defined on $(\Omega, \mathcal{F}, P)$. Since we use $W$-adapted control processes, we put

$$\mathcal{F}_t = \mathcal{F}_t^W \quad \text{and} \quad \boldsymbol{\Gamma} = \boldsymbol{\Gamma}^W.$$

Here we consider time-homogeneous controlled Zakai equations in our convenient assumptions; $(B_1')$ and $(B_2')$. By using results in Chap. 5, we investigate properties of solutions. In Sect. 6.1.1 we recall basic results, obtained in Chap. 5, and in Sect. 6.1.2 we study properties of solutions needed in the sequel.

© Springer Japan 2015

M. Nisio, *Stochastic Control Theory*, Probability Theory and Stochastic Modelling 72, DOI 10.1007/978-4-431-55123-2_6

### 6.1.1 Preliminaries

For $\gamma(\cdot) \in \boldsymbol{\Gamma}$, we consider the time-homogeneous controlled Zakai equation

$$\begin{aligned} du(t,x) =& \Big\{ \sum_{i=1}^{d} \partial_i \Big( \sum_{j=1}^{d} a^{ij}(x,\gamma(t)) \partial_j u(t,x) + b^i(x,\gamma(t)) u(t,x) \Big) \\ &+ \rho(x,\gamma(t)) u(t,x) \Big\} \, dt \\ &+ \sum_{k=1}^{m} \Big( \sum_{j=1}^{d} g_k^j(x,\gamma(t)) \partial_j u(t,x) + \hbar_k(x,\gamma(t)) u(t,x) \Big) \, dW^k(t). \end{aligned} \tag{6.1}$$

We assume that the following conditions are satisfied:

$(B_1')$

$$z := a^{ij}, \quad b^i, \quad \partial_i a^{ij}, \quad \partial_i b^i, \quad \rho, \quad g_k^j, \quad \partial_j g_k^j, \quad \text{and} \quad \hbar_k,$$
$$(i, j = 1, \ldots, d, k = 1, \ldots, m)$$

are Lipschitz continuous, say

$$|z(x_1,\gamma_1) - z(x_2,\gamma_2)|^2 \leq l(|x_1 - x_2|^2 + |\gamma_1 - \gamma_2|^2). \tag{6.2}$$

Further, there is a constant $K > 0$, such that

$$\begin{cases} \|z(\cdot,\gamma)\|_{C^3} \leq K, \ \forall \gamma \in \Gamma \ \text{for } z = a^{ij}, b^i, \\ \|z(\cdot,\gamma)\|_{C^2} \leq K, \ \forall \gamma \in \Gamma \ \text{for } z = g_k^i, \hbar_k. \end{cases} \tag{6.3}$$

$(B_2')$ ***Superellipticity*** (stronger than $(B_2)$)
There is a positive constant $\lambda_0$ such that

$$y^\top \Big( 2a(x,\gamma) - 3g(x,\gamma)g(x,\gamma)^\top \Big) y \geq \lambda_0 |y|^2, \ \forall y \in \mathbb{R}^d, \ \forall (x,\gamma) \in \mathbb{R}^d \times \Gamma.$$

We need $(B_2')$ in order to estimate $E\,\|u(t)\|^4$ (see Proposition 6.2).

In this chapter, we always assume $(B_1')$ and $(B_2')$. Hence $(B_4)$ holds, i.e., $z(t,x,\omega) := z(x,\gamma(t,\omega))$, for $z = a^{ij}, b^i, \rho, g_k^i, \hbar_k$, satisfies (5.104) with the constant $K_1$ independent of $\gamma(\cdot)$. We call the solution of (6.1) the response for $\gamma(\cdot)$ and denote it by $u^{\gamma(\cdot)}(\cdot)$.

Using Theorem 5.5 and Proposition 5.4 we obtain

**Theorem 6.1.** *Let $u_0$ be in $L^2(\Omega, \mathcal{F}_{T_0}; H_0)$. Then*

(i) *The controlled Zakai equation (6.1) with the initial condition $u(T_0) = u_0$ admits the unique solution $u(\cdot) := u^{\gamma(\cdot)}_{T_0 u_0}(\cdot)$ in $L^2([T_0, T] \times \Omega, (\mathcal{F}_t); H_1) \cap L^2(\Omega; C([T_0, T]; H_0))$ and obeys the estimate*

$$E_{T_0 u_0}\Big[\sup_{T_0 \le t \le T} \|u^{\gamma(\cdot)}(t)\|^2 + \int_{T_0}^{T} |||u^{\gamma(\cdot)}(t)|||^2\, dt\Big] \le N_0 E\|u_0\|^2, \tag{6.4}$$

*with a constant $N_0$ independent of $\gamma(\cdot)$, where the subscript $(T_0, u_0)$ refers to the initial condition.*

(ii) *If $u_0$ is in $L^2(\Omega, \mathcal{F}_{T_0}; H_1)$, then $u^{\gamma(\cdot)}_{T_0 u_0}(\cdot)$ is in $L^2([T_0, T] \times \Omega, (\mathcal{F}_t); H_2) \cap L^2(\Omega; C([T_0, T]; H_1))$ and satisfies*

$$E_{T_0 u_0}\Big[\sup_{T_0 \le t \le T} |||u^{\gamma(\cdot)}(t)|||^2 + \int_{T_0}^{T} \mathbb{I}u^{\gamma(\cdot)}(t)\mathbb{I}^2\, dt\Big] \le N_0 E|||u_0|||^2, \tag{6.5}$$

*with a constant $N_0$ independent of $\gamma(\cdot)$.*

Sometimes we omit the subscript $(T_0, u_0)$ for simplicity. We recall the notation

$$\mathbf{1}\gamma(\cdot) - \tilde{\gamma}(\cdot)\mathbf{1}_{T_0 T} := \|\gamma(\cdot) - \tilde{\gamma}(\cdot)\|_{L^2[T_0, T]}$$

and omit the subscript $T_0, T$, when there is no danger of confusion.

### 6.1.2 Basic Properties of Solutions

In this subsection, we will show that the solution depends continuously on the initial state, the time parameter and the control process. Here we deal with Eq. (6.1) in the Hilbert space framework, by putting

$$A^{\gamma(\cdot)}(t)\phi(x) = \sum_{i=1}^{d} \partial_i \Big(\sum_{j=1}^{d} a^{ij}(x, \gamma(t))\partial_j \phi(x) + b^i(x, \gamma(t))\phi(x)\Big) + \rho(x, \gamma(t))\phi(x),$$

$$G_k^{\gamma(\cdot)}(t)\phi(x) = \sum_{j=1}^{d} g_k^j(x, \gamma(t))\partial_j \phi(x) + \hbar_k(x, \gamma(t))\phi(x),$$

and

$$G^{\gamma(\cdot)}(t) = (G_1^{\gamma}(\cdot)(t), \ldots, G_m^{\gamma(\cdot)}(t)).$$

$(B_1')$ and $(B_2')$ ensure that the following coercivity condition is satisfied.

There are two constants, $\bar{\lambda} > 0$ and $\bar{\mu} \in \mathbb{R}^1$, such that, for any $\gamma(\cdot)$,

$$2\langle A^{\gamma(\cdot)}(t)\phi, \phi\rangle + 3\|G^{\gamma(\cdot)}(t)\phi\|_m^2 \leq -\bar{\lambda}|||\phi|||^2 + \bar{\mu}\|\phi\|^2, \quad \forall t, \quad P\text{-a.s.}, \quad \forall \phi \in H_1 \tag{6.6}$$

and

$$\begin{aligned} &2\langle A^{\gamma(\cdot)}(t)\phi, \phi\rangle_{(H_0 H_2)} + 3|||G^{\gamma(\cdot)}(t)\phi|||_m^2 \\ \leq &-\bar{\lambda}\mathbb{I}\phi\mathbb{I}^2 + \bar{\mu}|||\phi|||^2, \quad \forall t, \quad P\text{-a.s.}, \quad \forall \phi \in H_2 \end{aligned} \tag{6.7}$$

## 1. Dependence on the initial state

Let $u^{\gamma(\cdot)}(\cdot)$ and $\tilde{u}^{\gamma(\cdot)}(\cdot)$ be the responses for $\gamma(\cdot)$ with initial condition $u_0$ and $\tilde{u}_0$, respectively. Since (6.1) is a linear stochastic parabolic equation, $u^{\gamma}(\cdot)$ is the 0 process, if the initial state $u_0 = 0$, and Theorem 5.4 yields

**Proposition 6.1.**

$$E\Big[\sup_{T_0 \leq t \leq T} \|u^{\gamma(\cdot)}(t) - \tilde{u}^{\gamma(\cdot)}(t)\|^2 + \int_{T_0}^T |||u^{\gamma(\cdot)}(t) - \tilde{u}^{\gamma(\cdot)}(t)|||^2\, dt\Big] \leq N_1 E\|u_0 - \tilde{u}_0\|^2 \tag{6.8}$$

*and*

$$E\Big[\sup_{T_0 \leq t \leq T} |||u^{\gamma(\cdot)}(t) - \tilde{u}^{\gamma(\cdot)}(t)|||^2 + \int_{T_0}^T \mathbb{I}u^{\gamma(\cdot)}(t) - \tilde{u}^{\gamma(\cdot)}(t)\mathbb{I}^2\, dt\Big] \leq N_1 E|||u_0 - \tilde{u}_0|||^2 \tag{6.9}$$

*with a constant $N_1$ depending only on $\bar{\lambda}$, $\bar{\mu}$, and $K$ of* (6.3).

## 2. Estimation of the 4th moment

**Proposition 6.2.**

$$E_{T_0 u_0}\Big[\sup_{T_0 \leq t \leq T} \|u^{\gamma(\cdot)}(t)\|^4\Big] \leq N_2 E\|u_0\|^4, \tag{6.10}$$

$$E_{T_0 u_0}\Big[\sup_{T_0 \leq t \leq T} |||u^{\gamma(\cdot)}(t)|||^4\Big] \leq N_2 E|||u_0|||^4 \tag{6.11}$$

*with a constant $N_2$ depending only on $\bar{\lambda}$, $\bar{\mu}$, and $K$.*

*Proof.* For simplicity of notation, we omit $\gamma(\cdot)$ and $(T_0, u_0)$. $c_i, i = 1, 2, \ldots$ will denote constants independent of $t, \gamma(\cdot)$ and $u_0$.

From the energy equality

$$d\|u(t)\|^2 = (2\langle A(t)u(t), u(t)\rangle + \|G(t)u(t)\|_m^2)\, dt + 2(G(t)u(t)\, dW(t), u(t))$$

it follows that

$$
\begin{aligned}
d\|u(t)\|^4 &= \{4\|u(t)\|^2\langle A(t)u(t), u(t)\rangle + 2\|u(t)\|^2\|G(t)u(t)\|_m^2 \\
&\quad + 4|(G(t)u(t), u(t))_m|^2\}\,dt + 4\|u(t)\|^2(G(t)u(t), u(t))_m\,dW(t) \\
&\le (-2\bar{\lambda}\|u(t)\|^2|||u(t)|||^2 + 2\bar{\mu}\|u(t)\|^4)\,dt \\
&\quad + \text{stochastic integral,} \qquad (6.12)
\end{aligned}
$$

by (6.6). Let $\tau_n$ be the exit time of $\|u(t)\|$ from $[0, n]$. Then (6.12) yields

$$
E\|u(t\wedge\tau_n)\|^4 - E\|u_0\|^4 \le 2E\int_{T_0}^{t\wedge\tau_n}(-\bar{\lambda}\|u(s)\|^2|||u(s)|||^2 + \bar{\mu}\|u(s)\|^4)\,ds. \qquad (6.13)
$$

Therefore,

$$
E\|u(t\wedge\tau_n)\|^4 \le e^{2\bar{\mu}\cdot(t-T_0)}E\|u_0\|^4 \qquad (6.14)
$$

and

$$
E\int_{T_0}^{\tau_n}\|u(s)\|^2|||u(s)|||^2\,ds \le (2\bar{\lambda})^{-1}e^{2\bar{\mu}T}E\|u_0\|^4. \qquad (6.15)
$$

Next we compute the stochastic integral term. By the Burkholder–Davis–Gundy inequality, we have

$$
\begin{aligned}
&E\Big[\sup_{\theta<t\wedge\tau_n}\Big|\int_{T_0}^{\theta}\|u(s)\|^2\sum_{k=1}^{m}(G_k(s)u(s), u(s))\,dW^k(s)\Big|\Big] \\
&\le 6E\Big(\int_{T_0}^{t\wedge\tau_n}\|u(s)\|^6\|G(s)u(s)\|_m^2\,ds\Big)^{\frac{1}{2}} \\
&\le c_1E\Big(\int_{T_0}^{t\wedge\tau_n}\|u(s)\|^6|||u(s)|||^2\,ds\Big)^{\frac{1}{2}} \\
&\le c_1E\Big[\sup_{T_0\le s\le t\wedge\tau_n}\|u(s)\|^2\Big(\int_{T_0}^{t\wedge\tau_n}\|u(s)\|^2|||u(s)|||^2\,ds\Big)^{\frac{1}{2}}\Big] \\
&\le \frac{1}{8}J(t\wedge\tau_n) + c_2E\Big[\int_{T_0}^{t\wedge\tau_n}\|u(s)\|^2|||u(s)|||^2\,ds\Big], \qquad (6.16)
\end{aligned}
$$

where $J(\theta) = E[\sup_{T_0<s<\theta}\|u(s)\|^4]$.

Combining (6.12), (6.15), and (6.16) we obtain

$$
\begin{aligned}
J(t\wedge\tau_n) &\le 2\bar{\mu}E\int_{T_0}^{t}\|u(s\wedge\tau_n)\|^4\,ds + \frac{1}{2}J(t\wedge\tau_n) + c_3E\|u_0\|^4 \\
&\le \frac{1}{2}J(t\wedge\tau_n) + c_4E\|u_0\|^4 \quad \text{(by (6.14))}. \qquad (6.17)
\end{aligned}
$$

Letting $n \to \infty$, we conclude (6.10) by the monotone convergence theorem.

Finally, we prove (6.11) in the same way. Suppose that $E|||u_0|||^4 < \infty$. Then $u(\cdot)$ has continuous $H_1$-paths. Let $\tau_n$ be the exit time of $|||u(t)|||$ from $[0, n]$, and put

$$\xi(t) = |||u(t \wedge \tau_n)|||^2. \tag{6.18}$$

The energy equality (cf. Theorem 5.4) gives the dynamics of $\xi(t)$ and we obtain

$$\begin{aligned} d\xi^2(t) &= 2\xi(t)(2\langle A(t)u(t), u(t)\rangle_{(H_0 H_2)} + |||G(s)u(s)|||_m^2)\,dt \\ &\quad + 4\sum_{k=1}^{m}((G_k(t)u(t), u(t)))^2\,dt \\ &\quad + 4\xi(t)\sum_{k=1}^{m}((G_k(t)u(t), u(t)))\,dW^k(t) \\ &\leq 2\xi(t)(2\langle A(t)u(t), u(t)\rangle_{(H_0 H_2)} + 3|||G(s)u(s)|||_m^2)\,dt \\ &\quad + \text{stochastic integral}. \end{aligned} \tag{6.19}$$

From (6.19) and (6.7), it follows that

$$E[\xi(t)^2] \leq e^{2\bar{\mu}(t-T_0)}E|||u_0|||^4 \tag{6.20}$$

and

$$E\int_{T_0}^{\tau_n}\xi(s)\mathbb{I}u(s)\mathbb{I}^2\,ds \leq (2\bar{\lambda})^{-1}e^{2\bar{\mu}T}E|||u_0|||^4. \tag{6.21}$$

We estimate the stochastic integral term in the same way as (6.16), namely

$$E\Big[\sup_{T_0\leq\theta\leq t\wedge\tau_n}\Big|\int_{T_0}^{\theta}\xi(s)\sum_{k=1}^{m}((G_k(s)u(s), u(s)))\,dW^k(s)\Big|\Big] \leq c_5 E|||u_0|||^4. \tag{6.22}$$

Combining (6.19), (6.21), and (6.22) we conclude (6.11).

This completes the proof. □

## 3. Dependence on the time parameter

**Proposition 6.3.**

(i) *There is a constant $N_3$, depending only on $\bar{\lambda}$, $\bar{\mu}$ and $K$, such that, for any $\theta, t$ and $\gamma(\cdot)$,*

$$E_{T_0 u_0}\Big[\sup_{\theta\leq s\leq t}\|u^{\gamma(\cdot)}(s) - u^{\gamma(\cdot)}(\theta)\|^2\Big] \leq N_3|t-\theta|E|||u_0|||^2, \tag{6.23}$$

$$E_{T_0 u_0}\Big[\sup_{\theta\leq s\leq t}\|u^{\gamma(\cdot)}(s) - u^{\gamma(\cdot)}(\theta)\|_*^2\Big] \leq N_3|t-\theta|E\|u_0\|^2. \tag{6.24}$$

(ii) *Let* $u_0 = \phi$ $(\in H_0)$. *For* $\varepsilon > 0$, *there is* $\Delta(\varepsilon, \phi) > 0$, *such that for any* $t_1, t_2$ *and* $\gamma(\cdot)$,

$$E_{T_0\phi}\Big[\sup_{\theta,t\in[t_1,t_2]} \|u^{\gamma(\cdot)}(t) - u^{\gamma(\cdot)}(\theta)\|^2\Big] < \varepsilon \tag{6.25}$$

*whenever* $t_2 - t_1 < \Delta(\varepsilon, \phi)$.

(iii) ***Continuity at the initial time***

$$\lim_{\theta\downarrow T_0} E_{T_0\phi}\Big[\sup_{T_0<s<\theta} \|u^{\gamma(\cdot)}(s) - \phi\|^2\Big] = 0, \quad \textit{uniformly in } \gamma(\cdot), \tag{6.26}$$

$$\lim_{T_0\uparrow\theta} E_{T_0\phi}\Big[\sup_{T_0<s\le\theta} \|u^{\gamma(\cdot)}(s) - \phi\|^2\Big] = 0, \quad \textit{uniformly in } \gamma(\cdot). \tag{6.27}$$

*Proof.* (i) is immediate from (5.111) and (5.196).
(ii) Suppose $\psi \in H_1$. By the inequality

$$\sup_{\theta,t\in[t_1,t_2]} \|u(t) - u(\theta)\|^2 \le 2\sup_{t\in[t_1,t_2]} \|u(t) - u(t_1)\|^2 + 2\sup_{\theta\in[t_1,t_2]} \|u(\theta) - u(t_1)\|^2,$$

$$\bar{\Delta}(\varepsilon, \psi) := \varepsilon(4N_3|||\psi|||^2)^{-1} \tag{6.28}$$

satisfies (6.25) thanks to (6.23).
For $\phi \in H_0$, we fix $\tilde{\phi} \in H_1$, so that

$$\|\phi - \tilde{\phi}\|^2 < \frac{\varepsilon}{12N_1}. \tag{6.29}$$

Since for the response $\tilde{u}(t)$ with initial state $\tilde{\phi}$ (6.8) yields

$$E\Big[\sup_{T_0\le t\le T} \|u(t) - \tilde{u}(t)\|^2\Big] < \frac{\varepsilon}{12}, \tag{6.30}$$

we obtain

$$\begin{aligned} &E\Big[\sup_{\theta,t\in[t_1,t_2]} \|u(t) - u(\theta)\|^2\Big] \\ &\le 6E\Big[\sup_{T_0\le t\le T} \|u(t) - \tilde{u}(t)\|^2\Big] + 3E\Big[\sup_{\theta,t\in[t_1,t_2]} \|\tilde{u}(t) - \tilde{u}(\theta)\|^2\Big] \\ &< \varepsilon \quad \text{if } t_2 - t_1 < \bar{\Delta}\Big(\frac{\varepsilon}{6}, \tilde{\phi}\Big)(=: \Delta(\varepsilon, \phi)). \end{aligned}$$

This gives (ii).
Since (iii) is clear by (ii), we have completed the proof. □

**4. Dependence on control processes**

**Proposition 6.4.**

(i) *There is a constant $N_4$, depending only on $\bar{\lambda}, \bar{\mu}, K$, and $l$, such that, for $\psi \in H_1$,*

$$E_{T_0\psi}\Big[\sup_{T_0\le t\le T}\|u^{\gamma(\cdot)}(t)-u^{\tilde{\gamma}(\cdot)}(t)\|^2+\int_{T_0}^{T}|||u^{\gamma(\cdot)}(t)-u^{\tilde{\gamma}(\cdot)}(t)|||^2\,dt\Big]$$
$$\le N_4|||\psi|||^2(E[\mathbf{1}\gamma(\cdot)-\tilde{\gamma}(\cdot)\mathbf{1}^4])^{\frac{1}{2}}. \tag{6.31}$$

(ii) *For $\phi \in H_0$ and $\varepsilon > 0$, there is $\delta(\varepsilon,\phi) > 0$, such that*

$$E_{T_0\tilde{\phi}}\Big[\sup_{T_0\le t\le T}\|u^{\gamma(\cdot)}(t)-u^{\tilde{\gamma}(\cdot)}(t)\|^2+\int_{T_0}^{T}|||u^{\gamma(\cdot)}(t)-u^{\tilde{\gamma}(\cdot)}(t)|||^2\,dt\Big]<\varepsilon \tag{6.32}$$

*for any $\tilde{\phi} \in H_0$ with $\|\phi-\tilde{\phi}\|^2 < \frac{\varepsilon}{48N_1}$ whenever $E[\mathbf{1}\gamma(\cdot)-\tilde{\gamma}(\cdot)\mathbf{1}^4] < \delta(\varepsilon,\phi)$.*

*Proof.* For simplicity, we put $u = u^{\gamma(\cdot)}, \tilde{u} = u^{\tilde{\gamma}(\cdot)}$ and $v = u - \tilde{u}$. Then we have

$$dv(t) = A^{\gamma(\cdot)}v(t)\,dt + G^{\gamma(\cdot)}(t)v(t)\,dW(t) + Z(t)\,dt + \Psi(t)\,dW(t),\ t \in (T_0, T] \tag{6.33}$$

with

$$v(T_0) = 0,$$

where

$$Z(t) = (A^{\gamma(\cdot)}(t) - A^{\tilde{\gamma}(\cdot)}(t))\tilde{u}(t)$$

and

$$\Psi(t) = (G^{\gamma(\cdot)}(t) - G^{\tilde{\gamma}(\cdot)}(t))\tilde{u}(t).$$

Since $(B_1')$ provides a constant $c$, depending only on $l$, such that

$$\|Z(t)\|_*^2 \le c|\gamma(t)-\tilde{\gamma}(t)|^2|||\tilde{u}(t)|||^2, \quad \forall t, \quad P\text{-a.s.} \tag{6.34}$$

and

$$\|\Psi(t)\|_m^2 \le c|\gamma(t)-\tilde{\gamma}(t)|^2|||\tilde{u}(t)|||^2, \quad \forall t, \quad P\text{-a.s.} \tag{6.35}$$

by using the estimate

$$E\int_{T_0}^{T} |\gamma(t)-\tilde{\gamma}(t)|^2 |||\tilde{u}(t)|||^2\,dt \le E\Big[\sup_{T_0\le t\le T} |||\tilde{u}(t)|||^2 \mathbf{1}\gamma(\cdot)-\tilde{\gamma}(\cdot)\mathbf{1}^2\Big]$$

$$\le \Big(E\Big[\sup_{T_0\le t\le T} |||\tilde{u}(t)|||^4\Big]\Big)^{\frac{1}{2}} (E\mathbf{1}\gamma(\cdot)-\hat{\gamma}(\cdot)\mathbf{1}^4)^{\frac{1}{2}}$$

and (6.33)–(6.35), we can deduce (i) from (6.11).

(ii) For $\phi \in H_0$ and $\varepsilon > 0$, we fix $\phi^{\#} \in H_1$ such that

$$\|\phi-\phi^{\#}\|^2 < \frac{\varepsilon}{48N_1}. \tag{6.36}$$

Hence,

$$\|\tilde{\phi}-\phi^{\#}\|^2 < \frac{\varepsilon}{12N_1} \quad \text{for } \|\phi-\tilde{\phi}\|^2 < \frac{\varepsilon}{48N_1}. \tag{6.37}$$

Stressing the dependence on the initial state of response, we write

$$u_{T_0\tilde{\phi}}^{\gamma(\cdot)}(t) - u_{T_0\tilde{\phi}}^{\tilde{\gamma}(\cdot)}(t)$$
$$= (u_{T_0\tilde{\phi}}^{\gamma(\cdot)}(t) - u_{T_0\phi^{\#}}^{\gamma(\cdot)}(t)) + (u_{T_0\phi^{\#}}^{\gamma(\cdot)}(t) - u_{T_0\phi^{\#}}^{\tilde{\gamma}(\cdot)}(t)) + (u_{T_0\phi^{\#}}^{\tilde{\gamma}(\cdot)}(t) - u_{T_0\tilde{\phi}}^{\tilde{\gamma}(\cdot)}(t)). \tag{6.38}$$

Taking into account (6.8) and (6.31), we obtain

$$\text{LHS of (6.32)} \ < \frac{\varepsilon}{2} + N_4|||\phi^{\#}|||^2 (E\mathbf{1}\gamma(\cdot)-\tilde{\gamma}(\cdot)\mathbf{1}^4)^{\frac{1}{2}}.$$

Thus, putting $\delta(\varepsilon,\phi) := \varepsilon^2(2N_4|||\phi^{\#}|||^2)^{-2}$ we complete the proof. □

## 6.2 Formulations of Control Problems

In this section, we formulate optimization problems for stochastic systems described by Zakai equations, namely, we are concerned with optimization problems similar to those in Chap. 2, for Hilbert space-valued stochastic systems. When a control process $\gamma(\cdot)$ is chosen, the cost on the time internal $[T_0, t]$ is given by

$$\int_{T_0}^{t} r(u^{\gamma(\cdot)}(s), \gamma(s))\,ds + F(u^{\gamma(\cdot)}(t)),$$

where the integral part and $F(u^{\gamma(\cdot)}(t))$ denote a running cost and a terminal cost, respectively. We want to minimize (or maximize) the expectation of the cost, by taking a suitable $\gamma(\cdot)$. This optimization is a generalization of optimal control problem for partially observable diffusions, see Example 6.1. For the terminal cost, we introduce a Banach space $C^\#$ (see (6.40)). The basic properties of the terminal costs and running costs are studied in Sects. 6.2.2 and 6.2.3, respectively. Putting these results together, we state properties of the value function in Sect. 6.2.4.

### 6.2.1 Preliminaries

Firstly we define a Banach space $C^\#$. Put

$$F^\#(\phi) = \frac{F(\phi)}{1 + \|\phi\|} \quad \text{for } \phi \in H_0, \tag{6.39}$$

and define $C^\#$ by

$$C^\# = \{F \in C(H_0); F^\# \in C_{\mathrm{bu}}(H_0)\}. \tag{6.40}$$

Then $C^\#$ becomes a Banach space, equipped with norm $\| \quad \|_\#$ given by

$$\|F\|_\# = \sup_{\phi \in H_0} |F^\#(\phi)| \quad \text{for } F \in C^\#. \tag{6.41}$$

We write

$$F_1^\# \le F_2^\# \text{ if and only if } F_1^\#(\phi) \le F_2^\#(\phi), \ \forall \phi \in H_0, \tag{6.42}$$

or, equivalently $F_1(\phi) \le F_2(\phi), \forall \phi \in H_0$.

For the set $C^\#$, the following facts are clear. If $F$ is uniformly continuous, then $F \in C^\#$. If $F \in C^\#$, then $F$ is (at most) linearly growing and uniformly continuous on any bounded set.

From now on, we always take terminal costs from $C^\#$.

**Proposition 6.5.** *Let $F \in C^\#$ be given. We define $F^\S(\cdot) := F^\S(\cdot; T_0, t, \gamma(\cdot))$ by*

$$F^\S(\phi) = E_{T_0\phi} F(u^{\gamma(\cdot)}(t)), \quad \phi \in H_0. \tag{6.43}$$

*Then $F^\S$ is in $C^\#$. More precisely,*

$$\frac{|F^\S(\phi)|}{1 + \|\phi\|} \le \sqrt{1 \vee N_0}\|F\|_\#, \tag{6.44}$$

*with the constant $N_0$ of* (6.4), *and, for $\varepsilon > 0$, there is $\delta_1(\varepsilon; F) > 0$ independent of $t, T_0$ and $\gamma(\cdot)$, such that*

$$\left| \frac{F^{\S}(\phi)}{1 + \|\phi\|} - \frac{F^{\S}(\tilde{\phi})}{1 + \|\tilde{\phi}\|} \right| < \varepsilon \tag{6.45}$$

*whenever* $\|\phi - \tilde{\phi}\| < \delta_1(\varepsilon, F)$.

*Proof.* $c_i$ denotes a constant depending only on $\bar{\lambda}, \bar{\mu}, K$, and $l$. By (6.8), $F^{\S}$ is continuous. For simplicity, we put $u(t) = u^{\gamma(\cdot)}_{T0\phi}(t)$. Equation (6.44) follows from the inequality

$$\frac{|F(u(t))|}{1 + \|\phi\|} \leq \|F\|_{\#} \frac{1 + \|u(t)\|}{1 + \|\phi\|}. \tag{6.46}$$

Putting $\tilde{u}(t) = u^{\gamma(\cdot)}_{T0\tilde{\phi}}(t)$ and $v(t) = u(t) - \tilde{u}(t)$, let us show (6.45). We have

$$\begin{aligned}
&\left| F^{\#}(u(t)) \frac{1 + \|u(t)\|}{1 + \|\phi\|} - F^{\#}(\tilde{u}(t)) \frac{1 + \|\tilde{u}(t)\|}{1 + \|\tilde{\phi}\|} \right| \\
&\leq m_{F^{\#}}(\|v(t)\|) \frac{1 + \|u(t)\|}{1 + \|\phi\|} + \|F\|_{\#} \left| \frac{1 + \|u(t)\|}{1 + \|\phi\|} - \frac{1 + \|\tilde{u}(t)\|}{1 + \|\tilde{\phi}\|} \right| \\
&:= J_1 + \|F\|_{\#} J_2.
\end{aligned} \tag{6.47}$$

Since the modulus $m(\cdot)$ satisfies that, for any $\varepsilon > 0$, there is $C_\varepsilon > 0$, such that $m(a) \leq \varepsilon + C_\varepsilon a, a \geq 0$, we have

$$\begin{aligned}
E[m_{F^{\#}}(\|v(t)\|)^2] &\leq 2(\varepsilon^2 + C_\varepsilon^2 E\|v(t)\|^2) \\
&\leq 2(\varepsilon^2 + C_\varepsilon^2 N_1 \|\phi - \tilde{\phi}\|^2),
\end{aligned} \tag{6.48}$$

by (6.8). Thus, (6.4) and (6.48) yield

$$(EJ_1)^2 \leq c_1(\varepsilon^2 + C_\varepsilon^2 \|\phi - \tilde{\phi}\|^2). \tag{6.49}$$

For $EJ_2$, we observe that

$$J_2 \leq \frac{\|v(t)\|}{1 + \|\phi\|} + \frac{(1 + \|\tilde{u}(t)\|)\|\phi - \tilde{\phi}\|}{(1 + \|\phi\|)(1 + \|\tilde{\phi}\|)}. \tag{6.50}$$

Hence again (6.4) and (6.8) yield

$$EJ_2 \leq c_2 \|\phi - \tilde{\phi}\|. \tag{6.51}$$

which in conjunction with (6.47) and (6.49) gives (6.45). □

### 6.2.2 Properties of the Terminal Cost

Let us study how the terminal cost function, $\Lambda(t, T_0, \phi, \gamma(\cdot); F) := E_{T_0\phi} F(u^{\gamma(\cdot)}(t))$, depends on $T_0, \phi$, and $\gamma(\cdot)$.

**1. Dependence on the initial state**

**Proposition 6.6.** *For positive constants $\varepsilon$ and $n$, there is $\delta_2(\varepsilon, n; F) > 0$, such that*

$$E\Big[\sup_{T_0 \le t \le T} \Big|F(u^{\gamma(\cdot)}_{T_0\phi_1}(t)) - F(u^{\gamma(\cdot)}_{T_0\phi_2}(t))\Big|\Big] < \varepsilon, \quad \forall T_0 \text{ and } \gamma(\cdot), \tag{6.52}$$

*whenever $\|\phi_i\| < n$ $(i = 1, 2)$ and $\|\phi_1 - \phi_2\| < \delta_2(\varepsilon, n; F)$.*

The proof is easy, using $F(\psi) = F^{\#}(\psi)(1 + \|\psi\|)$.

**2. Dependence on time**

**Proposition 6.7.** *For $\phi \in H_0$ and $\varepsilon > 0$, there is $\delta_3(\varepsilon, \phi; F) > 0$, such that, for $t_1, t_2 \in [T_0, T]$,*

$$E_{T_0\phi}\Big[\sup_{s_1, s_2 \in [t_1, t_2]} |F(u^{\gamma(\cdot)}(s_2)) - F(u^{\gamma(\cdot)}(s_1))|\Big] < \varepsilon, \quad \forall \gamma(\cdot) \in \boldsymbol{\Gamma}, \tag{6.53}$$

*whenever $t_2 - t_1 < \delta_3(\varepsilon, \phi; F)$.*

*Proof.* By changing $\varepsilon$ to $\frac{\varepsilon}{2}$, we may assume $s_1 = t_1$ for the proof. We omit the superscript $\gamma(\cdot)$, for simplicity, and put $\xi = \sup_{s \in [t_1, t_2]} \|u(s) - u(t_1)\|$. Observing that

$$|F(u(s))| - F(u(t_1))| \le m_{F^{\#}}(\xi)\Big(1 + \sup_{T_0 \le t \le T} \|u(t)\|\Big) + \|F\|_{\#}\xi, \tag{6.54}$$

we have

$$\begin{aligned} &E_{T_0\phi}\Big[\sup_{t_1 \le s \le t_2} |F(u(s)) - F(u(t_1))|\Big] \\ &\le E_{T_0\phi}[\text{RHS of (6.54)}] \\ &\le 2(\varepsilon + C_\varepsilon \sqrt{E\xi^2})(1 + \sqrt{N_0}\|\phi\|) + \|F\|_{\#} E\xi. \end{aligned} \tag{6.55}$$

Now, applying Proposition 6.3 (ii) to $E\xi^2$, we complete the proof. □

**3. Dependence on $\gamma(\cdot)$**

**Proposition 6.8.**

(i) *For positive constants $\varepsilon$ and $n$, there is $\delta_4(\varepsilon, n; F) > 0$, such that*

$$E_{T_0\psi}\Big[\sup_{T_0\le t\le T}|F(u^{\gamma(\cdot)}(t)) - F(u^{\tilde\gamma(\cdot)}(t))|\Big] < \varepsilon \tag{6.56}$$

*for* $\psi \in H_1$ *with* $|||\psi||| \le n$, *whenever* $E\mathbf{1}\gamma(\cdot) - \tilde\gamma(\cdot)\mathbf{1}^4 < \delta_4(\varepsilon, n; F)$.

(ii) *For positive constants* $\varepsilon$ *and* $n$ *and* $\phi \in H_0$ *with* $\|\phi\| \le n$, *there are two positive constants* $\delta_5(\varepsilon, n; F)$ *and* $\delta_6(\varepsilon, n; F)$, *such that*

$$E\Big[\sup_{T_0\le t\le T}\Big|F(u^{\gamma(\cdot)}_{T_0\phi}(t)) - F(u^{\tilde\gamma(\cdot)}_{T_0\tilde\phi}(t))\Big|\Big] < \varepsilon \tag{6.57}$$

*whenever* $\|\phi - \tilde\phi\| < \delta_5(\varepsilon, n; F)$ *and* $E\mathbf{1}\gamma(\cdot) - \tilde\gamma(\cdot)\mathbf{1}^4 < \delta_6(\varepsilon, n; F)$.

*Proof.* (i) By using the estimate similar to (6.54) and (6.55) for $\xi := \sup_t \|u^{\gamma(\cdot)}_{T_0\psi}(t) - u^{\tilde\gamma(\cdot)}_{T_0\psi}(t)\|$ we can derive (6.56) from (6.31).

(ii) Put $\delta = \frac{1}{2}(1 \wedge \delta_2(\varepsilon, n+1; F))$. Then Proposition 6.6 leads to

$$E\Big[\sup_{T_0\le t\le T}\Big|F(u^{\gamma(\cdot)}_{T_0\phi}(t)) - F(u^{\gamma(\cdot)}_{T_0\tilde\phi}(t))\Big|\Big] < \varepsilon \tag{6.58}$$

for any $\gamma(\cdot) \in \boldsymbol{\Gamma}$, whenever $\|\phi - \tilde\phi\| < \delta$.

Next we fix $\psi \in H_1$, so that $\|\phi - \psi\| < \delta$ and $\|\psi\| < n$. Then

$$\|\tilde\phi - \psi\| < \delta_2(\varepsilon, n+1; F). \tag{6.59}$$

Taking into account the inequality

$$\begin{aligned}
&\sup_t |F(u^{\gamma(\cdot)}_{T_0\phi}(t)) - F(u^{\tilde\gamma}_{T_0\tilde\phi}(t))| \\
&\le \sup_t |F(u^{\gamma(\cdot)}_{T_0\phi}(t)) - F(u^{\gamma(\cdot)}_{T_0\psi}(t))| \\
&\quad + \sup_t |F(u^{\gamma(\cdot)}_{T_0\psi}(t)) - F(u^{\tilde\gamma(\cdot)}_{T_0\psi}(t))| \\
&\quad + \sup_t |F(u^{\tilde\gamma(\cdot)}_{T_0\psi}(t)) - F(u^{\tilde\gamma(\cdot)}_{T_0\tilde\phi}(t))|
\end{aligned}$$

and using (6.52), (6.59), and (6.56), we obtain (ii), with

$$\delta_5(\varepsilon, n; F) = \frac{1}{2}\Big(1 \wedge \delta_2\Big(\frac{\varepsilon}{3}, n+1; F\Big)\Big)$$

and

$$\delta_6(\varepsilon, \phi; F) = \delta_4\Big(\frac{\varepsilon}{3}, |||\psi|||; F\Big). \qquad \square$$

### 6.2.3 Running Cost

Suppose $r : H_0 \times \Gamma \mapsto \mathbb{R}^1$ is Lipschitz continuous, say

$$|r(\phi_1, \gamma_1) - r(\phi_2, \gamma_2)| \le l_r(\|\phi_1 - \phi_2\| + |\gamma_1 - \gamma_2|). \tag{6.60}$$

Put

$$r_0 = \sup_{\gamma \in \Gamma} |r(0, \gamma)|. \tag{6.61}$$

When a control process $\gamma(\cdot) \in \boldsymbol{\Gamma}$ is taken, the running cost $R$ on $[T_0, t]$ is given by

$$R(t, T_0, \phi, \gamma(\cdot)) = \int_{T_0}^{t} r(u_{T_0\phi}^{\gamma(\cdot)}(s), \gamma(s))\, ds. \tag{6.62}$$

Let us list several properties of the running cost function, by using results obtained in Sects. 5.4 and 6.1. By $K_i$, we denote a constant depending only on $\bar{\lambda}, \bar{\mu}, K, l, l_r$ and $r_0$.

**1. Bound**

$$\begin{aligned} E\Big[\sup_{T_0 \le t < T} |R(t, T_0, \phi, \gamma(\cdot))|\Big] &\le r_0 T + l_r E_{T_0\phi} \int_{T_0}^{T} \|u^{\gamma(\cdot)}(s)\|\, ds \\ &\le K_1(1 + \|\phi\|_*). \end{aligned} \tag{6.63}$$

**2. Dependence on the initial state**

$$E\Big[\sup_{T_0 \le t \le T} |R(t, T_0, \phi_1, \gamma(\cdot)) - R(t, T_0, \phi_2, \gamma(\cdot))|\Big] \le K_2 \|\phi_1 - \phi_2\|_*. \tag{6.64}$$

**3. Dependence on time**

$$E_{T_0\phi} \int_{\theta}^{t} |r(u^{\gamma(\cdot)}(s), \gamma(s))|\, ds \le K_3(1 + \|\phi\|_*)\sqrt{t - \theta}. \tag{6.65}$$

**4. Dependence on control processes**

$$\begin{aligned} &E\Big[\sup_{T_0 \le t \le T} |R(t, T_0, \phi, \gamma(\cdot)) - R(t, T_0, \phi, \tilde{\gamma}(\cdot))|\Big] \\ &\le \sqrt{\operatorname{diam} \Gamma}\, K_4(1 + \|\phi\|)(E[\mathbf{1}\gamma(\cdot) - \tilde{\gamma}(\cdot)\mathbf{1}^2])^{\frac{1}{4}}. \end{aligned} \tag{6.66}$$

*Proof.* For simplicity, we put $u(t) = u^{\gamma(\cdot)}(t), \tilde{u}(t) = u^{\tilde{\gamma}(\cdot)}(t)$ and $v(t) = u(t) - \tilde{u}(t)$. Then $v$ satisfies

$$\begin{cases} dv(t) = (A^{\gamma(\cdot)}(t)v(t) + Z(t))\,dt \\ \qquad\qquad + (G^{\gamma(\cdot)}(t)v(t) + \Psi(t))\,dW(t), \quad t \in (T_0, T], \\ v(T_0) = 0, \end{cases}$$

where

$$\begin{cases} Z(t) = (A^{\gamma(\cdot)}(t) - A^{\tilde{\gamma}(\cdot)}(t))\tilde{u}(t), \\ \Psi(t) = (G^{\gamma(\cdot)}(t) - G^{\tilde{\gamma}(\cdot)}(t))\tilde{u}(t). \end{cases}$$

From (5.188), we deduce that

$$E\int_{T_0}^{T} \|v(t)\|^2\,dt \le K_5 E\Big[\int_{T_0}^{T} (\|Z(s)\|_{H_{-2}}^2 + \|\Psi(s)\|_{*m}^2)\,dt\Big]. \tag{6.67}$$

Since $(B_1')$ yields

$$\begin{aligned} \|Z(t)\|_{H_{-2}}^2 &\le K_6|\gamma(t) - \tilde{\gamma}(t)|^2\|\tilde{u}(t)\|^2 \\ &\le K_6(\operatorname{diam}\Gamma)|\gamma(t) - \tilde{\gamma}(t)|\|\tilde{u}(t)\|^2 \end{aligned}$$

and

$$\|\Psi(t)\|_{*m}^2 \le K_6(\operatorname{diam}\Gamma)|\gamma(t) - \tilde{\gamma}(t)|\|\tilde{u}(t)\|^2,$$

inserting these estimates into (6.67), we obtain

$$\begin{aligned} E\int_{T_0}^{T} \|v(t)\|^2\,dt &\le K_7(\operatorname{diam}\Gamma)(E\mathbf{1}\gamma(\cdot) - \tilde{\gamma}(\cdot)\mathbf{1}^2)^{\frac{1}{2}}\Big(E\int_{T_0}^{T} \|\tilde{u}(s)\|^4\,ds\Big)^{\frac{1}{2}} \\ &\le K_8(\operatorname{diam}\Gamma)(E\mathbf{1}\gamma(\cdot) - \tilde{\gamma}(\cdot)\mathbf{1}^2)^{\frac{1}{2}}(1 + \|\phi\|^2). \end{aligned}$$

Hence, noticing that

$$\sup_t |R(t, T_0, \phi, \gamma(\cdot)) - R(t, T_0, \phi, \tilde{\gamma}(\cdot))| \le l_r \int_{T_0}^{T} (\|v(t)\| + |\gamma(t) - \tilde{\gamma}(t)|)\,dt,$$

we obtain (6.66). □

**5.** $R^{\dagger}(\phi) := \frac{ER(t,T_0,\phi,\gamma(\cdot))}{1+\|\phi\|}$ **is bounded and Lipschitz continuous uniformly in** $(t, T_0, \gamma(\cdot))$

Indeed, (6.63) and (6.64) imply

$$|R^{\dagger}(\phi)| \le K_1 \tag{6.68}$$

and

$$
\begin{aligned}
&|R^{\dagger}(\phi_1) - R^{\dagger}(\phi_2)| \\
&\leq E|R(t, T_0, \phi_1, \gamma(\cdot)) - R(t, T_0, \phi_2, \gamma(\cdot))| + \frac{E|R(t, T_0, \phi_2, \gamma(\cdot))|}{1 + \|\phi_2\|} \|\phi_1 - \phi_2\| \\
&\leq (K_1 + K_2)\|\phi_1 - \phi_2\|. \qquad (6.69)
\end{aligned}
$$

### 6.2.4 Value Functions

To emphasize the dependence on $F$, we write $F$ in the cost functional and the payoff:

$$
C(t, T_0, \phi, \gamma(\cdot); F) = \int_{T_0}^{t} r(u_{T_0\phi}^{\gamma(\cdot)}(s), \gamma(s))\, ds + F(u_{T_0\phi}^{\gamma(\cdot)}(t))
$$

and

$$
J(t, T_0, \phi, \gamma(\cdot); F) = EC(t, T_0, \phi, \gamma(\cdot); F).
$$

From (6.57) and (6.66), we deduce that, for $\varepsilon > 0$ and $\phi \in H_0$, there is a positive constant $\delta_C(\varepsilon, \phi; F)$ such that

$$
E\Big[\sup_{T_0 \leq t \leq T} |C(t, T_0, \phi, \gamma(\cdot); F) - C(t, T_0, \phi, \tilde{\gamma}(\cdot); F)|\Big] < \varepsilon \qquad (6.70)
$$

whenever $E\mathbf{1}\gamma(\cdot) - \tilde{\gamma}(\cdot)\mathbf{1}^2 < \delta_C(\varepsilon, \phi; F)$. Now we want to minimize the payoff by choosing a suitable control process from $\boldsymbol{\Gamma}$.

**Definition 6.1.** The minimum payoff

$$
V(t, T_0, \phi; F) := \inf_{\gamma(\cdot) \in \boldsymbol{\Gamma}} J(t, T_0, \phi, \gamma(\cdot); F) \qquad (6.71)
$$

is called the value function.

From (6.44) and (6.63), one immediately obtain the following remark.

*Remark 6.1.* There are constants $K_1$ and $K_2$, such that

$$
|J(t, T_0, \phi, \gamma(\cdot); F)| \leq K_1(1 + \|\phi\|_*) + K_2\|F\|_\#(1 + \|\phi\|), \quad \forall t, T_0, \gamma(\cdot). \qquad (6.72)
$$

Equation (6.72) provides a bound of the value function.

From the results in Sects. 6.2.2 and 6.2.3, the following properties of payoff and the value function are immediate.

**Proposition 6.9.** *Let $F \in C^{\#}$ be given.*

(i) $V(t, T_0, \cdot; F) \in C^{\#}$.
(ii) $V(t, T_0, \phi; F) = V(t - T_0, 0, \phi; F)$.
(iii) ***Dependence on the initial state***
*There are modulus functions, $m_n(\cdot; F), n = 1, 2, \ldots,$ such that*

$$\begin{cases} |J(t, T_0, \phi_1, \gamma(\cdot); F) - J(t, T_0, \phi_2, \gamma(\cdot); F)| \leq m_n(\|\phi_1 - \phi_2\|; F), \\ \forall t, T_0, \gamma(\cdot), \\ |V(t, T_0, \phi_1; F) - V(t, T_0, \phi_2; F)| \leq m_n(\|\phi_1 - \phi_2\|; F), \\ \forall t, T_0, \end{cases} \tag{6.73}$$

*whenever $\|\phi_1\|, \|\phi_2\| \leq n$,*
(iv) ***Dependence on time***
*For $\phi \in H_0$, there is a modulus function $m(\cdot; \phi, F)$, such that*

$$\begin{cases} |J(t_1, T_0, \phi, \gamma(\cdot); F) - J(t_2, T_0, \phi, \gamma(\cdot); F)| \leq m(|t_1 - t_2|; \phi, F), \\ \forall \gamma(\cdot) \in \boldsymbol{\Gamma}, \\ |V(t_1, T_0, \phi; F) - V(t_2, T_0, \phi; F)| < m(|t_1 - t_2|; \phi, F). \end{cases} \tag{6.74}$$

## 6.3 Dynamic Programming Principle (DPP)

As we said in Chap. 2, the DPP is a useful tool for analyzing control problems for finite-dimensional processes governed by SDEs. In this section, we focus on DPP for Hilbert space-valued processes governed by Zakai equations. We will formulate DPP via the semigroup constructed by means of the value function, in the same way as in Sect. 2.2. First, we show the DPP for the discrete-time case in Sect. 6.3.1. Then we deal with the continuous-time DPP in Sect. 6.3.2, by using a time discretization technique. We study properties of the semigroup and its generator in Sects. 6.3.3 and 6.3.4, respectively.

### *6.3.1 Discrete-Time DPP*

Let $\mathcal{D} = (t_1, \ldots, t_p)$ be a division of $[0, T]$. Put

$$\boldsymbol{\Gamma}^{\mathcal{D}} = \{\gamma(\cdot) \in \boldsymbol{\Gamma}; \gamma(t) = \gamma(t_i) \text{ for } t \in [t_i, t_{i+1}), i = 0, \ldots, p\}.$$

$\gamma(\cdot) \in \boldsymbol{\Gamma}^{\mathcal{D}}$ is called a switching control process for $\mathcal{D}$. Using switching control processes in $\boldsymbol{\Gamma}^{\mathcal{D}}$, we define the value function $V^{\mathcal{D}}$ by

$$V^{\mathcal{D}}(t, T_0, \phi; F) = \inf_{\gamma(\cdot)\in\boldsymbol{\Gamma}^{\mathcal{D}}} J(t, T_0, \phi, \gamma(\cdot); F). \tag{6.75}$$

The following proposition is easy.

**Proposition 6.10.** *Let $F$ be in $C^{\#}$.*

(i) $V^{\mathcal{D}}(t, T_0, \cdot; F) \in C^{\#}$

$$\|V^{\mathcal{D}}(t, T_0, \cdot; F)\|_{\#} \leq \sqrt{1 \vee N_0}((r_0 + l_r)T + \|F\|_{\#}), \tag{6.76}$$

*with the constant $N_0$ of* (6.4).

*By* (6.45) *and* (6.69), *there is a modulus $m(\cdot; F)$, such that*

$$\left|\frac{V^{\mathcal{D}}(t, T_0, \phi_1; F)}{1 + \|\phi_1\|} - \frac{V^{\mathcal{D}}(t, T_0, \phi_2; F)}{1 + \|\phi_2\|}\right| \leq m(\|\phi_1 - \phi_2\|; F) \tag{6.77}$$

*for any $t, T_0$ and $\mathcal{D}$,*

(ii) *There is a sequence of moduli $m_n(\cdot; F), n = 1, 2, \ldots,$ such that*

$$|V^{\mathcal{D}}(t, T_0, \phi_1; F) - V^{\mathcal{D}}(t, T_0, \phi_2; F)| \leq m_n(\|\phi_1 - \phi_2\|; F) \tag{6.78}$$

*for any $t, T_0$, and $\mathcal{D}$, whenever $\|\phi_1\|, \|\phi_2\| \leq n$,*

(iii) *For $\phi \in H_0$, there is a modulus $m_t(\cdot; \phi, F)$, such that*

$$|V^{\mathcal{D}}(t_1, T_0, \phi; F) - V^{\mathcal{D}}(t_2, T_0, \phi; F)| \leq m_t(|t_1 - t_2|; \phi, F) \tag{6.79}$$

*for any $T_0$ and $\mathcal{D}$.*

For $\theta < t$, we can define a monotone operator $V^{\mathcal{D}}_{\theta t}; C^{\#} \to C^{\#}$ by

$$V^{\mathcal{D}}_{\theta t} F(\phi) = V^{\mathcal{D}}(t, \theta, \phi; F), \quad \phi \in H_0. \tag{6.80}$$

Now we have;

*Remark 6.2.*

$$\mathcal{D} \subset \tilde{\mathcal{D}} \Longrightarrow V^{\mathcal{D}}_{\theta t} F(\phi) \geq V^{\tilde{\mathcal{D}}}_{\theta t} F(\phi), \quad \forall \phi \in H_0$$

and

$$F \leq \tilde{F} \Longrightarrow V^{\mathcal{D}}_{\theta t} F \leq V^{\mathcal{D}}_{\theta t} \tilde{F}.$$

Regarding $\gamma \in \Gamma$ as a constant control process, we introduce an auxiliary monotone operator $\nu_{\theta t}; C^{\#} \to C^{\#}$ by

$$\nu_{\theta t} F(\phi) = \inf_{\gamma \in \Gamma} J(t, \theta, \phi, \gamma; F) = \inf_{\gamma \in \Gamma} J(t - \theta, 0, \phi, \gamma; F). \tag{6.81}$$

For any $\theta, t$ and $F$, $J(t, \theta, \cdot; F)$ is continuous on $H_0 \times \Gamma$. Since $\Gamma$ is compact,

$$\Gamma_\phi := \{\gamma \in \Gamma; \nu_{\theta t} F(\phi) = J(t, \theta, \phi, \gamma; F)\} \tag{6.82}$$

is non-empty and compact. Hence there is a minimum selector $\hat{\gamma}_{\theta t F}$ (Borel mapping; $H_0 \mapsto \Gamma$) such that $\hat{\gamma}_{\theta t F}(\phi) \in \Gamma_\phi, \forall \phi \in H_0$. Thus

$$\nu_{\theta t} F(\phi) = J(t, \theta, \phi, \hat{\gamma}_{\theta t F}(\phi); F) \tag{6.83}$$

holds.

**Theorem 6.2.** *For* $0 \le i \le j \le p$,

$$V^{\mathcal{D}}_{t_i, t_{j+1}} F(\theta) = \nu_{t_i t_{i+1}} \cdots \nu_{t_j t_{j+1}} F(\phi) =: \prod_{k=i}^{j} \nu_{t_k t_{k+1}} F(\phi), \quad \forall \phi \in H_0. \tag{6.84}$$

*Outline of Proof.* Since we can apply the same arguments as in Theorem 2.2, we only sketch the proof.

Let $\gamma(\cdot) \in \boldsymbol{\Gamma}^{\mathcal{D}}$ be given. Since $(W(s) - W(t_j), s \ge t_j)$ is independent from $\mathcal{F}_{t_j}$, $(u^{\gamma(\cdot)}(t); t \in [t_j, t_{j+1}))$ is the response for the constant control $\gamma(t_j)$ with initial state $u^{\gamma(\cdot)}(t_j)$, under the conditional probability $P(\cdot|\mathcal{F}_{t_j})$. Hence, by freezing $\gamma(t_j)$ and $u^{\gamma(\cdot)}(t_j)$ we obtain

$$\begin{aligned} E(C(t_{j+1}, t_i, \phi, \gamma(\cdot); F)|\mathcal{F}_{t_j}) &\ge R(t_j, t_i, \phi, \gamma(\cdot)) + \nu_{t_j t_{j+1}} F(u^{\gamma(\cdot)}(t_j)) \\ &= C(t_j, t_i, \phi, \gamma(\cdot); \nu_{t_j t_{j+1}} F). \end{aligned} \tag{6.85}$$

Repeating the same arguments, we conclude that

$$J(t_{j+1}, t_i, \phi, \gamma(\cdot); F) \ge \prod_{k=i}^{j} \nu_{t_k t_{k+1}} F(\phi). \tag{6.86}$$

Taking the infimum over $\boldsymbol{\Gamma}^{\mathcal{D}}$ we obtain

$$V^{\mathcal{D}}_{t_i t_{j+1}} F(\phi) \ge \prod_{k=i}^{j} \nu_{t_k t_{k+1}} F(\phi). \tag{6.87}$$

Next we construct an optimal control process $\hat{\gamma}(\cdot)$ in $\boldsymbol{\Gamma}^{\mathcal{D}}$, by using the minimum selector. From (6.83) it follows that

$$\nu_{t_j t_{j+1}} F(u^{\gamma(\cdot)}(t_j)) = J(t_{j+1}, t_j, u^{\gamma(\cdot)}(t_j), \hat{\gamma}_j; F), \tag{6.88}$$

where $\hat{\gamma}_j := \hat{\gamma}_{t_j t_{j+1} F}(u^{\gamma(\cdot)}(t_j))$.

Put $F_m = \prod_{k=m}^{j} \nu_{t_k t_{k+1}} F, m = i, \dots, j$. On $[t_i, t_{i+1})$, we use the constant control $\hat{\gamma}_i := \hat{\gamma}_{t_i t_{i+1} F_{i+1}}(\phi)$ and denote its response with $\hat{u}(t_i) = \phi$, by $\hat{u}(t)$. Next we define $\hat{\gamma}_{i+1}$ by

$$\hat{\gamma}_{i+1} = \hat{\gamma}_{t_{i+1} t_{i+2} F_{i+2}}(\hat{u}(t_{i+1})) \quad \text{on } [t_{i+1}, t_{i+2})$$

and glue to $\hat{\gamma}_i$, namely

$$\hat{\gamma}(s) = \chi_{[t_i, t_{i+1})}(s)\hat{\gamma}_i + \chi_{[t_{i+1}, t_{i+2})}(s)\hat{\gamma}_{i+1}, \quad s \in [t_i, t_{i+2}).$$

$\hat{u}(t)$ denotes its response with $\hat{u}(t_i) = \phi$. Repeating this procedure, we construct $\hat{\gamma}(\cdot) \in \boldsymbol{\Gamma}^{\mathcal{D}}$ and its response $\hat{u}(\cdot)$. Then we have

$$J(t_{j+1}, t_i, \phi, \hat{\gamma}(\cdot); F) = \prod_{k=i}^{j} \nu_{t_k t_{k+1}} F(\phi). \tag{6.89}$$

Since (6.89) yields the opposite inequality for (6.87), we have completed the proof. □

**Corollary 6.1.** ***Discrete-time*** **DPP**

*Let $\mathcal{D} = (t_1, \dots, t_p)$ be a division of $[0, T]$. For $t_i \le t_l \le t_j$,*

$$V^{\mathcal{D}}_{t_i t_j} F(\phi) = V^{\mathcal{D}}_{t_i t_l}(V^{\mathcal{D}}_{t_l t_j} F)(\phi), \quad \forall \phi \in H_0. \tag{6.90}$$

## 6.3.2 *Continuous-Time* DPP

We will prove the continuous-time DPP by using time discretization approach (cf. [BN90]). Let us define $\mathrm{V}_{\theta t}; C^{\#} \to C^{\#}$ by

$$\mathrm{V}_{\theta t} F(\phi) = V(t, \theta, \phi; F), \quad \phi \in H_0. \tag{6.91}$$

We need to establish the semigroup property of $\{\mathrm{V}_{\theta t}; \theta \le t\}$, which formulates the DPP.

**Theorem 6.3.** *For* $0 \leq \theta_1 \leq \theta_2 \leq \theta_3 \leq T$,

$$\mathrm{V}_{\theta_1\theta_3} F(\phi) = \mathrm{V}_{\theta_1\theta_2}(\mathrm{V}_{\theta_2\theta_3} F)(\phi), \quad \forall \phi \in H_0. \tag{6.92}$$

*Proof.* We divide the proof into three steps.

**Step 1.** ***Approximation***
Let $\theta, t$ be given. Let $\mathcal{D}_n, n = 1, 2\ldots$ be a sequence of divisions satisfying

$$\theta, t \in \mathcal{D}_1, \quad \mathcal{D}_n \subset \mathcal{D}_{n+1}, \quad n = 1, 2\ldots, \quad \text{and} \quad \lim_{n\to\infty} |\mathcal{D}_n| = 0. \tag{6.93}$$

Then

$$\lim_{n\to\infty} V_{\theta t}^{\mathcal{D}_n} F(\phi) = \mathrm{V}_{\theta t} F(\phi), \quad \forall \phi \in H_0. \tag{6.94}$$

Indeed, by (6.93), $V_{\theta t}^{\mathcal{D}_n} F(\phi)$ is decreasing as $n \to \infty$, and

$$\mathrm{V}_{\theta t} F(\phi) \leq \lim_{n\to\infty} V_{\theta t}^{\mathcal{D}_n} F(\phi). \tag{6.95}$$

For the opposite inequality, we fix $\gamma(\cdot) \in \boldsymbol{\Gamma}$ arbitrarily and take $\gamma_{n'}(\cdot) \in \boldsymbol{\Gamma}^{\mathcal{D}_{n'}}$, so that

$$\begin{cases} \lim\limits_{n'\to\infty} E\mathbf{1}\gamma_{n'}(\cdot) - \gamma(\cdot)\mathbf{1}^2 = 0, \\ \lim\limits_{n'\to\infty} \gamma_{n'}(t) = \gamma(t), \quad \text{a.e. on } [0,T] \times \Omega. \end{cases} \tag{6.96}$$

Thus, (6.70) and (6.96) yield

$$\lim_{n'\to\infty} J(t, \theta, \phi, \gamma_{n'}(\cdot); F) = J(t, \theta, \phi, \gamma(\cdot); F). \tag{6.97}$$

Observing that

$$J(t, \theta, \phi, \gamma_{n'}(\cdot); F) \geq V_{\theta t}^{\mathcal{D}_{n'}} F(\phi) \geq \lim_{n\to\infty} V_{\theta t}^{\mathcal{D}_n} F(\phi),$$

we have

$$J(t, \theta, \phi, \gamma(\cdot), F) \geq \lim_{n\to\infty} V_{\theta t}^{\mathcal{D}_n} F(\phi), \quad \forall \gamma(\cdot) \in \boldsymbol{\Gamma},$$

from which the opposite of inequality (6.95) follows.
**Step 2.** Let us show that

$$\mathrm{V}_{\theta_1\theta_3} F(\phi) \geq \mathrm{V}_{\theta_1\theta_2}(\mathrm{V}_{\theta_2\theta_3} F)(\phi). \tag{6.98}$$

Let $\mathcal{D}_n, n = 1, 2, \ldots$ be divisions satisfying (6.93) with $\theta_1, \theta_2, \theta_3 \in \mathcal{D}_1$ instead of $\theta, t \in \mathcal{D}_1$. Then

$$\begin{aligned} V_{\theta_1\theta_3}^{\mathcal{D}_n} F(\phi) &= V_{\theta_1\theta_2}^{\mathcal{D}_n}(V_{\theta_2\theta_3}^{\mathcal{D}_n} F)(\phi) \\ &\geq V_{\theta_1\theta_2}^{\mathcal{D}_n}(\mathrm{V}_{\theta_2\theta_3} F)(\phi) \end{aligned} \tag{6.99}$$

follows from Corollary 6.1 and Remark 6.2. Letting $n \to \infty$ in (6.99), we obtain (6.98).

**Step 3.** ***Opposite of inequality* (6.98)**

For any $n$ and $m$, we have

$$\begin{aligned} V_{\theta_1\theta_3}^{\mathcal{D}_{n+m}} F(\phi) &\leq V_{\theta_1\theta_2}^{\mathcal{D}_{n+m}}(V_{\theta_2\theta_3}^{\mathcal{D}_m} F)(\phi) \\ &\leq J(\theta_2, \theta_1, \phi, \gamma(\cdot); V_{\theta_2\theta_3}^{\mathcal{D}_m} F), \quad \forall \gamma(\cdot) \in \boldsymbol{\Gamma}. \end{aligned} \tag{6.100}$$

Since (6.94) and Remark 6.1 yield

$$\lim_{m\to\infty} V_{\theta_2\theta_3}^{\mathcal{D}_m} F(u^{\gamma(\cdot)}(\theta_2)) = \mathrm{V}_{\theta_2\theta_3} F(u^{\gamma(\cdot)}(\theta_2)) \quad P\text{-a.s.}$$

and

$$|V_{\theta_2\theta_3}^{\mathcal{D}_m} F(u^{\gamma(\cdot)}(\theta_2))| \leq c(1 + \|F\|_\#)(1 + \|u^{\gamma(\cdot)}(\theta_2)\|) \quad P\text{-a.s.}$$

with a constant $c$ independent of $\theta_2, \theta_3, \mathcal{D}_m$, and $\gamma(\cdot)$, the dominated convergence theorem shows that

$$\lim_{m\to\infty} J(\theta_2, \theta_1, \phi, \gamma(\cdot); V_{\theta_2\theta_3}^{\mathcal{D}_m} F) = J(\theta_2, \theta_1, \phi, \gamma(\cdot); \mathrm{V}_{\theta_2\theta_3} F). \tag{6.101}$$

Thus, letting $m \to \infty$ in (6.100) and using (6.101), we have

$$\mathrm{V}_{\theta_1\theta_3} F(\phi) \leq J(\theta_2, \theta_1, \phi, \gamma(\cdot); \mathrm{V}_{\theta_2\theta_3} F), \quad \forall \gamma(\cdot) \in \boldsymbol{\Gamma} \tag{6.102}$$

from which the opposite of inequality (6.98) follows.

This completes the proof of the theorem. □

Since $\mathrm{V}_{\theta t} F = \mathrm{V}_{0 t-\theta} F$ thanks to the time homogeneity of the coefficients, we denote $\mathrm{V}_{0t}$ by $\mathrm{V}_t$, when there is no danger of confusion. Using this notation, we have a one-parameter semigroup $(\mathrm{V}_t, t \in [0, T])$ on $C^\#$.

*Remark 6.3.* **DPP**

$$\mathrm{V}_{t_1+t_2} F(\phi) = \mathrm{V}_{t_1}(\mathrm{V}_{t_2} F)(\phi). \tag{6.103}$$

Collecting previous results, we can easily deduce that $(\mathrm{V}_t) := (\mathrm{V}_t, t \in [0,T])$ is a monotone semigroup on $C^{\#}$, satisfying

(i) $\mu_0$-***contractiveness***

$$\|\mathrm{V}_t F - \mathrm{V}_t \tilde{F}\|_{\#} \le e^{\mu_0 t} \|F - \tilde{F}\|_{\#} \tag{6.104}$$

where $\mu_0 = 1 + \sqrt{N_1}$ with the constant $N_1$ of (6.8).

(ii) $t$-***continuity***

For any $\theta \in [0,T)$,

$$\lim_{t \to 0} \mathrm{V}_{\theta+t} F(\phi) = \mathrm{V}_\theta F(\phi), \quad \forall \phi \in H_0. \tag{6.105}$$

### 6.3.3 *Characterization of* $(\mathrm{V}_t)$

We will characterize $(\mathrm{V}_t)$ via payoffs. For $\gamma \in \Gamma$, we define $J_t^\gamma F(\phi)$ by

$$J_t^\gamma F(\phi) = J(t,0,\phi,\gamma;F) \; (= J(t+\theta,\theta,\phi,\gamma;F)), \quad \phi \in H_0. \tag{6.106}$$

Since $J_t^\gamma F$ is nothing but the value function for $\Gamma = \{\gamma\}$, $(J_t^\gamma, t \in [0,T])$ is a monotone semigroup on $C^{\#}$, satisfying (6.104) and (6.105). Further, for any $t$ and $\gamma$,

$$\mathrm{V}_t F \le J_t^\gamma F, \quad \forall F \in C^{\#}. \tag{6.107}$$

Put $\Sigma$ = the set of all $\mu_0$-contractive and $t$-continuous monotone semigroup $(\mathrm{S}_t) := (\mathrm{S}_t, t \in [0,T])$ on $C^{\#}$, satisfying

$$\mathrm{S}_t F \le J_t^\gamma F, \quad \forall F \in C^{\#}, \quad \forall t \in [0,T], \quad \forall \gamma \in \Gamma.$$

Now we can state

**Theorem 6.4.** $(\mathrm{V}_t)$ *is in* $\Sigma$. *Moreover for any* $(\mathrm{S}_t) \in \Sigma$,

$$\mathrm{S}_t F \le \mathrm{V}_t F, \quad \forall F \in C^{\#}, \quad \forall t \in [0,T], \tag{6.108}$$

*that is,* $(\mathrm{V}_t)$ *is the maximal element of* $\Sigma$, *called the envelope of* $((J_t^\gamma), \gamma \in \Gamma)$.

*Proof.* Let $t$ be given. Take a sequence of divisions, $\mathcal{D}_n, n = 1, 2, \ldots$, so that,

$$t \in \mathcal{D}_1, \quad \mathcal{D}_n \subset \mathcal{D}_{n+1}, \quad n = 1, 2, \ldots, \quad \lim_{n \to \infty} |\mathcal{D}_n| = 0.$$

Since, for any $F \in C^{\#}$ and any $\theta \le s$,

$$v_{\theta s} F(\phi) = \inf_{\gamma \in \Gamma} J_{s-\theta}^\gamma F(\phi) \ge \mathrm{S}_{s-\theta} F(\phi) \tag{6.109}$$

the monotonicity of $J_t^\gamma$ yields

$$\begin{aligned}\nu_{t_{j-1}t_j}(\nu_{t_j t}F)(\phi) &\geq \inf_{\gamma\in\Gamma} J^\gamma_{t_j-t_{j-1}}(\mathrm{S}_{t-t_j}F)(\phi)\\ &\geq \mathrm{S}_{t_j-t_{j-1}}(\mathrm{S}_{t-t_j}F)(\phi)=\mathrm{S}_{t-t_{j-1}}F(\phi).\end{aligned} \tag{6.110}$$

Repeating this procedure and using (6.84), we have

$$V_{0t}^{\mathcal{D}_n}F(\phi)\geq \mathrm{S}_t F(\phi),\quad n=1,2,\ldots. \tag{6.111}$$

Thus, taking into account (6.94) and letting $n \to \infty$ in (6.111), we obtain (6.108). □

### 6.3.4 Derivatives of $V_t F(\phi)$

Let us determine the generator of $(\mathrm{V}_t)$. Put

$$\mathcal{L}^\gamma F(\phi)=\langle A^\gamma\phi, DF(\phi)\rangle+\frac{1}{2}\sum_{k=1}^m\langle D^2F(\phi)G_k^\gamma\phi, G_k^\gamma\phi\rangle,\quad \gamma\in\Gamma. \tag{6.112}$$

**Theorem 6.5.** *For $F\in C^2_{\mathrm{bu}}(H_{-1})$, we have*

$$\lim_{t\to 0}\frac{1}{t}(\mathrm{V}_t F(\phi)-F(\phi))=\inf_{\gamma\in\Gamma}(\mathcal{L}^\gamma F(\phi)+r(\phi,\gamma))\quad \textit{for } \phi\in H_1. \tag{6.113}$$

*Proof.* Suppose that

$$|F(\phi)|+|||DF(\phi)|||+\|D^2F(\phi)\|_{L(H_{-1};H_1)}\leq K_0$$

and

$$\begin{aligned}|F(\phi)-F(\psi)|+|||DF(\phi)-DF(\psi)|||+\|D^2F(\phi)-D^2F(\psi)\|_{L(H_{-1};H_1)}&\\ \leq m_0(\|\phi-\psi\|_*).&\end{aligned}$$

Let $\phi\in H_1$ be given. First we fix $\gamma(\cdot)$ and evaluate

$$I(s):=\mathcal{L}^{\gamma(s)}F(u_{0\phi}^{\gamma(\cdot)}(s))-\mathcal{L}^{\gamma(s)}F(\phi). \tag{6.114}$$

We are going to show that, for $\varepsilon>0$, there is $\Delta(\varepsilon,\phi)>0$, independent of $\gamma(\cdot)$, such that

$$E\Big[\sup_{0\leq s\leq t}|I(s)|\Big]<\varepsilon\quad \text{for } t<\Delta(\varepsilon,\phi). \tag{6.115}$$

We divide the proof of (6.115) into four steps. For simplicity, we put $u(t) = u_{0\phi}^{\gamma(\cdot)}(t)$ and $c_i$ denotes a constant independent of $t, \phi$ and $\gamma(\cdot)$.

**Step 1.** There is $\Delta_1(\varepsilon, \phi) > 0$, independent of $\gamma(\cdot)$, such that

$$E\Big[\sup_{0\le s\le t} |\langle A^{\gamma(\cdot)}(u(s)-\phi), DF(u(s))\rangle|\Big] < \varepsilon, \quad \text{for } t < \Delta_1(\varepsilon, \phi). \tag{6.116}$$

Indeed, the expression inside [ ] is

$$\begin{aligned} &\le K \sup_{0\le s\le t} |||u(s)-\phi||| \sup_{0\le s\le t} |||DF(u(s))||| \\ &\le KK_0 \sup_{0\le s\le t} |||u(s)-\phi|||, \end{aligned} \tag{6.117}$$

where $K$ is the constant of $(B_1')$.

On the other hand, for $\tilde{\varepsilon} > 0$, we can take $\tilde{\phi} \in H_2$ so that $|||\phi - \tilde{\phi}||| < \tilde{\varepsilon}$. Thus, using the response $\tilde{u}(t) := \tilde{u}_{0\tilde{\phi}}^{\gamma(\cdot)}(t)$ we have

$$\begin{aligned} &E \sup_{0\le s\le t} |||u(s)-\phi||| \\ &\le E \sup_{0\le s\le T} |||u(s)-\tilde{u}(s)||| + E \sup_{0\le s\le t} |||\tilde{u}(s)-\tilde{\phi}||| + |||\tilde{\phi}-\phi||| \\ &\le (\sqrt{N_1}+1)\tilde{\varepsilon} + \sqrt{N_2}\sqrt{t}\,\mathbb{I}\tilde{\phi}\mathbb{I}, \end{aligned} \tag{6.118}$$

by (6.9) and (6.109). Now (6.117) and (6.118) yield (6.116).

**Step 2.** There is $\Delta_2(\varepsilon, \phi) > 0$, independent of $\gamma(\cdot)$, such that

$$E\Big[\sup_{0\le s\le t} |\langle A^{\gamma(\cdot)}(s)\phi, DF(u(s)) - DF(\phi)\rangle|\Big] < \varepsilon, \tag{6.119}$$

whenever $t < \Delta_2(\varepsilon, \phi)$.

Indeed, by (5.196), for any $\tilde{\varepsilon} > 0$,

$$\begin{aligned} \text{LHS of (6.119)} &\le K|||\phi|||Em_0\Big(\sup_{0\le s\le t} \|u(s)-\phi\|_*\Big) \\ &\le K|||\phi|||\Big(\tilde{\varepsilon} + C_{\tilde{\varepsilon}}E\Big[\sup_{0\le s\le t} \|u(s)-\phi\|_*\Big]\Big) \\ &\le K|||\phi|||(\tilde{\varepsilon} + C_{\tilde{\varepsilon}}\sqrt{c_0N}\sqrt{t}\|\phi\|). \end{aligned}$$

This yields (6.119).

**Step 3.** There is $\Delta_3(\varepsilon,\phi)>0$, independent of $\gamma(\cdot)$, such that

$$E\Big[\sup_{0\le s\le t}|\langle D^2F(u(s))G_k^{\gamma(\cdot)}(s)u(s),G_k^{\gamma(\cdot)}(s)(u(s)-\phi)\rangle|\Big]<\varepsilon,$$
$$k=1,2,\ldots,m, \tag{6.120}$$

whenever $t<\Delta_3(\varepsilon,\phi)$.
Indeed, we have

$$\begin{aligned}
&\text{LHS of (6.120)}\\
&\le E\Big[\sup_{0\le s\le t}|||D^2F(u(s))G_k^{\gamma(\cdot)}(s)u(s)|||\ \|G_k^{\gamma(s)}(s)(u(s)-\phi)\|_*\Big]\\
&\le K_0K^2E\Big[\sup_{0\le s\le t}\|u(s)\|\ \|u(s)-\phi\|\Big]\\
&\le K_0K^2\Big(E\sup_{0\le s\le t}\|u(s)\|^2\Big)^{\frac12}\Big(E\sup_{0\le s\le t}\|u(s)-\phi\|^2\Big)^{\frac12}\\
&\le c_1\|\phi\|\ |||\phi|||\sqrt{t}\quad\text{(by (5.111))},
\end{aligned}$$

with a constant $c_1$ independent of $k$ and $\gamma(\cdot)$. This yields (6.120).

**Step 4.** There is $\Delta_4(\varepsilon,\phi)>0$, independent of $\gamma(\cdot)$, such that

$$E\Big[\sup_{0\le s\le t}|\langle(D^2F(u(s))-D^2F(\phi))G_k^{\gamma(\cdot)}(s)\phi,G_k^{\gamma(\cdot)}(s)\phi\rangle|\Big]<\varepsilon,$$
$$k=1,2,\ldots,m, \tag{6.121}$$

whenever $t<\Delta_4(\varepsilon,\phi)$.
Indeed, for any $\tilde\varepsilon>0$,

$$\begin{aligned}
&\sup_{0\le s\le t}\|D^2F(u(s))-D^2F(\phi)\|_{L(H_{-1};H_1)}\|G_k^{\gamma(\cdot)}\phi\|_*^2\\
&\le\sup_{0\le s\le t}m_0(\|u(s)-\phi\|_*)K\|\phi\|^2\\
&\le K\Big(\tilde\varepsilon+C_{\tilde\varepsilon}\Big(\sup_{0\le s\le t}\|u(s)-\phi\|_*\Big)\Big)\|\phi\|^2.
\end{aligned}\tag{6.122}$$

Taking the expectation of (6.122) and using (5.196), we obtain (6.121).
Collecting the above results in Steps 1–4, we have (6.115).
Next we consider the running cost:

$$\begin{aligned}
E\Big[\sup_{0\le s\le t}|r(u(s),\gamma(s))-r(\phi,\gamma(s))|\Big]&\le l_rE\sup_{0\le s\le t}\|u(s)-\phi\|\\
&\le c_3\sqrt{t}|||\phi|||\quad\text{(by (5.111))},
\end{aligned}\tag{6.123}$$

with a constant $c_3$, independent of $\gamma(\cdot)$.

Thus, (6.115) and (6.123) yield that for any $\varepsilon > 0$, there is $\Delta^*(\varepsilon, \phi) > 0$, such that

$$E\Big[\sup_{0\le s\le t} |\mathcal{L}^{\gamma(s)} F(u(s)) + r(u(s), \gamma(s)) - \mathcal{L}^{\gamma(s)} F(\phi) - r(\phi, \gamma(s))|\Big] < \varepsilon, \quad \forall \gamma(\cdot) \tag{6.124}$$

whenever $t < \Delta^*(\varepsilon, \phi)$.
Finally, we note that

$$\begin{aligned}
&\inf_{\gamma(\cdot)\in\boldsymbol{\Gamma}} E\int_0^t (\mathcal{L}^{\gamma(s)} F(\phi) + r(\phi, \gamma(s)))\,ds \\
&\geq E\int_0^t \inf_{\gamma\in\Gamma}(\mathcal{L}^{\gamma} F(\phi) + r(\phi, \gamma))\,ds \\
&= t \inf_{\gamma\in\Gamma}(\mathcal{L}^{\gamma} F(\phi) + r(\phi, \gamma)) \quad \text{(because inside ( \quad ) is non random)} \\
&= \inf_{\gamma\in\Gamma} t(\mathcal{L}^{\gamma} F(\phi) + r(\phi, \gamma)) \\
&\geq \inf_{\gamma(\cdot)\in\boldsymbol{\Gamma}} E\int_0^t (\mathcal{L}^{\gamma(s)} F(\phi) + r(\phi, \gamma(s)))\,ds
\end{aligned} \tag{6.125}$$

because $\gamma \in \Gamma$ becomes a constant control process. Therefore, we have

$$\inf_{\gamma(\cdot)\in\boldsymbol{\Gamma}} E\int_0^t (\mathcal{L}^{\gamma(s)} F(\phi) + r(\phi, \gamma(s)))\,ds = t \inf_{\gamma\in\Gamma}(\mathcal{L}^{\gamma} F(\phi) + r(\phi, \gamma)). \tag{6.126}$$

Now we are ready to compute the LHS of (6.113). From Itô's formula of Theorem 5.7 it follows that

$$\begin{aligned}
&\mathrm{V}_t F(\phi) - F(\phi) \\
&= \inf_{\gamma(\cdot)\in\boldsymbol{\Gamma}} E\Big[\int_0^t r(u(s), \gamma(s))\,ds + F(u(t)) - F(\phi)\Big] \\
&= \inf_{\gamma(\cdot)\in\boldsymbol{\Gamma}} E\Big[\int_0^t (r(u(s), r(s)) + \mathcal{L}^{\gamma(s)} F(u(s)))\,ds\Big] \\
&= \inf_{\gamma(\cdot)\in\boldsymbol{\Gamma}} \Big\{E\Big[\int_0^t (r(u(s), \gamma(s)) + \mathcal{L}^{\gamma(s)} F(u(s)) - r(\phi, \gamma(s)) - \mathcal{L}^{\gamma(s)} F(\phi))\,ds\Big] \\
&\quad + E\int_0^t (r(\phi, \gamma(s)) + \mathcal{L}^{\gamma(s)} F(\phi))\,ds\Big\}.
\end{aligned}$$

Hence (6.124) and (6.126) lead to (6.113). □

Now Theorem 6.5 shows that $U(t, \phi) := \mathrm{V}_{T-t} F(\phi)$ has the backward dynamics

$$\partial_t U(t, \phi) + \inf_{\gamma \in \Gamma} (\mathcal{L}^\gamma U(t, \phi) + r(\phi, \gamma)) = 0.$$

This equation is called HJB equation.

## 6.4 Viscosity Solutions of HJB Equations

This section is devoted to the study of the HJB equations introduced at the end of Sect. 6.3. Let us consider HJB equation on $H_0$

$$\begin{aligned} &\partial_t U(t, \phi) \\ &+ \inf_{\gamma \in \Gamma} \Big( \langle A^\gamma \phi, DU(t, \phi) \rangle + \frac{1}{2} \sum_{k=1}^m \langle D^2 U(t, \phi) G_k^\gamma \phi, G_k^\gamma \phi \rangle + r(\phi, \gamma) \Big) = 0, \\ &\phi \in H_0, \quad t \in (0, T) \end{aligned} \tag{6.127}$$

with the lateral boundary condition

$$U(T, \phi) = F(\phi), \quad \phi \in H_0. \tag{6.128}$$

When the value function is smooth, the HJB equation provides the backward evolution equation by Theorem 6.5. However we can hardly expect that the value function will be regular. Here we will consider viscosity solution of the HJB equation according to [L89] and [GŚ00].

In Sect. 6.4.1, we give the definition of the viscosity solution. Since we treat the equation on $H_0$, the relation between the value function and the viscosity solution has already been stated in [GŚ00]. Thus we will sketch the proof of the assertion that the value function becomes a viscosity solution in Sect. 6.4.2 and the uniqueness theorem in Sect. 6.4.3. Applying these results, we revisit control problems for partially observable diffusions in Example 6.1.

### *6.4.1 Definitions*

Put

$$\mathbf{\Delta} = \Big\{ \delta(\cdot) \in C^1(0, T); \lim_{t \to 0} \delta(t) = \lim_{t \to T} \delta(t) = \infty \text{ and } \delta_* = \inf_t \delta(t) > 0 \Big\}. \tag{6.129}$$

Elements of $C^{12}_{\mathrm{bu}}((0, T) \times H_{-1})$ is called test functions. For a test function $\zeta(\cdot)$ and $\gamma \in \Gamma$, we put

$$\mathcal{L}^\gamma \zeta(t,\phi) = \langle A^\gamma \phi, D\zeta(t,\phi)\rangle + \frac{1}{2}\langle D^2\zeta(t,\phi)G^\gamma\phi, G^\gamma\phi\rangle_m. \tag{6.130}$$

where $\langle \psi, \xi\rangle_m = \sum_{k=1}^m \langle \psi_k, \xi_k\rangle$ .

**Definition 6.2.** Let $v(\cdot) \in C([0,T] \times H_0)$.

(a) $v(\cdot)$ is called a viscosity subsolution of (6.127)–(6.128), if for every test function $\zeta(\cdot)$ and for every $\delta(\cdot) \in \mathbf{\Delta}$, whenever $v - (\zeta + \frac{1}{2}\delta(t)\|\phi\|^2)$ attains a global maximum at $(\hat{t},\hat{\phi}) \in (0,T) \times H_0$, then $\hat{\phi} \in H_1$, and the subsolution inequality

$$\begin{aligned} 0 \le \partial_t \zeta(\hat{t},\hat{\phi}) &+ \frac{\delta'(\hat{t})}{2}\|\hat{\phi}\|^2 \\ &+ \inf_{\gamma\in\Gamma}\left\{\mathcal{L}^\gamma\zeta(\hat{t},\hat{\phi}) + \frac{1}{2}\delta(\hat{t})\mathcal{L}^\gamma\|\hat{\phi}\|^2 + r(\hat{\phi},\gamma)\right\} \end{aligned} \tag{6.131}$$

holds and

$$v(T,\phi) \le F(\phi), \quad \forall \phi \in H_0. \tag{6.132}$$

(b) $v(\cdot)$ is called a viscosity supersolution of (6.127)–(6.128), if for every test function $\zeta(\cdot)$ and for every $\delta(\cdot) \in \mathbf{\Delta}$, whenever $v - (\zeta - \frac{1}{2}\delta(t)\|\phi\|^2)$ attains a global minimum at $(\hat{t},\hat{\phi}) \in (0,T) \times H_0$, then $\hat{\phi} \in H_1$ and the supersolution inequality

$$\begin{aligned} 0 \ge \partial_t \zeta(\hat{t},\hat{\phi}) &- \frac{\delta'(\hat{t})}{2}\|\hat{\phi}\|^2 \\ &+ \inf_{\gamma\in\Gamma}\left\{\mathcal{L}^\gamma\zeta(\hat{t},\hat{\phi}) - \frac{1}{2}\delta(\hat{t})\mathcal{L}^\gamma\|\hat{\phi}\|^2 + r(\hat{\phi},\gamma)\right\} \end{aligned} \tag{6.133}$$

holds and

$$v(T,\phi) \ge F(\phi), \quad \forall \phi \in H_0. \tag{6.134}$$

(c) $v(\cdot)$ is called a viscosity solution, if it is both a viscosity subsolution and a viscosity supersolution.

### 6.4.2 Existence of Viscosity Solutions

Referring to [GŚ00], we prove

**Theorem 6.6.** *Let $F \in C^\#$ and put $U(t,\phi) = \mathrm{V}_{T-t}F(\phi)$. Then $U$ is a viscosity solution of (6.127)–(6.128).*

*Proof.* The proof is divided into three steps.

**Step 1.** Suppose that $(\hat{t},\hat{\phi}) \in (0,T)\times H_0$ is a maximizer of $U-(\zeta+\frac{1}{2}\delta(t)\|\phi\|^2)$. Then $\hat{\phi} \in H_1$.
Indeed, we have

$$|\zeta(t,\phi)| + |\partial_t\zeta(t,\phi)| + |||D\zeta(t,\phi)||| + \|D^2\zeta(t,\phi)\|_{L(H_{-1};H_1)} \le K_\zeta \quad \text{on } (0,T)\times H_0 \tag{6.135}$$

because $\zeta(\cdot) \in C^{12}_{\mathrm{bu}}((0,T)\times H_{-1})$.
By $u^{\gamma(\cdot)}(\cdot)$ we denote the response for $\gamma(\cdot)$ with the initial condition; $u^{\gamma(\cdot)}(\hat{t}) = \hat{\phi}$.
Since the semigroup property $\mathrm{V}_{\hat{t}t}\,\mathrm{V}_{tT} = \mathrm{V}_{\hat{t}T}\ (= \mathrm{V}_{T-\hat{t}})$ yields

$$0 = \inf_{\gamma(\cdot)\in\Gamma} E\Big[\int_{\hat{t}}^t r(u^{\gamma(\cdot)}(s),\gamma(s))\,ds + U(t,u^{\gamma(\cdot)}(t)) - U(\hat{t},\hat{\phi})\Big], \tag{6.136}$$

it follows that for any given $\gamma(\cdot)$,

$$-E\int_{\hat{t}}^t r(u^{\gamma(\cdot)}(s),\gamma(s))\,ds \le E[U(t,u^{\gamma(\cdot)}(t)) - U(\hat{t},\hat{\phi})]. \tag{6.137}$$

Fix $\gamma \in \Gamma$ arbitrarily and put

$$I(t) := E[U(t,u^\gamma(t)) - U(\hat{t},\hat{\phi})].$$

Noticing that $(\hat{t},\hat{\phi})$ is a global maximizer of $U - (\zeta + \frac{1}{2}\delta(t)\|\phi\|^2)$, we have

$$\begin{aligned} I(t) \le E\int_{\hat{t}}^t \Big\{&\partial_s\zeta(s,u^\gamma(s)) + \langle A^\gamma u^\gamma(s), D\zeta(s,u^\gamma(s))\rangle \\ &+ \frac{1}{2}\langle D^2\zeta(s,u^\gamma(s))G^\gamma u^\gamma(s), G^\gamma u^\gamma(s)\rangle_m \\ &+ \frac{1}{2}\delta'(s)\|u^\gamma(s)\|^2 + \delta(s)\langle A^\gamma u^\gamma(s), u^\gamma(s)\rangle \\ &+ \frac{1}{2}\delta(s)\|G^\gamma u^\gamma(s)\|_m^2\Big\}\,ds \end{aligned} \tag{6.138}$$

by Itô's formula (see Sect. 5.4).
Next we calculate the expression inside $\{\cdots\}$ in the RHS of (6.138). Take an integer $n_0$, so that

$$|\delta(s) - \delta(\hat{t})| + |\delta'(s) - \delta'(\hat{t})| \le 1 \quad \text{for } s \in \Big[\hat{t},\hat{t} + \frac{1}{n_0}\Big]. \tag{6.139}$$

By $c_i$ we denote positive constants, independent of $s$, $\hat{\phi}$ and $\gamma$. From (6.3), (6.6), and (6.135) it follows that

$$\begin{aligned} \text{inside } \{\cdots\} &\le c_1(1 + |||u^\gamma(s)||| + \|u^\gamma(s)\|^2) \\ &\quad - \bar{\lambda}\delta_* |||u^\gamma(s)|||^2 + \bar{\mu}(|\delta(\hat{t})| + 1)\|u^\gamma(s)\|^2 \\ &\le c_2(1 + \|u^\gamma(s)\|^2) - \frac{\bar{\lambda}}{2}\delta_* |||u^\gamma(s)|||^2 \\ &\le c_2\Big(1 + \sup_{\hat{t}\le s\le T} \|u^\gamma(s)\|^2\Big) - \frac{\bar{\lambda}}{2}\delta_* |||u^\gamma(s)|||^2. \end{aligned} \tag{6.140}$$

Thus, for $t \in [\hat{t}, \hat{t} + \frac{1}{n_0}]$,

$$I(t) \le c_3(t - \hat{t})(1 + \|\hat{\phi}\|^2) - \frac{\hat{\lambda}}{2}\delta_* E \int_{\hat{t}}^{t} |||u^\gamma(s)|||^2\, ds. \tag{6.141}$$

Noting that

$$\begin{aligned} E\int_{\hat{t}}^{t} |r(u^\gamma(s), \gamma)|\, ds &\le (t - \hat{t})c_4\Big(1 + E \sup_{\hat{t}\le s\le T} \|u^\gamma(s)\|\Big) \\ &\le c_5(t - \hat{t})(1 + \|\hat{\phi}\|), \end{aligned} \tag{6.142}$$

and combining (6.137), (6.141), and (6.142), we obtain

$$\frac{1}{t - \hat{t}} E\Big[\int_{\hat{t}}^{t} |||u^\gamma(s)|||^2\, ds\Big] \le c_6(1 + \|\hat{\phi}\|^2) =: a. \tag{6.143}$$

Put $t = \hat{t} + \frac{1}{n}$ $(n = n_0, n_0 + 1, \dots)$. Then there is $t_n \in [\hat{t}, \hat{t} + \frac{1}{n}]$, such that

$$E[|||u^\gamma(t_n)|||^2] \le a \quad (n = n_0, n_0 + 1, \dots), \tag{6.144}$$

by (6.143). Hence, along an appropriate subsequence $n'$, we have

$$u^\gamma(t_{n'}) \longrightarrow \tilde{u} \quad \text{weakly in } L^2(\Omega; H_1) \tag{6.145}$$

for some $\tilde{u} \in L^2(\Omega; H_1)$. Therefore $u^\gamma(t_{n'}) \to \tilde{u}$ weakly in $L^2(\Omega; H_0)$. But

$$\lim_{n'\to\infty} E[\|u^\gamma(t_{n'}) - \hat{\phi}\|^2] = 0, \tag{6.146}$$

because $u^\gamma \in L^2(\Omega; C([\hat{t}, T]; H_0))$. Consequently $\tilde{u} = \hat{\phi}$ $P$-a.s., which show that $\hat{\phi} \in H_1$.

**Step 2.** Suppose that $(\hat{t},\hat{\phi}) \in (0,T) \times H_0$ is a global minimizer of $U - (\zeta - \frac{1}{2}\delta(t)\|\phi\|^2)$. Then $\hat{\phi} \in H_1$.

Indeed, we take $\frac{1}{n}$-optimal control process $\gamma_n(\cdot)$ at $t_n := \hat{t} + \frac{1}{n} (n = n_0, n_0 + 1, \ldots)$, namely

$$0 \le E_{\hat{t}\hat{\phi}}\Big[\int_{\hat{t}}^{t_n} r(u^{\gamma_n(\cdot)}(s), \gamma_n(s))\, ds + U(t_n, u^{\gamma_n(\cdot)}(t_n)) - U(\hat{t},\hat{\phi})\Big] < \frac{1}{n}. \tag{6.147}$$

Since $(\hat{t},\hat{\phi})$ is a global minimizer of $U - (\zeta - \frac{1}{2}\delta(t)\|\phi\|^2)$, referring to (6.143) and omitting the subscript $\hat{t},\hat{\phi}$ for simplicity, we have

$$n \int_{\hat{t}}^{t_n} E|||u^{\gamma_n(\cdot)}(s)|||^2\, ds \le c_7(1 + \|\hat{\phi}\|^2) =: a. \tag{6.148}$$

Since the LHS $= \int_0^1 E|||u^{\gamma_n(\cdot)}(\hat{t} + \frac{\theta}{n})|||^2\, d\theta$, there is $\theta_n \in [\hat{t}, \hat{t} + \frac{1}{n}]$, such that

$$E|||u^{\gamma_n(\cdot)}(\theta_n)|||^2 \le a \quad (n = n_0, n_0 + 1, \ldots).$$

Thus, along an appropriate subsequence $n'$,

$$u^{\gamma_n(\cdot)}(\theta_{n'}) \longrightarrow \tilde{u} \quad \text{weakly in } L^2(\Omega; H_1) \tag{6.149}$$

for some $\tilde{u} \in L^2(\Omega; H_1)$. On the other hand, (6.26) shows that

$$\lim_{n' \to \infty} E[\|u^{\gamma_{n'}(\cdot)}(\theta_{n'}) - \hat{\phi}\|^2] = 0. \tag{6.150}$$

Now, (6.149) and (6.150) imply that $\tilde{u} = \hat{\phi}$ $P$-a.s., i.e., $\hat{\phi} \in H_1$.

**Step 3.** ***Subsolution inequality***

By Step 1, we suppose that $(\hat{t},\hat{\phi}) \in (0,T) \times H_1$ is a global maximizer of $U - (\zeta + \frac{1}{2}\delta(t)\|\phi\|^2)$. Then Itô's formula together with (6.137) yields

$$\begin{aligned}
0 &\le \inf_{\gamma(\cdot)\in\Gamma} E\Big[\int_{\hat{t}}^{t} r(u^{\gamma(\cdot)}(s), \gamma(s))\, ds \\
&\quad + \zeta(t, u^{\gamma(\cdot)}(t)) - \zeta(\hat{t},\hat{\phi}) + \frac{1}{2}(\delta(t)\|u^{\gamma(\cdot)}(t)\|^2 - \delta(\hat{t})\|\hat{\phi}\|^2)\Big] \\
&= \inf_{\gamma(\cdot)\in\Gamma} E\int_{\hat{t}}^{t} \Big(r(u^{\gamma(\cdot)}(s), \gamma(s)) + \partial_t\zeta(s, u^{\gamma(\cdot)}(s)) + \frac{\delta'(s)}{2}\|u^{\gamma(\cdot)}(s)\|^2 \\
&\quad + \mathcal{L}^{\gamma(s)}\zeta(s, u^{\gamma(\cdot)}(s)) + \frac{1}{2}\delta(s)\mathcal{L}^{\gamma(s)}\|u^{\gamma(\cdot)}(s)\|^2\Big)\, ds.
\end{aligned} \tag{6.151}$$

We claim that, for $\varepsilon > 0$, there is $\Delta_1(\varepsilon, \hat{\phi}) > 0$, independent of $\gamma(\cdot)$, such that

$$E\Big[\sup_{\hat{t}\le s\le t} |||u^{\gamma(\cdot)}(s) - \hat{\phi}|||^2\Big] < \varepsilon, \quad \forall \gamma(\cdot), \tag{6.152}$$

whenever $t - \hat{t} < \Delta_1(\varepsilon, \hat{\phi})$.

Indeed, for $\tilde{\varepsilon} > 0$, we take $\psi \in H_2$, so that $|||\hat{\phi} - \psi|||^2 < \tilde{\varepsilon}$. Let $v^{\gamma(\cdot)}$ denote the response for $\gamma(\cdot)$, with $v^{\gamma(\cdot)}(\hat{t}) = \psi$. Then, the inequality

$$|||u^{\gamma(\cdot)}(s) - \hat{\phi}||| \le |||u^{\gamma(\cdot)}(s) - v^{\gamma(\cdot)}(s)||| + |||v^{\gamma(\cdot)}(s) - \psi||| + |||\psi - \hat{\phi}|||$$

and relations (5.107) and (5.112) yield

$$E\Big[\sup_{\hat{t}\le s\le t} |||u^{\gamma(\cdot)}(s) - \hat{\phi}|||^2\Big] \le c_6\tilde{\varepsilon} + c_7 \mathbb{I}\psi\mathbb{I}^2 (t - \hat{t}). \tag{6.153}$$

from which (6.152) follows.

By using (6.152), we can take $\Delta_2(\varepsilon, \hat{\phi}) > 0$, independent of $\gamma(\cdot)$, so that

$$E\Big[\sup_{\hat{t}\le s\le t} \delta(s)|\mathcal{L}^{\gamma(s)}\|u^{\gamma(\cdot)}(s)\|^2 - \mathcal{L}^{\gamma(s)}\|\hat{\phi}\|^2|\Big] < \varepsilon, \quad \forall \gamma(\cdot) \tag{6.154}$$

whenever $t - \hat{t} < \Delta_2(\varepsilon, \hat{\phi})$.

We treat $\mathcal{L}^{\gamma(s)}\zeta(s, u^{\gamma(\cdot)}(s))$ and other terms of (6.151) in the same way as the proof of Theorem 6.5. Then (6.151) shows that the subsolution inequality holds.

Since we can establish the supersolution inequality by the same arguments, this completes the proof. □

### 6.4.3 Uniqueness of the Viscosity Solution

To show the uniqueness of the viscosity solution, we state the comparison theorem [GŚ00], Th. 6.1. Denote by $\mathscr{C}$ the set of all $U \in C([0, T] \times H_0)$ such that

(a)

$$\lim_{\|\phi\|\to\infty} \sup_t \frac{|U(t, \phi)|}{\|\phi\|^2} = 0,$$

(b) for any bounded set $\Lambda$ of $H_0$, $U_{/[0,T]\times\Lambda}$ is in $C_u([0, T] \times H_{-1})$.

**Comparison Theorem**

Let $U$ and $V$ be in $\mathscr{C}$. If $U$ and $V$ are a viscosity subsolution and a viscosity supersolution of (6.127)–(6.128), respectively, then $U \le V$ on $[0, T] \times H_0$.

From this theorem, we deduce;

**Theorem 6.7.** *Let $F$ be in $C_u(H_{-1})$. Under $(B_1')$ and $(B_2')$, $U(t,\phi) = \mathrm{V}_{T-t}F(\phi)(= \mathrm{V}_{0\,T-t}F(\phi))$ is the unique viscosity solution of (6.127)–(6.128) in $\mathscr{C}$.*

*Proof.* It is enough to show that $U$ is in $\mathscr{C}$. Since $U$ satisfies (a) by (6.72), we only need to prove (b).

Let $\gamma(\cdot)$ be given. Let $u(\cdot)$ and $\tilde{u}(\cdot)$ denote the response for $\gamma(\cdot)$ with $u(t) = \phi$ and $\tilde{u}(t) = \tilde{\phi}$, respectively. We first consider the terminal cost. For $\varepsilon > 0$, we have

$$\begin{aligned} E|F(u(T)) - F(\tilde{u}(T))| &\le E m_F(\|u(T) - \tilde{u}(T)\|_*) \\ &\le \varepsilon + C_\varepsilon E\|u(T) - \tilde{u}(T)\|_* \\ &\le \varepsilon + C_\varepsilon \sqrt{N}\|\phi - \tilde{\phi}\|_* \quad \text{(by (5.189))}. \end{aligned} \tag{6.155}$$

For the running cost, (6.64) leads to

$$\begin{aligned} E\int_t^T |r(u(s),\gamma(s)) - r(\tilde{u}(s),\gamma(s))|\,ds &\le l_r E\int_t^T \|u(s) - \tilde{u}(s)\|\,ds \\ &\le l_r\sqrt{NT}\|\phi - \tilde{\phi}\|_*. \end{aligned} \tag{6.156}$$

Thus, (6.155) and (6.156) show that $U(t,\cdot) \in C_u(H_{-1})$ uniformly on $[0,T]$, namely there is a modulus $m(\cdot)$, such that

$$|U(t,\phi) - U(t,\tilde{\phi})| \le m(\|\phi - \tilde{\phi}\|_*) \quad \forall t. \tag{6.157}$$

Next we consider time continuity. Let $u(\cdot)$ be the response for $\gamma(\cdot)$, with $u(0) = \phi$. Since $U(t,\phi) = \mathrm{V}_{0\,T-t}F(\phi)$, we see that, for $\varepsilon > 0$,

$$\begin{aligned} E|F(u(t_1)) - F(u(t_2))| &\le E m_F(\|u(t_1) - u(t_2)\|_*) \\ &\le \varepsilon + C_\varepsilon\sqrt{N_3}\|\phi\|\sqrt{(t_1 - t_2)} \end{aligned} \tag{6.158}$$

by (6.24). Now (6.67) and (6.158) yield

$$\begin{aligned} &\sup_{\gamma(\cdot)\in\Gamma} |J(t_1,0,\phi;\gamma(\cdot);F) - J(t_2,0,\phi;\gamma(\cdot);F)| \\ &< \varepsilon + c_7(1 + C_\varepsilon)\|\phi\|\sqrt{|t_1 - t_2|}, \end{aligned} \tag{6.159}$$

with a constant $c_7$ independent of $\phi$.

Now (6.158) and (6.159) conclude (b). □

*Example 6.1.* ***Control of partially observable diffusion processes***

Let us consider a controlled diffusion $X$ and its observation process $Y$, according to 5.5.1. For a policy $\zeta$ (see Definition 5.5.1), $X$ and $Y$ evolve according to the SDEs

$$\begin{cases} dX(t) = \alpha(X(t), \zeta(t, Y))\, dB(t) + \sigma(X(t), \zeta(t, Y))\, dW(t) \\ \qquad\qquad + b(X(t), \zeta(t, Y))\, dt, \quad t \in (T_0, T], \\ X(T_0) = X_0, \end{cases} \tag{6.160}$$

and

$$\begin{cases} dY(t) = h(X(t), \zeta(t, Y))\, dt + dW(t), \quad t \in (T_0, T], \\ Y(T_0) = 0, \end{cases} \tag{6.161}$$

where $B$, $W$, and $X_0$ are mutually independent and $B$ and $W$ are $d_0$- and $m$-dimensional Wiener processes on $(\tilde{\Omega}, \tilde{\mathcal{F}}, \tilde{P})$. The problem is to minimize the payoff;

$$J(T_0, X_0; \zeta) := E_{T_0 X_0}\Big[\int_{T_0}^{T} l(X(t), \zeta(t, Y))\, dt + f(X(T))\Big], \tag{6.162}$$

by choosing an appropriate policy $\zeta$. We denote the set of all policies by $\Gamma^*$.

Recalling Proposition 5.9, we will formulate the problem as a control problem with full observation by using the unnormalized conditional probability density, given by the Zakai equation. Let $\beta$ be an $m$-dimensional Wiener process, defined on $(\Omega, \mathcal{F}, P)$. Suppose that $X_0$ has a probability density function $\phi \in H_0$. For $\gamma(\cdot) \in \boldsymbol{\Gamma}^{\beta}$, we consider the Zakai equation

$$\begin{cases} dq(t) = A^{\gamma(\cdot)} q(t)\, dt + \sum_{k=1}^{m} G_k^{\gamma(\cdot)} q(t)\, d\beta^k(t), \quad t \in (T_0, T], \\ q(T_0) = \phi, \end{cases} \tag{6.163}$$

where

$$A^{\gamma}\psi = \sum_{i=1}^{d} \partial_i \Big[\sum_{i=1}^{d} K^{ij}(x, \gamma)\partial_j \psi - \Big(b^i(x, \gamma) - \sum_{j=1}^{d} \partial_j K^{ij}(x, \gamma)\Big)\psi\Big],$$

with $K = \frac{1}{2}(\alpha\alpha^{\top} + \sigma\sigma^{\top})$, and

$$G_k^{\gamma}\psi = -\sum_{i=1}^{d} \sigma_k^i(x, \gamma)\partial_i \psi + \Big(h^k(x, \gamma) - \sum_{i=1}^{d} \partial_i \sigma_k^i(x, \gamma)\Big)\psi, \quad k = 1, \ldots, m.$$

We assume the smoothness condition $(B_1')$ and uniform positive definiteness condition for $\alpha\alpha^\top$:

$$y^\top \alpha\alpha^\top(x,\gamma)y \geq \lambda_0|y|^2, \quad \forall y \in \mathbb{R}^d, \quad \forall x, \gamma.$$

Further, we assume that $l(\cdot,\gamma) \in H_0$ with $\sup_\gamma \|l(\cdot,\gamma)\| < \infty$ and $f \in H_1$.

Since $\zeta(t,\beta)$ is in $\boldsymbol{\Gamma}^\beta$ and, for $\gamma(\cdot) \in \boldsymbol{\Gamma}^\beta$, we can take $\zeta \in \Gamma^*$ so that $\gamma(t) = \zeta(t,\beta)$ almost everywhere on $[0,T] \times \Omega$, Proposition 5.9 asserts

$$J(T_0, X_0; \zeta) = E_{T_0\phi}\Big[\int_{T_0}^T (l(\cdot,\gamma(t)), q^{\gamma(\cdot)}(t))\,dt + (f, q^{\gamma(\cdot)}(T))\Big], \tag{6.164}$$

where $q^{\gamma(\cdot)}(\cdot)$ is the solution of (6.163), and $J(T_0, X_0; \zeta)$ depends on $T_0$, $\phi$ and $\zeta$.

Further, the value function $V(T_0,\phi) := \inf_{\zeta\in\Gamma^*} J(T_0, X_0; \zeta)$ is nothing but the infimum of the RHS of (6.164) over $\gamma(\cdot) \in \boldsymbol{\Gamma}^\beta$. Since the terminal cost $F(\psi) := (f,\psi)$ is in $C(H_{-1})$, Theorem 6.7 says that the value function $V(\cdot)$ is the unique viscosity solution of the HJB equation

$$\partial_t V(t,\phi) + \inf_{\gamma\in\Gamma}\Big\{\langle A^\gamma\phi, DV(t,\phi)\rangle + \frac{1}{2}\sum_{k=1}^m \langle D^2V(t,\phi)G_k^\gamma\phi, G_k^\gamma\rangle + (l(\cdot,\gamma),\phi)\Big\}$$
$$= 0, \quad t \in (0,T), \quad \phi \in H_1,$$

with the lateral boundary condition

$$V(T,\phi) = (f,\phi).$$

# References


[A03] R.A. Adams, *Sobolev Spaces*, 2nd edn. (Academic, Amsterdam/Boston, 2003)

[BC09] A. Bain, D. Crisan, *Fundamentals of Stochastic Filtering* (Springer, New York/London, 2009)

[Be52] R. Bellman, On the theory of dynamic programming. Proc. Nat. Sci. U.S.A. **38**, 716–719 (1952)

[Be57] R. Bellman, *Dynamic Programming* (Princeton University Press, Princeton, 1957)

[Be75] V.E. Benes, Composition and invariance methods for solvinig some stochastic control problems. Adv. Appl. Prob. **7**, 299–329 (1975)

[BSW80] V.E. Venes, L.A. Shepp, H.S. Witsenhaussen, Some solvable stochastic control problems. Stochastics **4**, 39–83 (1980)

[Be92] A. Bensoussan, *Stochastic Control of Partially Observable Systems* (Cambridge University Press, Cambridge/New York, 1992)

[BN90] A. Bensoussan, M. Nisio, Nonlinear semigroup arising in the control of diffusions with partial observation. Stoch. Stoch. Rep. **30**, 1–45 (1990)

[BP99] T.R. Bieleckii, S.R. Pliska, Risk sensitive dynamic asset management. Appl. Math. Optim. **39**, 337–366 (1999)

[BS73] F. Black, M. Scholes, The pricing of options and corporate liabilities. J. Polit. Econ. **81**, 637–659 (1973)

[BM07] R. Buckdahn, J. Ma, Pathwise stochastic control problems and stochastic HJB equations. SIAM J. Control Optim. **45**, 2224–2256 (2007)

[Br06] S. Brendle, Portfolio selection under incomplete information. Stoch. Proc. Appl. **116**, 701–723 (2006)

[CPY09] M.H. Chsng, T. Pang, J. Yong, Optimal stopping problem for stochastic differential equations with random coefficients. SIAM J. Control Optim. **48**, 941–971 (2009)

[CI90] M.G. Crandall, H. Ishii, The maximum principle for semicontinuous functions. Differ. Integral Equ. **3**, 1001–1014 (1990)

[CIL92] M.G. Crandall, H. Ishii, P.L. Lions, A user's guide to viscosity solutions. Bull. AMS NS **27**, 1–67 (1992)

[DaPZ92] G. Da Prato, J. Zabczyk, *Stochastic Equations in Infinite Dimenmensions* (Cambridge University Press, Cambridge, 1992)

[ElKPQ97] N. El Karoui, S. Peng, M.C. Quenez, Backward stochastic differential equations in finance. Math. Finance **7**, 1–71 (1997)

[ElKQ95] N. El Karoui, M.C. Quenez, Dynamic programming and pricing of contingent claims in an incomplete market. SIAM. J. Control Optim. **33**, 29–66 (1950)


© Springer Japan 2015

M. Nisio, *Stochastic Control Theory*, Probability Theory and Stochastic Modelling 72, DOI 10.1007/978-4-431-55123-2

[E83] L.C. Evans, Classical solutions of Hamilton-Jacobi-Bellman equation for uniformly elliptic operators. Trans. AMS **275**, 245–255 (1983)

[E98] L.C. Evans, *Partial Differential Equations*. GSM19 (AMS, Providence, 1998)

[FH11] W.H. Fleming, D. Hernandez-Hernandez, On the value of stochastic differential games. Commun. Stoch. Anal. **5**, 341–351 (2011)

[FKSh10] W.H. Fleming, H. Kaise, S.J. Sheu, Max-plus stochastic control and risk sensitivity. Appl. Math. Optim **62**, 81–144 (2010)

[FMcE95] W.H. Fleming, W.H. McEneaney, Risk sensitive control on infinite time horizon. SIAM J. Control Optim. **33**, 1881–1915 (1995)

[FR75] W.H. Fleming, R.W. Rishel, *Deterministic and Stochastic Optimal Control* (Springer, Berlin/New York, 1975)

[FSh99] W.H. Fleming, S.J. Sheu, Optimal long term growth rate of expected utility of wealth. Ann. Appl. Prob. **9**, 871–903 (1999)

[FSh00] W.H. Fleming, S.J. Sheu, Risk sensitive control and an optimal investment model. Math. Finance **10**, 197–213 (2000)

[FSh02] W.H. Fleming, S.J. Sheu, Risk sensitive control and an optimal investment model II. Ann. Appl. Prob. **12**, 730–767 (2002)

[FS06] W.H. Fleming, H.M. Soner, *Controlled MarkovProcesses and Viscosity Solutions*, 2nd edn. (Springer, New York 2006)

[FSo89] W.H. Fleming, P.E. Souganidis, On the existence of value function of two-plsyer, zero-sum stochastic differential games. Indiana Math. J. **38**, 293–314 (1989)

[FKK72] M. Fujisaki, G. Kallianpur, H. Kunita, Stochastic differential equations for the nonlinear filtering problem. Osaka J. Math. **9**, 19–40 (1972)

[GŚ00] F. Gozzi, A. Świech, Hamilton-Jacobi-bellman equations for the optimal control of the Duncan-Mortensen-Zakai equation. J. Funct. Analy. **172**, 466–510 (2000)

[HP81] J.M. Harrison, S.R. Pliska, Martingales and stochastic integrals in the theory of continuous ytading. Stoch. Proc. Appl. **11**, 215–260 (1981)

[HP83] J.M. Harrison, S.R. Pliska, Stochastic calculus model of continuous trading; complete markets. Stoch. Proc. Appl. **15**, 313–316 (1983)

[HNSh10] H. Hata, H. Nagai, S.I. Sheu, Asymptotics of the probability minimizing a down-side risk. Ann. Appl. Prob. **20**, 52–89 (2010)

[HS10] H. Hata, J. Sekine, Explicit solution to a certain nonELQG risk-sensitive stochastic control problem. Appl. Math. Optim. **62**, 341–380 (2010)

[IW81] N. Ikeda, S. Watanabe, *Stochastic Differential Equations and Diffusion Processes* (North Holland, Amsterdam/New York, 1981)

[Is92] H. Ishii, Viscosity solutions for a class of Hamilton-Jacobi equations in Hilbert spaces. J. Funct. Anal. **105**, 301–341 (1992)

[I42] K. Itô, Differential equations determining Markov processes. Zenkoku Shijo Sugaku Danwakai **244**, 1352–1400 (1942). (In Japanese)

[I51] K. Itô, *On Stochastic Differential Equations*. Memoirs of the American Mathematical Society, vol. 4 (AMS, New York City, 1951)

[JLL90] P. Jaillet, D. Lamberton, B. Lapeyer, Variational inequalities and the pricing of American options. Acta Appl. Math. **21**, 263–289 (1990)

[KS91] I. Karatzas, S.E. Sheve, *Brownian Motion and Stochastic Calculus*, 2nd edn. (Springer, New York, 1991)

[KS98] I. Karatzas, S.E. Sheve, *Methods of Mathematical Finance* (Springer, New York, 1998)

[Ko04] S. Koike, *A Biginner's Guide to the Theory of Viscisity Solutions*. MSJ Memoirs, vol. 13 (JMS, Tokyo, 2004)

[KN02] K. Kuroda, H. Nagai, Risk sensitive portfolio optimization and infinite time horizon. Stoch. Stoch. Rep. **73**, 309–331 (2002)

[Kr09] N.V. Krylov, *Controlled Diffusion Processes*, 2nd edn. (Springer, Berlin, 2009)

[Ku67] H.J. Kushner, Dynamical equations for optimal nonlinear filtering. J. Differ. Equ. **3**, 179–190 (1967)

[La83] S. Lang, *Real Analysis*, 2nd edn. (Addison-Wesley, New York, 1983)

[L83] P.L. Lions, Optimal control of diffusion processes and Hamilton-Jacobi-Bellman equations I. J. Commun. PDE. **8**, 1101–1134 (1983)

[L88] P.L. Lions, Viscosity solutions of fully nonlinear second-order equations and optimal stochastic control in infinite dimensions. Part I. The case of bounded stochastic evolution. Acta Math. **161**, 243–278 (1988)

[L89] P.L. Lions, Viscosity solutions of fully nonlinear second-order equations and optimal stochastic control in infinite dimensions. Part II. Optimal control of Zakai equation, in *Stochastic Partial Differential Equations and Applications II*, ed. by G. Da Prato, L. Tubaro. Lecture Notes in Mathematics, vol. 1390 (Springer, Berlin/Heidelberg, 1989), pp. 147–170. Part III. Uniqueness of viscosity solutions for general second order equations. J. Funct. Anal. **86**, 1–18 (1989)

[LN83] P.L. Lions, M. Nisio, A uniqueness result for the semigroup associated with HJB equations. Proc. Jpn. Acad. **58**, 273–276 (1983)

[LS01] R.S. Liptser, A.N. Shiryayev, *Statistics of Random Processes I, II*, 2nd edn. (Springer, New York/Berlin, 2001)

[Ma00] C. Martini, American option prices as unique viscosity solutions to degenerate HJB equations, Rapport de rech, INRIA, 2000

[Me71] R.C. Merton, Optimal consumption and portfolio rules in continuous time. J. Econ. Theory **3**, 373–413 (1971)

[Me73] R.C. Merton, Theory of rational option pricing. Bell J. Econ. Manag. Sci. **4**, 141–183 (1973)

[M88] M. Metivier, *Stochastic Partial Differential Equations in Infinite Dimensional Spaces* (Scuola Superiore, Pisa, 1988)

[My66] P.A. Meyer, *Probability and Potentials* (Blaisdell, Waltham, 1966)

[Mo10] H. Morimoto, *Stochastic Control and Mathematical Modeling, Applications in Economics* (Cambridge University Press, Cambridge/New York, 2010)

[Na03] H. Nagai, Optimal strategies for risk sensitive portfolio optimization problems for general factor models. SIAM J. Control Optim. **41**, 1779–1800 (2003)

[N76] M. Nisio, On a nonlinear semigroup attached to stochastic optimal control. Pull. RIMS **12**, 513–537 (1976)

[N78] M. Nisio, On nonlinear semigroups for Markov processes associated with optimal stopping. Appl. Math. Optim. **4**, 143–169 (1978)

[N81] M. Nisio, *Lecture on Stochastic Control Theory*. ISI Lecture Notes, vol. 9 (McMillan India, Delhi 1981)

[N88] M. Nisio, Stochastic differential games and viscosity solutions of Isaacs equations. Nagoya Math. J. **110**, 163–184 (1988)

[P79] E. Pardoux, Stochastic partial differential equations and filtering of diffusion processes. Stochastics **3**, 127–167 (1979)

[P93] E. Pardoux, Stochastic partial differential equations, a review. Bull. Sc. Math. **117**, 29–47 (1993)

[PP90] E. Pardoux, S. Peng, Adapted solution of backward stoshastic equation. Syst. Control Lett. **14**, 55–61 (1990)

[Pe92] S. Peng, Stochastic Hamilton-Jacobi-Bellman equations. SIAM J. Control Optim. **30**, 284–304 (1992)

[Ph02] H. Pham, Smooth solutions to optimal investment models with stochastic volatilities and portfolio constraints. Appl. Math. Optim. **46**, 55–78 (2002)

[R90] B.L. Rozovskii, *Stochastic Evolution Systems* (Kluwer Academic, Dordrecht/Boston, 1990)

[S08] A.N. Shiryayev, *Optimal Stopping Rules*, 2nd edn. (Springer, Berlin/Heidelberg 2008)

[St11] L. Stettner, Penalty method for finite horizon stopping problems. SIAM J. Control Optim. **49**, 1078–1099 (2011)

[SV79] D.V. Strook, S.R.S. Varadhan, *Multidimensional Diffusion Processes* (Springer, Berlin/New York, 1979)

[W71] J.C. Willems, Least squares stationary optimal control and the algebraic Ricatti equation. IEEE. Trans. Auto. Control **16**, 621–635 (1971)

[Wo68] W.M. Wonham, on a matrix Ricatti equation of stochastic control. SIAM J. Control Optim. **6**, 681–697 (1968)

[Y80] K. Yosida, *Functional Analysis*, 6th edn. (Springer, Berlin/New York, 1980)

[YZ99] J. Yong, X.Y. Zhou, *Stochastic Controls, Hamiltonian Systems and HJB Equations* (Springer, New York, 1999)

# Index



© Springer Japan 2015

M. Nisio, *Stochastic Control Theory*, Probability Theory and Stochastic Modelling 72, DOI 10.1007/978-4-431-55123-2

GPSR Compliance
The European Union's (EU) General Product Safety Regulation (GPSR) is a set of rules that requires consumer products to be safe and our obligations to ensure this.

If you have any concerns about our products, you can contact us on

ProductSafety@springernature.com

In case Publisher is established outside the EU, the EU authorized representative is:

Springer Nature Customer Service Center GmbH
Europaplatz 3
69115 Heidelberg, Germany

www.ingramcontent.com/pod-product-compliance
Ingram Content Group UK Ltd.
Pitfield, Milton Keynes, MK11 3LW, UK
UKHW021522300726
14060UKWH00012B/630

* 9 7 8 4 4 3 1 5 5 1 2 2 5 *